# Seeing Trees

# *Seeing Trees*

A History of Street Trees in
New York City and Berlin

SONJA DÜMPELMANN

Yale UNIVERSITY PRESS

NEW HAVEN AND LONDON

Published with assistance from The Foundation for Landscape
Studies and a Dean's Annual Research Grant from the Harvard
University Graduate School of Design.

Yale University Press books may be purchased in quantity for
educational, business, or promotional use. For information, please
e-mail sales.press@yale.edu (U.S. office) or sales@yaleup.co.uk
(U.K. office).

Designed by Sonia L. Shannon
Set in Electra type by Integrated Publishing Solutions
Printed in the United States of America.

Frontispiece (detail of Figure 5.19): Chargesheimer's view of
Stalinallee from the terrace of Café Warschau, ca. 1953. (Repro:
Rheinisches Bildarchiv Köln, Chargesheimer, rba_L010538_32)

page 19 (see Figure 3.2): A scene along Madison Square, 1901.
(Photo by the Byron Company. Geographic File, box 39, folder:
Fifth Avenue 3, image 91237d, New-York Historical Society.
Collection of the New-York Historical Society)

page 123 (detail of Figure 8.4): Newly planted silver lindens and
flexible seating on Unter den Linden near the intersection with
Friedrichstrasse, 1949. (bpk Bildagentur/Kunstbibliothek
Staatliche Museen Berlin/Willy Römer/Art Resource, NY)

Library of Congress Control Number: 2018938623
ISBN 978-0-300-22578-5 (hardcover : alk. paper)

A catalogue record for this book is available from the
British Library.

This paper meets the requirements of ANSI/NISO Z39.48-1992
(Permanence of Paper).

10 9 8 7 6 5 4 3 2 1

*To my parents*

# CONTENTS

# ACKNOWLEDGMENTS

IN THE COURSE OF RESEARCHING and writing this book, I received the help and support of many institutions and individuals to whom I am very grateful. The Foundation for Landscape Studies facilitated this book's publication through its David R. Coffin Publication Grant. The Harvard University Graduate School of Design (GSD) and a Dean's Research Grant from the school also offered generous financial support for this project. The book was further advanced by the opportunity to develop and offer a variety of seminar and lecture courses at the GSD that revolved around trees. I would like to thank the students in these courses for their engaged discussions and work, and my colleague Gary Hilderbrand for co-teaching a class with me in spring 2015 that enabled us to bring together his knowledge and perspectives on urban trees as a practicing landscape architect with mine as a historian.

I presented some of the first forays of this work in 2014 at the Twelfth International Conference on Urban History in Lisbon, where my paper was part of a panel organized by historians Dorothee Brantz and Peter Clark, to whom I am very grateful. Ever the enthusiast when it comes to nature in the city, Dorothee was among the first to offer encouragement and support when I embarked on this project. I presented papers and lectures related to the book at the Society of Architectural Historians conferences in 2015 and 2016, Princeton University, the University of Oregon, Dumbarton Oaks, the New York Botanical Garden, the Technical University Berlin, the University of Kassel, Leibniz University Hannover, and Albert-Ludwigs University Freiburg: thanks to all who invited me.

I also thank the many archivists and librarians who answered questions and made materials available at Frances Loeb Library and Special Collections of the Harvard University Graduate School of Design, Harvard University Herbaria & Libraries, Harvard University Archives, the Huntington Library, Patricia Klingenstein Library of the New-York Historical Society, Archives at the American Museum of Natural History Research Library, New York City Municipal Archives, La Guardia and Wagner Archives, New York Public Library, New York City Department of Parks and Recreation, Columbia University Archives, Rockefeller Archive Center, Brooklyn Public Library, Brooklyn Museum, Brooklyn Historical Society, Cooper Union Library Archives, Syracuse University Moon Library, DC Public Library Special Collections, Department of Rare Books & Special Collections at Rush Rhees Library of the University of Rochester, Cushing Memorial Library & Archives at Texas A & M University, the State Historical Society of Missouri, Davey Tree Expert Company, Staatsbibliothek Berlin, Stadtbibliothek Berlin, Papiertiger Archiv und Bibliothek der sozialen Bewegungen in Berlin, Archives of Leibniz Institute for Research on Society and Space, Museum Pankow Archiv, Studienarchiv Umweltgeschichte at Hochschule Neubrandenburg, Bundesarchiv Berlin-Lichterfelde, and Universitätsarchiv Dresden. Special thanks go to Monika Bartzsch, Kerstin Bötticher, and Andreas Matschenz at Landesarchiv Berlin; to Kerstin Ehlebracht and Klaus Wichert at Berlin's Senatsverwaltung für Umwelt, Verkehr und Klimaschutz; and to Stefanie Sentner at Humboldt-Universität zu Berlin. At Landesdenkmalamt Berlin Leonie Glabau and Heike Waschner gave me access to and entrusted me with materials still stored in boxes in the catacombs of Altes Stadthaus. My research and experiences in Berlin during this project would not have been the same without my conversations with Manfred Butzmann, Hans Georg Büchner, Almut Jirku, Wolfgang Krause, Siegfried Sommer, and Stephan Strauss, who generously offered their time and insights into life, urban trees, and municipal open-space planning during and after Berlin's division. Wolfgang Krause found the keys to open storage rooms in Pankow and Prenzlauer Berg filled with documents that may soon land in dumpsters. Not only did he lead me to these hidden spaces that in themselves are part of Berlin's landscape history, he also took the time to show me some of the street trees planted under his direction of Prenzlauer Berg's district Department of Gardens before and after the fall of the wall. I also thank Phyllis Anderson, who generously passed on to me her personal collection of street tree brochures from the late 1970s to the mid-1980s; Jürgen Hübner-Kosney, who shared documents from his personal archive; and Joachim Wolschke-Bulmahn for the brochure on trees from his personal antiquarian collection.

Over the years I worked on this project, Dalal Alsayer, Leandro Couto de Alameida, Zach Crocker, Clemens Finkelstein, and Farhad Mirza assisted me in various tasks, especially with the scanning of documents. I thank them for their patience and good humor.

My colleagues Anita Berrizbeitia, Peter Del Tredici, Richard Forman, Peter Galison, Kenneth Helphand, Susan Herrington, Jerold Kayden, Niall Kirkwood, Marcus Köhler, Norbert Kühn, Keith Morgan, Joshua Shannon, and John Stilgoe have been supportive of this project in many ways, some they may not even be aware of. Christine Smith has been a mentor and friend throughout the years of work on this book, making those other parts of academic life easier to handle. Similarly, a big thank-you goes to my office neighbors Stephen Gray, Michael Hooper, and Jane Hutton for creating that legendary fourth-floor spirit!

At Yale University Press I thank Jean Thomson Black, who was most generous with her time and expertise and ensured a smooth production process. Michael Deneen, Jeffrey Schier, and the production team provided crucial assistance in transforming the manuscript and images into this printed book. It would not be the same without the masterful copyediting of Robin DuBlanc. I am very thankful to all of them.

Toward the end of finalizing this book, Felix Bärwald provided hope, trust, and support; during the time it took to research and write it, my father planted his favorite shrub, a sea buckthorn, teasing it to grow like a tree. And while my mother and I watched his experiment proceed with loving familiar ridicule, my mother, as always, read my manuscript before anybody else did, offering that first litmus test that is so valuable. Thank you for everything.

# Seeing Trees

# Introduction

## *Seeing the Urban Forest*

"Forest character and the urban hustle and bustle; holy dignity and se-renity, and mundane noise are no doubt opposites," German land-scape gardener G. A. Schulze wrote in 1881. Schulze had put a finger on the arguable paradox of planting trees in areas that had in many cases been cleared of trees for urban development in the first place. At the time, more and more cities had begun to systematically plant trees along their streets. But did trees belong in the city? Few gardeners doubted the aesthetic and climatological use of street trees in the metropolis. Some decades later, in 1920, American horticulturalist Furman Lloyd Mulford made clear, "If it is impossible to grow trees on a street, . . . that street should be closed for human use until conditions are so improved that it will support trees." No trees, no humans. Early street tree experts did not fail to stress the intimate connection between humans and trees. They were complementary. As one commentator put it, in city planning "the most impor-tant factor involved besides ourselves is the tree." As this book shows, tree planting has always been as much about humans as trees. By caring for street trees, park executives would care for citizens, Wichita director of parks and forestry Alfred MacDonald pointed out in 1923. Later, in the 1970s, when North American cities were troubled by disinvestment and decay, urban foresters continued to emphasize the role of trees in urban society. Although various qualifiers were used to describe trees in populated areas—*shade, amenity, ornamental, street,* and *landscape*—one tree expert thought that the clearest nomination would simply be "trees for people." Another professional explained that the urban forester was a "street smart forester" and "a people manager, not a tree manager."[1]

As much as trees have naturally belonged in the city, trees along city streets oc-cupy contested terrain, and some of the biggest urban conflicts have therefore revolved around them. In fact, both tree planting and tree felling in cities have often been conflictual. Street trees have stood at the very core of events that have shaped and

1

characterized the evolution of the modern city. In the nineteenth and twentieth centuries they contributed to turning cities into both local and global places. This book is
about street trees' literal and figurative entanglement not only in the built urban structure but also in urban social, cultural, and political life as a whole.

In 1932 at the Eighth National Shade Tree Conference in the United States, tree
expert and landscape engineering professor Laurie D. Cox observed the increasing
interest that street trees had elicited in U.S. cities in the previous decades. Indeed, beginning with the tree-planting activities led by nineteenth-century village-improvement
societies and promoted by Arbor Day, street tree planting had taken off across the
country. But Cox also noted that although much time and money had been spent on
street trees, their planting and care lacked notable progress. In general, he reported,
street tree planting had probably not improved in the last fifty to one hundred years,
neither in the United States nor abroad. Writing in Germany in the late 1970s, garden
and landscape historian Dieter Hennebo made a similar observation in a seminal
essay on the history of street trees. According to him, general insights concerning street
trees had not changed much since the late nineteenth century, when most cities had
begun to more systematically line their streets with trees for reasons of hygiene and
aesthetics.[2]

While it is true that many of the general principles and practices of street tree
planting and care in our cities are old, even going back centuries, a closer look at the
planting, maintenance, and management of street trees as well as their reception and
representation by experts and nonexperts also reveals changes in politics, technology,
science, and design, especially in the period between the late nineteenth and early
twenty-first centuries, the time covered in this book. Increasing urbanization and the
reform movements of the Progressive Era marked a shift in the treatment of trees in
cities on both sides of the Atlantic, including New York City and Berlin, the twin focus
of this book. Urban planners and designers, tree experts and arborists, city officials
and citizens had to resolve the growing tension between the modernization of the city
and the conservation of trees. They had to deal with the tensions between nativism and
globalism; the vernacular and modernity; nature's agency and human actors; and technology, science, economy, and aesthetics. The history of street tree planting therefore
offers insights into how our relationships with both nature and the city have changed
over time. The stories in this book reveal how street trees have been considered variously as aesthetic objects, creators of space, territorial markers, instruments of emancipation and empowerment, sanitizers, air-conditioners, nuisances, upholders of moral
values, economic engines, scientific instruments, and ecological habitats. Looking at
street trees enables us to see both the forest and the trees.

## Urban Forestry

Street tree planting is part of what in the United States was first called city forestry and
has since the 1960s become more widely known as urban forestry. Erik Jorgensen,
professor of forest pathology at the University of Toronto, has often been credited with

coining the term *urban forestry*, which he defined as "a specialized branch of forestry" that had "as its objectives the cultivation and management of trees for their present and potential contribution to the physiological, sociological and economic wellbeing of urban society." But while the term's international adoption beginning in the 1970s can be traced back to Jorgensen's research activities in Toronto, it was already being used in the late nineteenth century to describe the systematic planting, management, and care of street trees. Urban forestry was, as George R. Cook, the general superintendent of parks in Cambridge, Massachusetts, explained in 1894, "an art requiring special knowledge, cultivated taste and a natural sympathy with plant life." With city governments recognizing their new responsibility, urban forestry quickly turned into what became more widely known as city forestry before assuming the original nomenclature again in the 1960s. This latter shift marked the increased recognition that all trees administered by a municipal government agency were parts of a larger environmental system, a forest. The relentless spread of Dutch elm disease had taught foresters that it was crucial to see the forest as well as the trees if serious damage was to be avoided. The urban forest therefore also stretched beyond municipal boundaries into areas that have been described as suburban, peri-urban, and rural. In the 1980s this holistic understanding of the urban forest began to be adopted in Europe, where town forestry had a long tradition and street trees were managed by municipal park and garden departments.[3]

Even if "urban forestry" was in 1960s North America first understood as an oxymoron, many early tree experts had not seen an opposition between "the urban" and the forest. Not all trees, however, were considered suitable to urban conditions, given their various physiological needs, growth habits, and aesthetics. Their planting demanded careful curation, as both Schulze and Mulford pointed out as early as the late nineteenth and early twentieth centuries. Tree experts found that some trees had a habit that was ill adapted to the space constraints of inner cities. Landscape gardener Schulze, for example, warned in the 1880s that horse chestnut, elm, poplar, and black locust were not suitable for city streets. He proposed instead that false acacia 'Bessoniana,' Mediterranean hackberry, and common hackberry should be used more often. By 1901, the Association of German Garden Artists had established four street tree classes, categorized by the trees' height and canopy size. Each class was to be treated differently with regard to the distance between trees, the distance of planting from the building line, or front yard, and the density of buildings in the area. Similarly, in the northeastern United States, landscape architect Elbert Peets thought of ailanthus, Oriental plane, and Norway maple as "city trees," whereas elms and sugar maples were better adapted to village streets. But decades later, sugar maples, with their wide canopies, proved ill suited as roadside trees in general, being particularly susceptible to drought and road salts. American elms were decimated by Dutch elm disease beginning in the 1930s and were replaced in many locations by green ash, which in turn was destroyed even more extensively by the emerald ash borer. Ultimately, the replacement of one species with another did not suffice. Tree experts argued that new street trees had to be cultivated: those that could resist air pollution and road salt, and that had small and narrow crowns. What was needed was what Schulze had already in the 1880s called a *Musterbaum*, and what early American city foresters called the "ideal" or "model tree." To this effect, German dendrologist and nursery owner Johann Gerd Krüssmann, for example, bred the narrow-crowned *Platanus x hybrida* 'Dortmund' and the columnar *Ginkgo biloba* 'Tremonia.' More than in Europe, nurserymen and institutions in the United States undertook systematic efforts to cultivate new ideal street trees that would meet the social and economic requirements of the city.[4]

In 1950 Oakland's superintendent of parks William Penn Mott Jr. had demanded the production of "a hybrid tree for street tree planting" that could survive in the modern city and that would cost under 25¢ per year to maintain. To him it appeared only logical that nurserymen and plant breeders should produce a tree that met specifications like an ascending habit, a tap root, a long life, pest resistance, and tolerance of poor soil conditions. Ideally, that tree was to be non-fruit-bearing yet flowering and colorful, and it would feature bright green shiny foliage. The plant industry began to take increasing notice in the 1960s and 1970s. After the popular red maple was discovered in the 1960s to have a high failure rate and large variations in fall color and growth, in 1971 a systematic breeding program was begun to select appropriate cultivars with steady characteristics, thereby industrializing biological innovation in the realm of street trees. To identify pollution-resistant trees, cultivars of common street tree species like red maple, Norway maple, ginkgo, honey locust, and London plane tree were subjected to air-pollution tests in special fumigation chambers and on field sites in the greater New York City area. Individual plant breeders and nurseries as well as breeding programs at larger institutions in the United States did not focus their search for new cultivars only on native species but included those from other continents as well. For example, in 1979 the University of Pennsylvania began studying twenty European tree species and cultivars, hoping to expand the range of suitable street trees in the northeastern United States and to replace the American elm with an elm of similar grandeur that was resistant to Dutch elm disease. Attempts to produce the latter included growing seedlings from irradiated seed. This new form of techno-scientific production relied on atomic energy, for which new peacetime uses had been sought in the postwar years. Many tree cultivars produced through these efforts in the United States can also be found along European streets today. However, as helpful as these breeding programs were, they also had negative aspects, encouraging monocultures along city streets that could easily spread harmful insects and fungi.[5]

Like insects and fungi, trees had their own life force that could be tamed but not entirely controlled. Trees' agency and relative permanence, often outlasting human lives, have contributed to their frequent anthropomorphization, which in playful ways further blurs distinctions between nature and culture (figure I.1).

Tree cartoons that play with the analogy between humans and trees abound. At the beginning of the twentieth century, trees were considered shrewd and "cunning as a rat," with "a degree of intelligence worthy of the animal kingdom." Tragically, by developing girdling roots they could "commit suicide" through root strangulation. Advising tree wardens, the Massachusetts Forestry Association wrote in 1900, "To be

FIGURE I.1.
Cartoon of a man walking his
dog past a tree that holds on
to the bars of the protective
fencing, 2013. (Sam Gross/
The New Yorker Collection/
The Cartoon Bank)

perfectly healthy a tree, like a human being, must have a sound body covered with an unbroken and healthy skin." To achieve this purpose, tree doctors and tree surgeons thought to save trees by filling their cavities. Later in the twentieth century, chemotherapy, endoscopy, and finally computer tomography were applied to determine tree health and find out more about trees' inner lives.[6]

As the application of these technologies and the search for and production of new tree cultivars show, the systematic introduction of street trees in cities has been an example of both innovation and preservation. As urban designers planned with trees in mind and nurserymen created new trees, as cities began to plant and protect trees, and as citizens began to protest against their destruction, nature in the form of urban trees was—and, of course, still is—an important part of culture in the city.

## Rural-Urban Transformation

The chapters in this book reveal how in the process of cultivating trees and planting them along streets, cities were naturalized and trees were urbanized. In the late nineteenth and early twentieth centuries, tree planting became an integral part of the urbanization process, transcending the artificial distinctions between rural and urban, natural and cultural. Practices commonly applied to trees outside of cities were transferred to the city. If tree guards had been used to protect trees in pastures against the appetites of livestock, in the city they saved street trees from nibbling horses and urinating dogs. Whereas the cavities of notable old trees had for centuries been boarded up or filled with various natural materials, by the early twentieth century "modern" concrete fillings had become a standard practice ostensibly prolonging street trees' lives. Tree and bird conservation efforts also began outside of urban areas, leading to enthusiastic responses in cities by the turn of the twentieth century. Street trees were planted to provide habitats for birds, and bird sanctuaries, baths, feeding stations, and nesting boxes were established throughout public urban parks. Street tree planting also became a means for women to transcend the domestic sphere and enter public spaces as they moved from the peripheral suburbs to the center stage of the formerly male-dominated business districts. Although street trees were planted first and foremost for aesthetic and climatological purposes, at various moments throughout the twentieth century, particularly in times of crisis, street trees became a life-saving resource, turning the city into a forest or orchard that could be harvested. Street trees provided lumber for firewood, fruit for a vitamin-rich diet, and flowers and leaves for beverages. Leaves were also composted to produce mulch for inner-city planting areas. New York City began composting in 1935 to substitute for the decreasing animal manure resulting from the city's diminishing number of horses and the region's agricultural mechanization and consequent reduction of livestock. In Berlin fallen foliage and pruned limbs from street trees became a highly sought-after commodity for composting in the postwar years when topsoil in the city was lacking. Street trees therefore turned the city into a territory in which natural resources were produced and consumed.[7]

## Internationalism and Nativism

As shown in various chapters in this book, street trees were used to achieve autarky, and they were weapons of defense. They literally and figuratively demarcated and delimited more or less porous boundaries, and they were used as instruments of identity politics. Political conflict and nativist ideologies were projected onto urban trees, which became carriers and signifiers of xenophobic sentiment. On occasion, political conflict showed merely in the rhetoric used around trees. During World War II, for example, Long Island resident Mrs. M. Cooper asked the Department of Parks and Recreation to replace a tree that had died as a result of her son cutting the tree's bark while playing "killing the Japs." And New York City Department of Parks landscape architect Francis Cormier asked the Queens director of parks to "order a blitzkrieg" on the maple wilt that had affected large sugar maples on Twenty-Ninth Avenue. Some years later, during the early years of the Cold War, street trees appeared not as victims but as weapons. In his 1958 keynote address to the Thirty-Fourth National Shade Tree Conference in Asheville, North Carolina, Erle Cocke, decorated war veteran and president of Delta Airlines, emphasized the importance of trees as keepers of American virtues. "We can meet the challenge of the Soviet threat only if we return to the moral and spiritual ideals of our forefathers," he posited, forefathers who had always made major decisions under shade trees that were an "umbrella for peace itself." According to Cocke, trees, arborists, and tree experts could therefore help the United States maintain world leadership by deterring and standing up to Russia.[8]

Street tree care and planting had both conservative and progressive dimensions, and their use in identity politics could be positive or negative. Street trees were planted with the intention of evoking feelings of neighborhood and citywide pride and belonging. For example, New York City's 1943 planning lists for postwar street tree planting showed the predominance of different species in the various boroughs: Oriental plane in Manhattan; Norway maple and Oriental plane in Brooklyn; pin oak, linden, and American elm in the Bronx; pin oak in Queens; and a mixture of Norway maple, pin oak, and red oak in Richmond (today Staten Island). Berlin's citywide tree-planting concept after the fall of the wall ensured the city's renown as the city of lindens. As discussed in chapter 8, Berlin's linden trees planted along its most famous boulevard and its many radial streets had a centripetal and unifying effect at various moments in the city's history.[9]

Yet at times, as illustrated in chapter 1, certain species also evoked xenophobic reactions that demanded their eradication and replacement. The topic of trees attracted attention from both progressives and conservatives, but particularly the latter. On both sides of the Atlantic, many early nationalist-conservative landscape designers like Frank A. Waugh, Harry Maasz, Gustav Allinger, Heinrich Friedrich Wiepking-Jürgensmann, and Alwin Seifert embraced trees, in some cases publishing entire books on trees and their use in urban and rural environments.[10]

In the years after World War II, when landscape designers sought to use large-scale street tree planting to protect Berlin from desertification—as shown in chapter 5—they uncritically continued the association of cities with steppes, an association common in anti-urban conservative and nationalist circles in the previous decades. Eugenics and race theories had bolstered this connection already in the early twentieth century. Many conservatives believed that urbanization threatened the superior Germanic race, which was considered to have evolved in the wooded areas of the temperate zone now deforested and threatened by desertification. Tree planting was therefore perceived as salvation. Not only were trees thought to cleanse the air and provide shade and lumber, the argument was that by providing a healthy living environment for the "Germanic race," trees could also prevent the "southern race" from taking over. In this vein in 1937, Frankfurt's director of gardens Max Bromme wrote that urban trees and green open space were necessary for a physically and mentally healthy productive, forceful, and industrious people, and that new tree cultivars therefore needed to be produced for the "steppe climate of city streets."[11]

Just as street trees were used to protect and segregate, they were also used to overcome physical and mental boundaries. Various citizen groups planted street trees to reclaim their right to the city, as shown in chapters 3, 4, and 7. Tree seeds and cultivars, practices of street tree planting and care, and technologies and machines were exchanged between people, cities, and countries.

## Street Tree Exchange

Beginning in the late nineteenth century and continuing throughout the twentieth, street trees produced a "global cultural pattern" in cities around the world. Central to the spread of the street tree idea was the increasing mobility of tree experts and the public, and the exchange of knowledge in the popular press and specialist publications. North American city forestry built upon the earlier movement to establish a professional forestry in the United States. Ties to forestry experts in Germany and France proved particularly important in this endeavor. American scientists and foresters like William F. Fox, Gifford Pinchot, Hugh Potter Baker, and G. Frederick Schwarz traveled to Germany and France to take classes and obtain degrees in the prominent schools of forestry there. German foresters in turn emigrated or temporarily moved to the United States as soon as they saw the opportunities awaiting them on the other side of the Atlantic. Such was the case of Carl Alwyn Schenck, who was hired by George W. Vanderbilt as estate forester at Biltmore, North Carolina, where he founded the Biltmore Forest School in 1898. Bernhard Eduard Fernow, who came to the United States to be with his American wife, held degrees from the Münden Forestry Academy near Hannover and the University of Königsberg; he soon became the third chief of the USDA's Division of Forestry and then influenced the development of North American forestry in academic positions at Cornell, Penn State, and the University of Toronto.[12]

Some German-born or German-inspired foresters became interested in city forestry, and American professionals like the young landscape architects Elbert Peets and Wayne C. Holsworth as well as early city foresters like William Solotaroff in East Orange, New Jersey; Jacob H. Prost in Chicago; and O. C. Charlton in Dallas looked to both Germany and France, especially Berlin and Paris, for models of successful urban street tree planting. Whether in the realm of soil conditions, irrigation practices, or methods to combat insects, Germany and France featured prominently in the descriptions and experiences American experts referenced. Peets, who intended to write a book on street trees, was even hoping to embark on a research trip to Europe, but his plan was thwarted by the outbreak of World War I. Holsworth, in contrast, used his time spent in the American Expeditionary Forces in France during the war to observe French forestry methods, and in the course of a period of convalescence he visited many French cities to study their urban tree plantings. Later in the twentieth century the situation would reverse. Beginning in the postwar period, North American street tree planting would be reported on in Europe, especially in West Germany, and machines like the tree mover, tree-care and pruning practices, and tree cultivars like "Greenspire" littleleaf linden (figure I.2) would be imported from the United States.[13]

## Model Tree Cities

New York City and Berlin were among the cities that early urban planners and designers looked to for comparison, seeking grounds for analysis and an exchange of ideas. The two cities were both important cultural and industrial centers in the late nineteenth and early twentieth centuries, when each undertook initiatives to consolidate. By 1920 they had become metropolitan centers, with 5.6 million inhabitants in New York and 3.9 million in Berlin. Berlin's forty-four thousand street trees, cared for by the Department of Gardens (figure I.3), drew the attention of New York City landscape architect John Y. Cuyler, who was serving as consulting landscape architect to

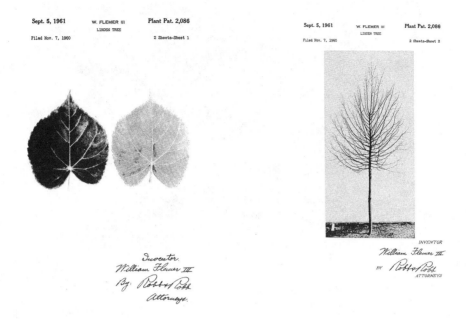

FIGURE I.2.
United States Plant Patent for the new linden tree cultivar known as 'Greenspire'
littleleaf linden, bred by William Flemer III at his Princeton nurseries in New Jersey,
filed in 1960 and patented in 1961. (U.S. Patent and Trademark Office)

the city's private Tree Planting Association. In 1908 he recommended that New York City follow Berlin in its extensive municipal tree plantings.[14]

It was not only the experts who commented on Berlin as an exemplar of a tree and park city and a model of civic improvement for American cities. A Harlem resident who criticized the "bare hard streets" that "sadly need the softening touch of green leaves" wished that "a New York millionaire or even a city official" would someday be inspired by a "walk through the miles of plane trees, chestnuts and lindens in Paris and Berlin." In the 1920s, a seventh-grade pupil from Manhattan praised Berlin's Tiergarten for its neatness and lack of litter in a school essay, and Manhattan resident Robert J. Caldwell pointed to Berlin's riverside parklands as examples of what could be done with Manhattan's river fronts. Berlin's tree planting and its shady streets were looked at favorably by New Yorkers and Americans in general as well as by tree experts from within Germany itself, where Berlin was considered a model tree city. But another European city and its American cousin also aroused attention in New York and other U.S. cities when it came to trees: Paris and Washington, DC.[15]

In the American specialist literature and in popular newspaper accounts of the late nineteenth and early twentieth centuries, the French and American capitals were presented as models to be followed in issues ranging from the number of trees planted

FIGURE I.3.
Men working on Berlin street tree, ca. 1920. (Photo: akg-images, 896297)

to the planting methods and materials employed. The architecture, parks, gardens, and urban design of Paris held great fascination for American designers and planners during the City Beautiful movement between 1893 and the first decades of the twentieth century. Department of the Seine prefect George-Eugène Haussmann's urban renewal plans for Second Empire Paris set a precedent for park-system planning and systematic street tree planting, among other things. The late nineteenth-century manuals on street tree planting by French arboriculturists Jules Nanot and Adolphe Chargueraud—themselves inspired by the work undertaken under Haussmann— were cited by American city foresters and their German colleagues, and the Parisian street tree ordinance was used as a model for the draft of a bill submitted to the New York legislature in the 1870s. But already earlier, when the United States was erecting its own capital at the end of the eighteenth century, France had provided a model and expertise. The French-born architect and engineer Pierre Charles L'Enfant delivered a plan for Washington, DC, that was modeled on Versailles. From the very beginning, therefore, the young nation's capital was equipped with wide boulevards and avenues as well as numerous "reservations," open spaces that were reserved for parks and gardens enclosing public buildings.[16]

Washington was the first city in the United States to plan and implement a comprehensive street tree–planting program, beginning in the fall of 1872 under the administration of Governor Alexander R. Shepherd (plate 1). From 1884, tree planting and care was in the hands of Trueman Lanham until his death in 1916, when his son Clifford took over. The city quickly became known as the "city of trees." Its trees were "almost the first thing" visitors remarked upon, and it was their absence that Washingtonians first noticed in other cities, the *Washington Star* insisted in 1889. Expert and public opinion did not differ much. Street trees were what made Washington one of the most beautiful cities in America, according to New York state forester William F. Fox. By 1912, nearly 280 miles of streets in Washington were lined with trees on both sides, summing up to around one hundred thousand trees, of which around twenty-five hundred to three thousand were replaced annually. They turned the city into a model for urban tree planting and care, and by the 1920s into a "city forest" and a capital that appeared green to aviators.[17]

Already in the late nineteenth century Washington had been styling itself to overtake European capitals when it came to trees. Whereas Berlin's central linden boulevard Unter den Linden was less than a mile in length and, according to a 1886 newspaper article, "more appreciable in history than in reality," the new American Unter den Linden—Washington's Massachusetts Avenue—was "four miles of vigorous and stately young lindens, twenty to thirty feet high." The fresh New World Unter den Linden would soon be connected to twelve further miles of streets and avenues planted with lindens and to an entire street tree network consisting of various tree species running along the main avenues. Already in 1884 the *Chicago Daily Tribune* had reported that the "finest single street" was "Massachusetts Avenue, where linden are planted four abreast throughout four miles, recalling famous Unter den Linden in Berlin," again referring to the other European capital that informed American urban tree planting at the time. In New York City, "the magnificent central strip of trees in the Boulevard," that is, today's Broadway north of Columbus Circle, was considered to far surpass in beauty the trees of Berlin's Unter den Linden until the subway was built and hampered tree growth.[18]

But despite this pride, American city foresters were concerned about their cities lagging behind Berlin, Paris, and other European cities. New York landscape architect John Y. Culyer lamented in 1909, "We are woefully behind all the large cities of the Old World and of many in our own country." By the early twentieth century the competition was on. When it came to street trees, Newark's Shade Tree Commission stated in 1913, there was no question "that an American city should not be outdone by foreign cities."[19]

## Competition and Comparison

Competition, comparison, and knowledge transfer played important roles in street tree development on both sides of the Atlantic. Despite the different circumstances in New York City and Berlin and the various stories told in this book, the cities' street tree histories also show similarities. Many conflicts have been persistent and recurrent in one or another form throughout the twentieth century and into the twenty-first in both cities.

For example, various crises led to the use of street tree lumber as firewood. Already in the late nineteenth century the park departments in both cities had given away or sold lumber to other municipal units, private citizens, and commercial lumbermen. But during World War I and the economic crises in the 1920s and 1930s, increasing provisions were made to supply poor households with firewood from street and park trees. In New York City, the chestnut blight had led to an unprecedented large-scale removal of trees from the city's streets and parks in 1912. The coal shortage that hit the city in 1917 upon U.S. entry into the war lasted into the early 1920s, again turning street and park trees into a welcome timber supply. In Manhattan and the Bronx, the Department of Parks collected, dried, sawed, and chopped wood into small pieces suitable for small stoves and furnaces, and the wood was given away for free to the needy at distribution points throughout the city during the coldest months of the year.[20]

Other crises also turned New York City street trees into lumber. After 1933, Dutch elm disease and the great 1938 and 1944 September hurricanes required additional pruning and removal of trees. After the 1944 Great Atlantic Hurricane, 16,800 street trees had to be removed from the city's streets, and 12,000 had to be pruned. Given the Department of Parks' labor shortage because of the war, residents had to assist in removing and cutting up the fallen trees. By October, the department was inviting citizens to pick up wood from fallen trees in various locations in each borough.[21]

As elaborated in chapter 5, in Berlin in these years it was no natural disaster that struck down trees. Instead it was the firebombs of the Second World War. But Berlin's number of street trees had already been decreasing in the decades before the war as Dutch elm disease ravaged city trees and provided much-needed firewood. A police order issued on 9 September 1931 required affected trees to be felled and their lumber used for construction or as firewood. From a certain point of view, therefore, Dutch elm disease during the world economic crisis could be considered a blessing in disguise, but not all wood was used wisely. The Wilmersdorf district councilor pointed out in the winter of 1930 how shameful it was that wood from pruned trees was burned openly on the streets when it could warm suffering human bodies. He requested that the Berlin Magistrate inform the Labor Office and Bureau of Charities so that the jobless and poor could collect the branches and use them as firewood.[22]

"The life of the modern tree," as the journal *Parks & Recreation* titled an article in 1935, was always one of conflict, contention, and contestation on both sides of the Atlantic. Above- and belowground electrical and telegraph wires, subways, leaking gas pipes, air and soil pollution, and the widening of streets caused distress for street trees. Sometimes they were stolen or vandalized. One form of vandalism was trees' use as advertising columns. In the late nineteenth century, New York City's and Brooklyn's city ordinances forbade the use of trees as notice boards and advertising pillars. In October 1917, the city of Boston organized a "tear-down week" to protect trees from nails and pamphlets that, according to scientists at the time, provided easy breeding grounds for insects besides often disorienting travelers. In Berlin, trees were turned into notice boards particularly in the postwar years, when places to append notices were lacking in the ruined inner city. Authorities sought to stifle the habit only a few months after German capitulation, when the Berlin police president issued an order prohibiting the use of trees as notice boards.[23]

Over time, damage from horses was replaced by damage from motor vehicles, and ruin through de-icing salt took the place of ruin through rock salt water from ice-cream wagons. Dogs, on the other hand, provided a continual matter of contention. In New York City, the Board of Health had already declared in 1918 that dogs had to be controlled so that they would not "commit any nuisance upon any sidewalk of any public street." Nevertheless, dogs' use of trees as "watering posts" began to cause serious concern in the 1940s. By then, dog urine had seriously damaged London plane trees on the Upper West and East Sides, for example, along Manhattan and Morningside Avenues, and along Seventy-Seventh and Sixty-Second Streets. The only way that the city's tree men saw to alleviate the situation was to shield the trunks with metal or other types of covers.[24]

In Berlin in 1912 the Department of Gardens similarly complained of dogs injuring young plane trees. In the German capital, where dogs were explicitly allowed in unpaved areas, numerous street trees suffered urine burns that rendered the trees more subject to fungi infections and disease. In 1957 West Berlin's daily *Tagesspiegel* reported humorously that there were two and a half trees for every dog in the city: 90,258 dogs and 220,000 street trees. Some years later, the Association for the Protection of Berlin Trees drew attention to the fact that West Berlin's 93,000 dogs produced 35 million liters of urine that contributed to damaging street trees. By 1985 the topic of trees and dogs had gained enough traction for the ruling West Berlin Christian Democrats to feature a dachshund on their election campaign poster, accompanied by the slogan "Berlin is back! We let new trees grow." But the challenge of preventing dogs from using street trees as watering posts has persisted. By 1990 West Berlin counted around 100,000 dogs, which, besides liters of urine, also produced twenty tons of feces per day.[25]

However, not only dog urine threatened trees—human urine was a problem as well. Between 1996 and 2003 Berlin's Straße des 17. Juni running through Tiergarten was the venue of the Loveparade, a big techno music festival that turned the tree-covered areas on either side into large open-air restrooms, a situation heavily criticized by conservationists. A similar phenomenon had been under attack already in the second half of the nineteenth century when there were only a few public urinals. Human urine was found to be one of the damaging factors affecting Berlin's street trees, especially the older and larger trees near bars and drinking establishments. Street trees near the breweries along Schönhauser Allee, for example, were observed to be affected particularly badly.[26]

The trees themselves were often perceived by citizens as nuisances. They were considered hazards housing bees that endangered children. Caterpillars falling from trees onto passersby and people sitting on benches led to frequent complaints and sometimes even to requests to fell trees. Trees attracted roosting starlings that made noise and covered sidewalks with their droppings. When Brooklyn residents complained, the Department of Parks inquired at the Zoological Garden for advice on "bird psychology" and how the birds might be prevented from roosting in the trees. On the other hand, as shown for New York City in chapter 3, trees were considered a way to attract birds to the city in an effort to protect both birds and trees. The city followed a movement that had begun in German cities. In Berlin, the distribution of birdbaths and nesting boxes had been undertaken as early as the 1880s, and by the 1920s nesting boxes, Berlepsch bird feeders, and birdbaths had been distributed in many parks. The belief that these measures would not only protect and increase bird life but also encourage human convalescence in hospitals led the city's Commission for Nature Protection to suggest to the Department of Public Health the installation of birdbaths, feeders, and nesting boxes in hospital gardens. In 1929 the Magistrate advised the district Departments of Gardens to undertake all available measures to promote bird life in the city, which had suffered bird and egg hunting by children and teenagers; an increase of cats; and the modernization of parks and gardens that had purportedly led to the removal of shrubs. For the purpose, it was proposed that side and minor streets be planted with globular varieties of maple, locust, and red hawthorn as their dense canopies provided nesting opportunities for songbirds.[27]

Trees also provided more intangible and ephemeral nuisances. The fruit of female ginkgo trees smelled, and in Germany tree experts warned that the fine hairs on the underside of plane trees' leaves, released during pruning activities, could cause respiratory allergies. This observation had been publicized already around the turn of the century and received renewed attention in the early 1950s.[28]

By far the largest tree concern in New York City and Berlin, however, was security and public safety. Street trees could obstruct views and light, and falling branches could kill people. While business owners complained about their shop windows or advertising being obstructed from view, residents complained about street trees obstructing daylight penetration into their apartments and covering street lighting at nighttime. The former was a particularly common resident complaint, but in Berlin during the postwar years it also became an argument used by the city to explain why it was removing densely planted trees: removing tree cover and enabling light penetration into apartments would save energy. Citizen complaints regarding streetlight obstruction also involved conflicts between different municipal departments. New York residents often complained because they felt that trees obstructing lighting rendered their streets unsafe and prone to burglaries, and turned them into potential crime scenes. On occasion the Department of Parks could show, however, that the issue at stake was not, or not only, lack of street tree pruning but lack of sufficient lighting altogether.[29]

Obstruction of traffic signals was also bemoaned in Berlin. In an effort to ensure road safety, the various affected city departments communicated about the issue directly. More than in New York City, in Berlin, the pruning of trees was undertaken to ensure the safety of traffic. But on occasion, other security concerns also required attention. For example, in 1949, when street lighting had not yet been restored to its "peacetime density," the Berlin police considered the lack of lighting along Sonnenallee a threat to public safety as it could facilitate muggings. The police requested that the Department of Gardens fell the street trees adjacent to lanterns and prune all others. In more ways than one, therefore, throughout history it has been important to see both the forest and the trees.[30]

## Seeing Trees

Street tree planting in New York City is the subject of the first part of this book. Four chapters tell the story of the city's street trees. Despite—or perhaps rather because of—its unique character, the practices New York City adopted were observed by many cities in the United States and abroad, like Cleveland and Madrid. Chapters 1 and 2 uncover the roles of New York City's urban government, philanthropists, businessmen, social and public health reformers, tree experts, and design professionals in the struggle to establish city forestry in the United States. They show how in this process trees became one of the building materials of the Taylorized American city, governed and administered according to the principles of Frederick Winslow Taylor: street tree planting contributed to the establishment of this system of comprehensive scientific urban management based upon streamlined processes.[31]

Chapters 3 and 4 reveal how women and African Americans respectively used street tree planting to (re)claim their right to the city. Chapter 3, "Tree Ladies," draws attention to the activities of female New York philanthropists, social reformers, activists, and early design professionals who played an important role in urban tree planting and preservation. Concentrating on the period between 1900 and 1930, this chapter shows how New York women embraced street trees as a means and symbol of empowerment, emancipation, and even resistance. Initiating street tree–planting campaigns, the women transgressed the separation of private and public spheres, and the binary of male-coded architecture and female-coded nature. Street trees both marked and created elite and female space in the city (figure I.4).

FIGURE I.4.
Female street tree workers in Berlin during World War I, ca. 1916.
(Photo: akg-images, 558798)

Some decades later, in the 1960s and 1970s, street tree planting became a means for African American citizens in the rundown neighborhoods of Harlem and Bedford-Stuyvesant to claim their right to the city. Chapter 4, "Planting Civil Rights," focuses on the activities of the Neighborhood Tree Corps, a community tree-planting and environmental education program in Bedford-Stuyvesant that built identity, community pride, and opportunities for children's nature education. The chapter situates the events not only in a citywide and national context, where similar initiatives abounded at the time, but also within contemporary discourses in American sociology and in the nascent field of environmental psychology. As the chapter shows, street tree planting filled the visual and phenomenological gap between top-down policies and actual experience in rundown neighborhoods. It was a means and method of encouraging both public and private activism in the spirit of civil rights.

The second part of the book is dedicated to street trees in Berlin. Chapter 5, "Burning Trees," tells the story of how during and after World War II Berlin's street trees were used in political diplomacy and warfare, providing physical protection, material resources, rhetorical instruments, and symbols of the city's resurrection and rebirth. In the postwar years, street trees offered protection against the cold in winter and against the dust clouds of the rubble and ruins in summer, and they became a fulcrum of contention in some of the first ideological battles of the Cold War during the Berlin Blockade. Street trees were important elements not only in the mitigation of the local urban climate and in the early scientific endeavors of urban climatology, but also in the global political climate at the time.

The sixth chapter, "Greening Trees," focuses on the competition between East and West Berlin in reestablishing and restoring Berlin's green heritage in the postwar years. It shows how interest in the various functions of street trees peaked in both parts of the city at this time and uncovers the various means, methods, and ideologies that the two city governments employed in their urban greening efforts and their legitimization. Research projects were undertaken in both parts of the city to scientifically prove previous pseudoscientific assumptions regarding the health and aesthetic functions of street trees, for example, their capacity to buffer sound, a function that was increasingly important given the growing automobile traffic. Seeking to base their practice in Marxist theory and socialist ideology, East Berlin and East German landscape architects emphasized street trees' materialist functions and their efficient production, planting, and maintenance in "socialist city greening." In West Berlin, as in other parts of West Germany, the opportunities offered by reconstruction inspired landscape architects to discuss the street tree aesthetic and consider alternate street tree–planting paradigms. These recalled prewar and even late nineteenth-century aesthetic ideals that often built on nationalist-conservative sentiment.

Chapter 7, "Shades of Red," explores the relationship between top-down and bottom-up initiatives for the protection and management of street trees in East and West Berlin in the 1970s and 1980s. At the time, street trees were dying because of increasing traffic and the use of road salts. For East and West Berliners alike, street tree planting, care, and protection was a means to oppose the state and its environmental politics. The new genre of street tree art and various citizen initiatives for street tree protection in East and West Berlin forced the city governments to more effectively care for their street trees. Experts devised methods for the value and risk assessment of trees, and they used aerial infrared photography to facilitate the assessment of tree damage and disease as well as to monitor trees' spatial distribution and temporal development. Seeing the larger picture from above helped to identify, address, and ultimately remedy some of the problems on the ground.

Chapter 8, "Unity and Variety," shows how street trees were used after the fall of the Berlin wall to unify a heterogeneous urban fabric, level out an uneven tree canopy throughout the city, and reestablish Berlin's image as a green linden city. Tree planting was a displacement activity, an activity that sought to heal the fissures of the German psyche as well as the physical fissures in the urban fabric. It was an activity filled with the desire to make Berlin whole again and to overcome the different cultural traditions, frameworks, and collective memories in East and West Germany. Throughout their history, the linden trees of Berlin's iconic boulevard Unter den Linden have played a decisive role in the city's various unification efforts. Their inherent malleable nature has turned the famous boulevard into one of the longest-lasting elements of Berlin's urban design. With German unification, however, the municipal government of street trees also began to erode, so that by 2017 the city's street tree legacy in part depended on public-private partnerships.

The epilogue picks up on the role of public government and private initiative in street tree planting in New York City and Berlin. It offers an outlook on the street tree landscape of the future, which, more than in the past, will have to come to terms with climate change.

Throughout the twentieth century, New York City and Berlin, like many other cities, have searched for the ideal street tree, a search that continues to this day. And as the chapters in this book show, not only do trees change—we change along with them.

# PART ONE
# New York City

# 1
# Tree Doctor vs. Tree Butcher
*Standardized Trees and the Taylorization of New York City*

I n December 1897, the *New York Times* reported an incident in Brooklyn that involved a shade tree, its three protectors, and "the hirelings of a corporation." The location was Brooklyn's twenty-ninth ward, once the location of the village of Parkville. Only a few weeks after the incident, Brooklyn was to become part of one of New York City's five new boroughs. The scene described by the *Times* was one of war, a battle between the three members of the Matthews family and the workers of a telephone company. The object of contention was a shade tree standing in front of the Matthews family home that "was threatened with the axe to make way for telephone wires." The *Times* humorously reported: "On Sunday morning Mr. Matthews arose to find that three holes for telephone poles had been dug in front of his premises. He objected to the proceeding and jumped into one of the holes. . . . The son Leo just then came charging down the path from the house with a revolver in his hand. He gave it to the father and intrenched himself in one of the other holes. Mr. Matthews waved the weapon, promising faithfully to kill all comers until Mrs. Matthews made a sortie from the house and joined the line of battle after successfully taking the third excavation." According to the report, all policemen and laborers in the ward "turned out to assault the Matthewses' position, which was accomplished after a great strategic display." The family was reported to have retreated in an orderly fashion, "gathering around the big shade tree," declaring that they would "never see the tree go down" but "die fighting in the shade of its branches first."[1]

The Matthews family was quite literally up in arms and down in the hole, trying to defend their shade tree against the "tree butchers," the laborers of the telephone company that was threatening to cut the tree to make room for its wires. The conservation of the Brooklyn shade tree stood against innovation and modernization. Alexander Graham Bell had placed the first New York to Chicago call in October 1892, only five years before the reported Brooklyn battle. Once Bell's patents expired in 1893

FIGURE 1.1.
Street tree pruning in the Bronx,
ca. 1913. (New York City Parks
Photo Archive, ANR734)

and 1894, a telecommunications boom gripped the country. The new telephone companies that entered a free telecommunications market expanded the network and significantly changed the streetscape of large cities and small towns through the addition
of myriad overhead wires.[2]

The conflict relayed in the *New York Times* was typical for a time when the city
was expanding, engulfing and destroying fields, forests, wildlife, and trees in the process. Telegraph and telephone lines as well as electrical cables disturbed trees above-
ground (figure 1.1). After 1884, when state legislation had begun to force such wires
underground in New York City and Brooklyn, street trees were cut down to lay subterranean conduits. In addition, only a few trees survived the construction of the subway.
Quite literally, then, street trees and modern facilities stood in competition for space
and resources.[3]

Street trees have always stood in conflict with a variety of actors and agents,
human and otherwise; they have always been objects of contention and marked bat-
tlegrounds, often quite literally, as in the Brooklyn incident. In the battle over space
and resources, trees have become urbanized and the city has become naturalized.
Although street trees have consistently been signifiers of nature in the city—and, as
such, have often been considered as standing in opposition to culture and artifice—
they have also stood for the fusion of nature and culture as well as for the entanglement of nature, culture, science, and technology. The term *arboriculture* itself, which
by the early twentieth century had come to be used in English to describe the planting

and maintenance of street trees, encapsulates this fusion of nature and culture by means of science and technology.[4]

Although trees as living beings have always been understood to stand in contrast to the inert hard surfaces of the city's urban fabric, they became one of the building materials of the Taylorized American city. Their introduction along urban streets contributed to the establishment of scientific management, leading in some cases even to the application of principles of scientific forestry to what since the 1960s has come to be called the urban forest. Through various practices, including planting and pruning, that were used by the new tree wardens, shade tree commissioners, city foresters, arboriculturists, tree doctors, and tree surgeons, trees were standardized, domesticated, and quite literally pressed into shape and "straightened," the term for pruning activities that remedied the effects of storms and hurricanes. While the planting of street trees gave rise to a new urban professional—the city forester—it also led to various tensions: between forestry and landscape professionals; between urban governments and the public; between living trees and the inert hard surfaces, materials, and technologies of the city; and between the utility and aesthetics of trees.[5]

## The Heat of Treeless Streets

By the middle of the nineteenth century, New York City's urban center lacked trees. Urban development had created treeless streets that appeared to some as an unvaried sterile environment in which "the uncomely rock" was exposed "to the gaze of every passer." As early as 1822, the *New-York Evening Post for the Country* had argued for more trees in the city, singling out the "public walk running easterly down Chamber-street, north of the City Hall" and "the Battery, which presents such a beautiful prospect to the passenger entering the harbor from sea" as the first two thoroughfares that should be planted with "trees of lofty growth." Yet the newspaper's opening optimism was soon dampened. Street tree planting was a private affair, and the newspaper had little faith in the city's inhabitants. It concluded, "What is every body's business is nobody's, and though [tree planting] is admitted to be very desirable by every body, yet we have reason to fear it will not be undertaken by any body." Some fifty years later, at a time when the nation's capital was embarking on its first systematic tree-planting efforts, the nation's largest city was criticized as the metropolis with the fewest street trees. So few trees were lining New York City streets—mostly ailanthus, elms, and buttonwoods—that they could almost all be individually listed. As landscape architect John Y. Culyer commented in the early twentieth century, trees downtown had "met an unfortunate fate, . . . the expansion of our business enterprise and the demands of a rapidly increasing population."[6]

To instigate change, it took crises that challenged human survival itself. The increased attention to urban trees was triggered by two related developments: increasing urbanization and related health problems on the one hand, and deforestation on the other. Between 1832 and 1866, repeated cholera outbreaks killed altogether 12,230

FIGURE 1.2.
Arbor Day in Lower Manhattan, 1904. The tree planting in front of Public School
No. 177 was sponsored by the Tenement Shade Tree Committee.
(*Annual Report of the Tree Planting Association of New York City, February 1905*
[New York: Gillis, 1905]. General Research Division, The New York Public Library,
Astor, Lenox and Tilden Foundations.)

people in New York City. By the turn of the twentieth century, living conditions had
deteriorated further due to overcrowding and poor sanitary and housing conditions.
Outside of cities, the country's timber resources—the extensive forests stretching west—
were being depleted and it became clear that they were not infinite.

Forestry and what became known around the turn of the century as city forestry
were closely connected. Rural tree-planting and conservation efforts were extended into
the cities, and foresters became concerned about urban trees. Arbor Day, first established in 1872 by J. Sterling Morton as a response to wide-ranging deforestation, soon
became an instrument to reafforest not only the rural plains and mountainsides but also
city streets. For example, on Arbor Day in 1904 the New York City Tree Planting Association's Tenement Shade Tree Committee supported the planting of thirteen trees in
front of Public School No. 177 at the corner of Market and Monroe Streets (figure 1.2).
In Brooklyn in 1911 the American Association for the Planting and Preservation of City
Trees used Arbor Day to promote children's tree clubs and their care for street trees.

Although his proper domain was conservation of the country's vast forested lands,
Gifford Pinchot, chief of the U.S. Forest Service, was often cited in these early urban
tree-planting efforts. Having lived in the city, he and his wife were two of the first
members of the Tree Planting Association of New York City, founded in 1897. By 1907
Pinchot had become one of its eight trustees.[7]

But despite these close connections between scientific and city forestry, their objectives were very different. Scientific forestry, as developed since the seventeenth century in France and in the German states, was geared toward the production of timber. Although monoculture and the depletion of timber resources in the late nineteenth century led German foresters to pay greater attention to forest ecology and aesthetics, their aim continued to be a sustainable timber yield. Street trees, in contrast, were by this time planted in cities first and foremost because of their climatic, sanitary, and aesthetic functions. They were both "shade trees" that provided cooling effects in the summer heat and "ornamental trees" that were pleasing to look at and thought to reduce modern nervousness in the process. Street trees could enhance urban space and its aesthetic appearance. They brought nature into the city and were, as Pinchot put it, "the only form through which the residents of the city can come in daily contact with nature as we know it in the woods and fields." In all these different capacities, trees were thought to promote public health.[8]

Public health reformers like New York's eminent physician Stephen Smith, who in 1911 at the age of eighty-eight became president of the city's Tree Planting Association, argued that trees could save lives. Smith, who pioneered New York City's sanitary reforms and founded the Metropolitan Board of Health and later the American Public Health Association, was the author of a groundbreaking study that correlated children's mortality from "diarrheal diseases" with high temperatures. The study made the rounds in the 1870s and continued to be referenced into the early twentieth century, for example, by tree experts William F. Fox and William Solotaroff. On the basis of his data collection from the early 1870s, Smith found a clear increase in child mortality between the months of February and July. In general, the death rate in the summer months, July, August, and September, was the highest. Smith's data suggested further that a correlation existed between the highest average weekly temperatures and the highest death rates. For example, in 1872, 618 children under five years died from diarrheal diseases in the week ending 6 July. In contrast, Smith counted only 17 children who died in the week ending 3 February, the week with the lowest average temperature that year. From his statistics, the physician concluded that climate mitigation in the city could save five thousand to eight thousand lives per year. To "equalize the temperature" in the city and "purify its atmosphere," Smith recommended planting street trees. While trees were, as scientists and health reformers did not fail to point out repeatedly in the nineteenth century, "evaporating machines" that had a cooling effect in the summer and supposedly a warming effect in the winter, supporters of the miasma theory also thought that trees could disinfect the air and neutralize malaria. They believed that trees absorbed poisonous gases and produced "health-giving oxygen." Thus, "trees in a crowded city [were] a self-acting sanitarium."[9]

To illustrate the enormous effectiveness attributed to trees' climatological and cooling functions, Smith drew attention to what became known as the Washington elm study. Subsequently referenced repeatedly in the popular press, the study had first been mentioned by Harvard professor of botany Asa Gray in his 1857 *First Lessons in Botany and Vegetable Physiology* and was attributed in later publications to Harvard

FIGURE 1.3.
The Washington elm on the Cambridge Common, 1875. (Photo by J. W. Black and
Company. Houghton Library, Harvard University, MS Am 2861)

College mathematics professor Benjamin Peirce. His calculations of the surface area
of tree foliage of the famous Washington elm standing on the Cambridge Common
found that it had produced two hundred thousand square feet, or five acres, on the
basis of an estimated crop of 7 million leaves (figure 1.3). These numbers helped to
substantiate and rationalize the shade and cooling effect of trees that individuals commonly perceived only subjectively. Now scientific data also proved that trees could
provide enormous relief.[10]

## Fighting for Trees

Tirelessly working for the improvement of public health in his city, in 1873 Smith
drafted and introduced a bill into the legislature for the cultivation of street trees. He
hoped that a more uniform and systematic treatment of street tree planting would

improve health conditions, particularly in the tenement districts. He also hoped that shaded streets would improve the city's overall appearance. But the bill stalled, and it took three more attempts and significant amendments before Mayor Low finally passed it in 1902. The amendments, however, turned the requirements to establish a Bureau of Forestry, appoint a city forester, and provide a budget for the planting and care of street trees into mere recommendations. Thus the bill was rendered almost useless to the Department of Parks.[11]

Without the appropriate personnel and budget, tree planting continued to remain a largely private affair in New York City throughout the first decades of the twentieth century. Opinions on this matter had always varied. Many followed Smith and tree experts like William Solotaroff and William F. Fox who argued that "the planting and care of street trees belong to the City Government as much as street paving." Others saw property owners' rights curtailed. Brooklyn engineer and multimillionaire Samuel B. Duryea, who in 1882 would be among the incorporators and benefactors of the Tree Planting and Fountain Society of Brooklyn and became its president in 1890, had held in 1881 that a property owner should have the right to decide whether he wanted to plant a tree in front of his property and which species. But in the minds of City Beautiful advocates and Progressive Era social and public health reformers, New York City lagged behind cities like Minneapolis, St. Louis, Buffalo, and Newark that had appointed and empowered commissions or special bureaus to plant and care for street trees. In New York City, therefore, private enterprise had to make amends for the missed opportunity.[12]

While his bill was stalled in the legislature, Stephen Smith became one of the first members of the Tree Planting Association of New York City. In 1896 a group of men led by Colonel Cornelius B. Mitchell had resolved to found an association to promote street tree planting, care, and protection. They were inspired on the one hand by the consolidation of Greater New York and the civic pride it engendered. On the other hand, they were motivated by the dearth of street trees and the poor condition of those that existed. When the association was incorporated in January of the following year, its first list of members read like a New York who's who. Besides Smith, it included philanthropist and housing reformer Robert de Forest; art dealer Samuel P. Avery; painter Lockwood de Forest; sculptor Augustus Saint-Gaudens; industrialist and former mayor Edward Cooper; churchmen William Wilmerding Moir and Henry C. Potter; U.S. Navy captain A. T. Mahan; chief justice of the New York Court of Common Pleas Charles P. Daly; and financiers J. Pierpoint Morgan, W. Bayard Cutting, and William Collins Whitney. Forester Gifford Pinchot and Samuel Parsons Jr., landscape architect in the Department of Parks, as well as landscape architect John Y. Culyer, who served as the association's first secretary and advisory forester, provided the organization with professional legitimacy. Of the 173 initial members, 28 were women. They included wives of the New York merchant aristocracy as well as women like Margaret Blaine Damrosch, wife of the famous conductor and composer Walter Damrosch.

The Tree Planting Association produced and circulated pamphlets on tree plant-ing and care, encouraged Arbor Day celebrations, raised funds for planting trees to serve as object lessons, and lobbied for legislation promoting and protecting street trees. In its attempt to raise general awareness, in the first decade of the twentieth cen-tury the association expanded its range of activities to include assistance to the police department in the prosecution of tree vandalism, and advocacy for tree and forest conservation not only within the city but throughout the nation. But in its first years, it was its assistance to residents interested in tree planting and the work of the associ-ation's Tenement Shade Tree Committee, founded in 1902, that was most in evidence (see chapter 4). First president Cornelius B. Mitchell reported that 2,176 trees had been planted under the auspices of the association in 1901–2, and once the Tenement Shade Tree Committee had begun its work, more street trees were planted along ten-ement blocks and in front of public schools.[13]

The Tree Planting Association's relationship with the Department of Parks turned out to be rather complicated, leading to much frustration and contention, for example, with regards to the choice of tree species. While the Department of Parks criticized the association's decision to plant the fast-growing, fragile, care-intensive Carolina poplar as short-sighted and little conducive to the growth of a long-lasting canopy, the association accused the city of neglecting street trees in general. They were neither planted in sufficient numbers, nor were they cared for appropriately.[14]

Indeed, good intentions were seldom carried through. For example, when the city was making plans to widen Delancey Street approaching the new Williamsburg Bridge, the Tree Planting Association in 1905–6 submitted to the Department of Parks planting designs prepared by its advisory landscape architect and forester John Y. Culyer (figure 1.4). Although the submissions were encouraged and approved, a few years later, the situation was far from what Culyer had envisioned. Instead of trees lining the sidewalks and a double row of trees enclosing a central streetcar line, the only trees planted in two rows above what by then had become a subway line were struggling (figure 1.5). The poplars had been planted on top of the new subway along the center of the street in 1912, but the four to six feet of soil between street level and the subway tunnel was too shallow a depth for the poplars to thrive. In 1914 the Depart-ment of Parks therefore had to remove fifty-two trees and plant a hundred new ones.[15]

The department's relative inertia on the one hand and the city's continual devel-opment on the other inspired the ceaseless, energetic Smith to enlarge the Tree Plant-ing Association's scope once he had assumed its presidency in 1911. The organization moved its headquarters to a new office further downtown in Manhattan and adapted to the needs of Greater New York by nominating vice presidents for all five boroughs. Smith's presidency marked a further shift. In his forceful declaration of intent, he envisioned a citywide tree survey, for which he finally sought assistance from the New York State College of Forestry at Syracuse University. Against all odds and despite the lack of governmental cooperation, in the summer of 1914 Henry R. Francis, who had joined the College of Forestry as assistant professor of landscape extension only the year before, began the street tree survey for the Tree Planting Association.[16]

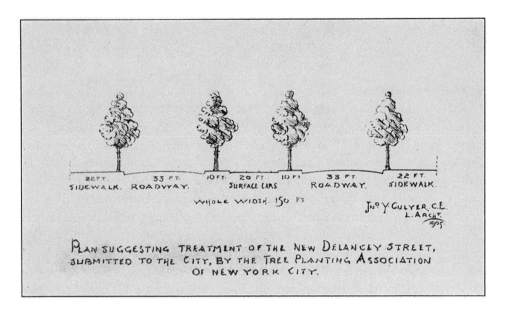

FIGURE 1.4.
Envisioned street tree planting along Delancey Street, 1905. Section drawing by
John Y. Culyer for the Tree Planting Association of New York City. (*Annual Report of
the Tree Planting Association of New York City, 1905.* General Research Division,
The New York Public Library, Astor, Lenox and Tilden Foundations)

FIGURE 1.5.
Delancey Street Parkway (with school garden in the foreground) shortly after the
poplars were planted along the median, ca. 1912. (New York City Parks Photo
Archive, ANR703)

## Saving Trees

Sent to New York City by the dean of his college and hosted by the Tree Planting Association, Francis set up headquarters in the American Museum of Natural History and went to work. His findings were excruciating. Many trees were in bad shape. They showed large wounds that had not been treated correctly and concrete fillings that were damaged or crumbling. Unprotected by tree guards, many trees had been de-barked by horses that had been hitched to their trunks (figures 1.6 and 1.7). They were growing in inadequate openings of the concrete paving and suffering from under-ground gas leaks. Trees were also unevenly distributed throughout the city, and where they were planted at all, they often lacked a consistent and uniform planting pattern as most planting was done by private property owners. Given the lack of a systematic plan, streets were planted with different, often ill-adapted species and with "little or no regard to the width of streets, nature of soil, nor to the artistic effect of certain trees in relation to the height of buildings upon those streets." New York's unsystematic and irregular tree-planting efforts had led to "complete denudation of large areas since trees have been removed continuously and none have been planted in replacement." The fear of a treeless city was not unrealistic if no significant changes in the planting, care, and management of street trees were undertaken. The Department of Parks' statistics testified to this observation. In Manhattan in 1908 and 1909, 535 trees were removed and none were planted; in Richmond, 755 trees were removed and only 62

TREE USED AS A HITCHING POST, ALBANY, N. Y.

AN UNPROTECTED TREE.
BARK GNAWED BY HORSES.

FIGURE 1.6.
Trees debarked and damaged by horses. (William F. Fox, *Tree Planting on Streets and Highways* [Albany: J. B. Lyon, 1903])

FIGURE 1.7.
Warning against the misuse of street trees as hitching posts, ca. 1914. (Published in a
brochure by the Shade Tree Commission of the City of Newark, New Jersey, and in
*A Plea for Street Trees in the Borough of the Bronx* [New York: J. J. Little and Ives, 1916])

were planted; in Queens and Brooklyn, 2,837 trees were taken down and 90 planted;
data for the Bronx was lacking.[17]

According to Francis, the only remedy was the establishment of a central Bureau of Tree Culture and the employment of a city forester in each borough who
would organize and supervise all tree work undertaken by a group of trained arboriculturists, foremen, and workmen. Each city forester would conduct and keep a tree
census, specify the material for a central municipal nursery, collaborate with the borough's engineering department, issue permits, consider complaints, determine work
methods, and control contractual work and specifications. Francis hoped that this centralized and coordinated management could put "all matters pertaining to the vegetation of the City of Greater New York for the first time on a thoroughly systematic and
businesslike basis, and the transformation of the present unsatisfactory and expensive
tree situation into one of economy and systematic beautification."[18]

Francis's work received wide-ranging attention and praise in the press, and the
Department of Parks could not fail to notice. The president of the Park Board, the
Honorable Cabot Ward, became interested in the matter and requested further studies and surveys from the College of Forestry the following year, a project that would
be funded by John D. Rockefeller. In summer 1915 Francis was pursuing fieldwork
elsewhere, so the task was passed on to Laurie Davidson Cox, who had joined the
faculty only a year after Francis, in 1914.[19]

In New York City, Cox refined the overview and general recommendations
made by his colleague the year before. Focusing on Manhattan, where dense and tall
buildings that blocked light and sun created exceptionally adverse conditions for tree
growth, Cox developed a plan for a street tree system accommodating twenty-five
thousand street trees. According to his survey, which covered the island south of 110th
Street, fifteen thousand of these trees were already planted (plate 2). Cox provided

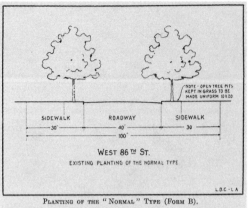

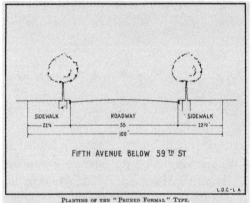

FIGURE 1.8.
Two different prototypes for street
tree planting in Manhattan, 1915,
proposed by Laurie D. Cox.
(Laurie Davidson Cox, *A Street
Tree System for New York City,
Borough of Manhattan: Report to
Honorable Cabot Ward, Com-
missioner of Parks, Boroughs of
Manhattan and Richmond, New
York City* [Syracuse: Syracuse
University, 1916])

specifications for tree pits, guards, and stakes as well as a prototypical planting plan determining that the ideal distance between all species of trees in all locations was fifty feet. He developed eight different prototypes of tree planting that could be applied to the streets throughout the borough (figure 1.8). His proposals did not shy away from tackling some of the densest and busiest areas of the city, including Fifth Avenue below Fifty-Ninth Street; Second Avenue between Seventh and Fifteenth Streets; West Eleventh and Twenty-Third Streets; East Thirty-Sixth Street; and Tenth Avenue between Fifty-Seventh and Seventy-Second Streets. Along these streets he proposed the "Pruned Formal Type," a variation of the "Normal Type" consisting of rows of trees along the sidewalk. Inspired by examples in European cities, street trees of the "Pruned Formal Type" were kept small and compact through regular pruning, and their aesthetic was thought to add "dignity to the architecture of the street," which

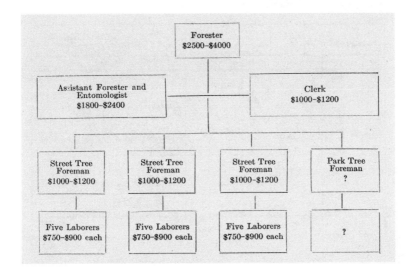

FIGURE 1.9.
Cox's proposal for the organization of a Bureau of City Forestry, 1915. (Laurie Davidson Cox,
*A Street Tree System for New York City, Borough of Manhattan: Report to Honorable Cabot Ward,
Commissioner of Parks, Boroughs of Manhattan and Richmond, New York City*
[Syracuse: Syracuse University, 1916])

Cox considered especially appropriate on streets like Fifth Avenue. The trees regarded as the best adapted to conditions in Manhattan were Oriental plane, ginkgo, European linden, Norway maple, pin oak, Carolina poplar, and, in locations where no other tree would grow, the tree of heaven, or ailanthus.[20]

To realize his street tree system, an appropriate management system had to be instituted. Building upon Francis's suggestions the year before, Cox recommended the creation of the Bureau of City Forestry in the Department of Parks led by a city forester (figure 1.9), advice that would be followed in 1917. At the end of the summer he provided the president of the Park Board with prototypical street tree–planting plans for various locations throughout the borough, with specifications for planting and management and a budget for the realization of his proposed street tree system. Cox also delivered a street tree census for most of the borough that he had already partially transcribed to a card system specifically developed for Manhattan.

Cox adapted the street tree census suggested by his colleague William Solotaroff to the conditions in New York City. Solotaroff had proposed a combined method of field books, maps, and index cards to record the location, species, diameter, planting date, size when planted, nursery of origin, life history, and condition of existing and newly planted trees. Whereas Solotaroff had suggested one card for every tree, New York City's size and its density of buildings made it more practical to list the data for all trees of one block on one card (figure 1.10). Cox also added information on whether trees were equipped with guards, grates, and stakes.[21]

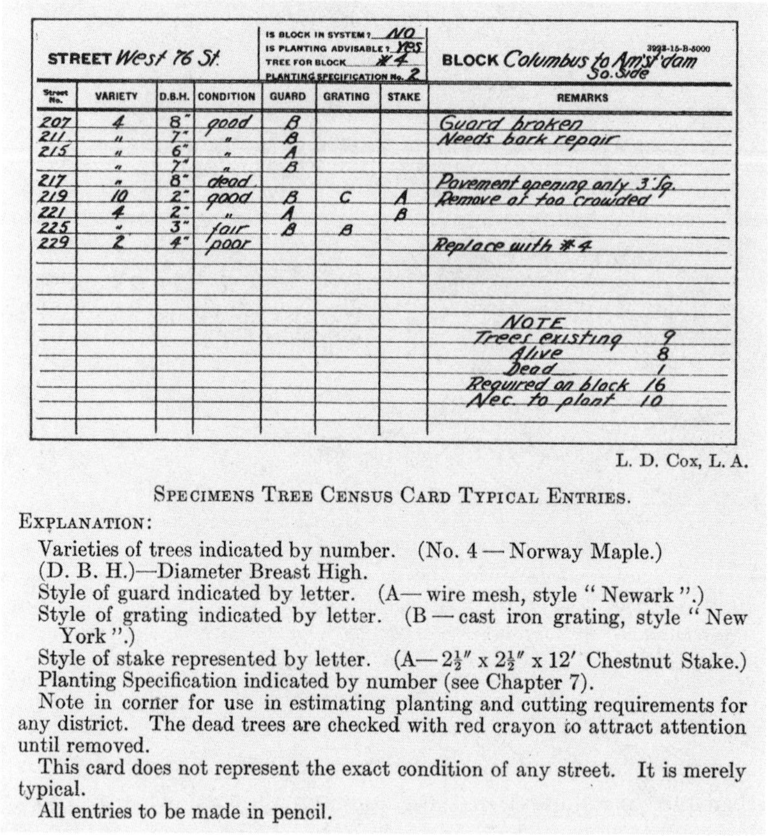

| STREET West 76 St. | | | | IS BLOCK IN SYSTEM? NO / IS PLANTING ADVISABLE? YES / TREE FOR BLOCK #4 / PLANTING SPECIFICATION No. 2 | | | BLOCK Columbus to Am'st'dam So. Side    3093-15-B-8000 |
|---|---|---|---|---|---|---|---|
| Street No. | VARIETY | D.B.H. | CONDITION | GUARD | GRATING | STAKE | REMARKS |
| 207 | 4 | 8" | good | B | | | Guard broken |
| 211 | " | 7" | " | B | | | Needs bark repair |
| 215 | " | 6" | " | A | | | |
| | " | 7" | | B | | | |
| 217 | " | 8" | dead | | | | Pavement opening only 3 sq. |
| 219 | 10 | 2" | good | B | C | A | Remove at too crowded |
| 221 | 4 | 2" | " | A | | B | |
| 225 | " | 3" | fair | B | B | | |
| 229 | 2 | 4" | poor | | | | Replace with #4 |

NOTE
Trees existing    9
Alive    8
Dead    1
Required on block    16
Nec. to plant    10

L. D. Cox, L. A.

SPECIMENS TREE CENSUS CARD TYPICAL ENTRIES.

EXPLANATION:

Varieties of trees indicated by number. (No. 4 — Norway Maple.)
(D. B. H.)—Diameter Breast High.
Style of guard indicated by letter. (A— wire mesh, style " Newark ".)
Style of grating indicated by letter. (B — cast iron grating, style " New York ".)
Style of stake represented by letter. (A— $2\frac{1}{2}''$ x $2\frac{1}{2}''$ x 12' Chestnut Stake.)
Planting Specification indicated by number (see Chapter 7).
Note in corner for use in estimating planting and cutting requirements for any district. The dead trees are checked with red crayon to attract attention until removed.
This card does not represent the exact condition of any street. It is merely typical.
All entries to be made in pencil.

FIGURE 1.10.

Cox's proposal for an index card record system of street trees, 1915. (Laurie Davidson Cox, A Street Tree System for New York City, Borough of Manhattan: Report to Honorable Cabot Ward, Commissioner of Parks, Boroughs of Manhattan and Richmond, New York City [Syracuse: Syracuse University, 1916])

Protective guards, wires, grates, and frames were designed to limit soil compaction, enable water and air to infiltrate into the ground, and protect the tree stems from mechanical damage by horses and collisions with carriages and automobiles. With regards to the wires and frames protecting tree stems, centuries-old practices from rural environments, where wooden frames had kept grazing animals from debarking young trees, were transferred into an urban context, giving expression on a micro scale to the larger rural-urban transformations that were occurring at the time (figures 1.11 and 1.12).

For New York City's street tree system, Cox developed specifications for different guards that came with names like Newark for a wire mesh tree guard, Albany and Manhattan for two different kinds of iron tree guards, and New Haven, New York, Knickerbocker, and Manhattan for different tree grates (figure 1.13).[22]

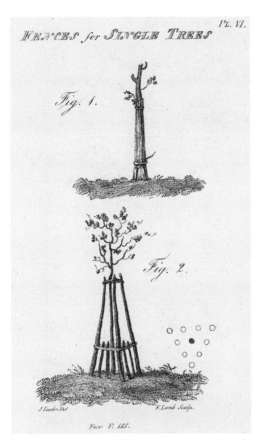

FENCES for SINGLE TREES

Fig. 1.

Fig. 2.

J. Loudon Del.

F. Lamb Sculp.

Face P. 155.

FIGURE 1.11.
Nineteenth-century tree guards used in meadows as protective measures against pasturing animals and deer. (John Claudius Loudon, *Observations on the Formation and Management of Useful and Ornamental Plantations* [Edinburgh: Archibald Constable, 1804])

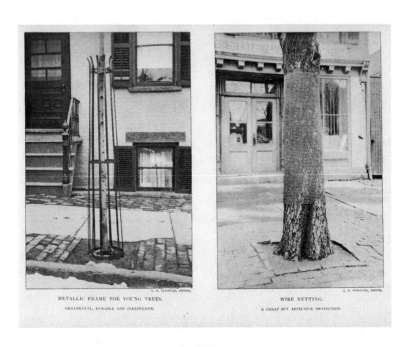

METALLIC FRAME FOR YOUNG TREES.

ORNAMENTAL, DURABLE AND INEXPENSIVE.

WIRE NETTING.

A CHEAP BUT EFFECTIVE PROTECTION.

FIGURE 1.12.
Early twentieth-century street tree guards. (William F. Fox, *Tree Planting on Streets and Highways* [Albany: J. B. Lyon, 1903])

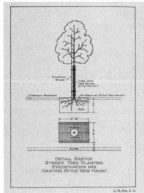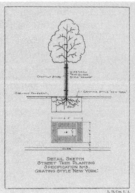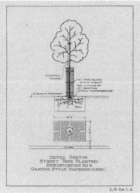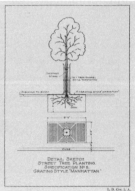

FIGURE 1.13.
The tree grates New Haven, New York, Knickerbocker, and Manhattan, developed
for street trees in Manhattan. (Laurie Davidson Cox, *A Street Tree System for
New York City, Borough of Manhattan: Report to Honorable Cabot Ward,
Commissioner of Parks, Boroughs of Manhattan and Richmond, New York City*
[Syracuse: Syracuse University, 1916])

## A New Profession?

It became clear that besides tree laws, ordinances, surveys, planting specifications, and standardized protective elements, systematic street tree management and care required trained professionals and workers. City forestry included elements of city planning, landscape architecture, and forestry, three fields that were in the process of professionalization at the time. Landscape architects were heavily involved in the new fields of city planning and city forestry, and many early city foresters quickly became members of the early planning organizations. But despite these personal and professional overlaps, Cox voiced concern that city foresters often neglected design, engineering, and administration, concentrating instead on the horticultural and entomological issues of tree planting and care. He was drawing attention to one of the struggles affecting early city forestry: its formation as a profession, a process that went hand in hand with the systematization of street tree management. As the founder of the nation's pioneering four-year professional program in city forestry that provided the foundation for Syracuse University's later landscape architecture program, Cox played an important role in this development.[23]

The city forestry program offered in the Department of Landscape Engineering at the New York State College of Forestry at Syracuse University beginning in 1914 was an attempt to contribute to forming a new discipline and profession and addressing the noticeable lack of urban professionals educated in the new field. The course was aimed at training men for public service and in "the expert arboricultural knowledge of the forester and the artistic appreciation of the landscape architect." To this end, the curriculum included courses in plant materials, arboriculture, landscape construction and design, and city planning.[24]

Although the program addressed the higher education of future city foresters and Cox admitted in 1932 that much had been learned, at the Eighth National Shade Tree Conference he also deplored that "the art has not made notable progress. . . . On . . . average there is probably no better street tree planting being done today either in this country or abroad than was being done 50 or 100 years ago. . . . In spite of all our publicity and apparent public interest, there is no division of our city government where money is spent more niggardly." At the Syracuse University College of Forestry only three students graduated with a master's degree in city forestry in 1916 and 1917, and this advanced degree was no longer listed after 1931. In short, city forestry and city foresters were underappreciated, and while advanced academic training came to appear unnecessary for city foresters, the arboricultural training of their foremen and workmen was still unresolved.[25]

Outside of the university, this situation was caused partly by the fact that the boundaries between theory and practice in street tree management had remained unclear and confused. Despite the city forestry program at Syracuse, tree experts and workmen still lacked certified titles and formal training. Already in the early twentieth century, some considered the title "city forester" inappropriate, as the "esthetical nature" of dealing with street trees had nothing to do "with the care and rational treatment of . . . forest areas." William Solotaroff, the New Jersey authority on city forestry sought out by the cities of Harrisburg, Pittsburgh, Chicago, and New Orleans to consult on how to place street trees under municipal control, explained further that the use of "city forester" was unsuitable because it implied that the city was growing trees for lumber. Although the title was applied in cities throughout the nation in the early twentieth century, replacing "tree warden," a term introduced by an 1899 Massachusetts state law, "arboriculturist" appeared to many in the early twentieth century a more appropriate description.[26]

But unease about naming remained, leading Wayne C. Holsworth, who graduated from Harvard's School of Landscape Architecture in 1921 with a thesis entitled "Street Tree Planting in Its Relation to City Planning," to argue that "street tree commissioner" might be the most appropriate term as neither was he a forester nor did he deal exclusively with the care of individual trees like the arboriculturist. At the National Shade Tree Conferences in the 1930s discussions still revolved around questions of title, standards, and certification. Many agreed that the time was ripe for a professionalization of the field that would distinguish it from commercial tree services through an ethics based upon concerns for tree life and beauty and for the societal values trees expressed. In 1936 forester and entomologist Orville W. Spicer argued for the need of legislation following the models of engineering and architecture that would finally set standards and protect a professional title to be carried only by qualified and experienced individuals. Despite these efforts, however, today urban forestry, not unlike in the early twentieth century, continues to be led predominantly by experts who, due to the lack of urban forestry degrees, hold college and university degrees in forestry, landscape architecture, horticulture, or plant science.[27]

## "Tree Butchers" or "Tree Doctors"?

Similar to the confusion revolving around the title, standards, and duties of street tree managers and "diagnosticians," the training of workmen—the "arboriculturists," "tree doctors," "tree surgeons," and "dendricians," as they were variously called in the early twentieth century—was also a fraught issue. Given that trees were neither human beings nor animals, Elbert Peets warned of the confusion embedded in terms like *tree surgery* and *tree doctoring*. It would take tree experts until the 1930s to settle on the title "arborist" for tree workers and even longer, until 1964, to found the first independent Society of Municipal Arborists. Even then, training and certification were handled unevenly across the country and in the public and private realms, with some of the big tree expert companies offering their own certificate programs. Not surprisingly, then, in 1913 the young Elbert Peets—inspired in his own work by the Arts and Crafts movement—had described arboriculture as a trade without formal training. There was, according to him, "no regular way to learn how to repair trees"; the arboriculturist was "a man who develops a 'tree sense' rapidly, masters the tricks of the cambium, and is endowed with a moderate amount of native gumption." But to leave this field of occupation to men with a "tree sense" also opened doors to those without one.[28]

The Tree Planting and Fountain Society of Brooklyn found in the 1890s that "tree butchers" and "irresponsible men" had earned money offering their services under false pretenses in the name of the society, injuring and destroying trees by pruning them badly and applying pretended insect remedies. In Chicago, city forester Jacob H. Prost warned that "tree peddlers should be held in suspicion." In Dallas, his colleague O. C. Charlton warned of "itinerant stranger[s] peddling a wagon load of trees whose roots may have been exposed for hours or days to drying by wind and sun." William F. Fox, superintendent of New York state forests, advised caution when assigning tree planting and care: "This work should not be entrusted to ignorant, inexperienced persons, as is too often the case. Men of this class frequent our cities, and solicit employment as tree pruners. With glib tongues they describe the defects, real or otherwise, in street or lawn trees, and obtain permission to do some work. As a result, beautiful specimens have been disfigured or irremediably injured." Forester Bernhard E. Fernow warned of the "quacks," and Pennsylvania State University zoologist H. A. Surface of the "tree fakers," whose prescriptions for tree care were not based on scientific knowledge and "have not proved themselves effective" but were instead harming the trees. "We are in need of tree doctors, but not of tree butchers," the F. A. Bartlett Tree Expert Company proclaimed, albeit certainly not without self-interest.[29]

The public also accused untrained men of butchering trees. A case in point was the treatment of street trees in the Bronx in 1914. A member of the Woman's Municipal League denounced the Department of Parks for unfair politics and labor practices, "making poplars look like clothes-horses and maples like elms, and leaving them bleeding in the Spring time after the sap had begun to flow." In support of the enraged lady, another city resident called attention to this "mutilation of street trees in the

Bronx by the tree butchers" of the Department of Parks. "Tree Lover" observed that limbs were cut off regardless of whether they were healthy or decayed and that the cut wounds were not treated according to the latest state of the art. It was clear that the trees were not in the hands of skilled laborers but were being "slowly slaughtered by those who should protect them." In his defense, Bronx park commissioner Thomas W. Whittle explained that the men carrying out the work came to the department via the Civil Service Commission and had to be trained on the job. Given the miles of street trees in the Bronx, it was impossible for the head gardener to personally supervise all pruning and trimming. Furthermore, Whittle drew attention to the fact that his department was not responsible for the borough's street tree legacy, which largely consisted of brittle and fragile silver maples and poplars that now required extensive pruning to maintain tree health and stability. There was no doubt, then, that attention had to be paid to the treatment and care of trees and to their quality.[30]

## Model Trees

The comprehensive street tree system that Cox proposed for New York City and that the tree-loving public was requesting was based upon regularity and uniformity. It involved the production, planting, and care of standardized street trees and systematized work processes and practices governed by a central agency. Street tree systems both resulted from and contributed to the Taylorization of the American city. It appeared that street trees had to be systematically and scientifically managed. Informal plantings by residents and private organizations were no longer sufficient. At the very least, they had to be controlled by the Department of Parks. Planting and care based upon the latest scientific knowledge and carried out with the most innovative tools and skill was to replace a practice formerly largely grounded in traditional craft knowledge.

In developing efficiency principles for street tree planting and care, both the relatively long lives and stasis of trees as well as their changeable nature and temporary existence had to be taken into account. Street trees began to be understood in this dichotomy. On the one hand, they were natural entities with their own agency and life force. On the other hand, they were controlled, shaped, and protected by humans.

Tree experts generally understood the ideal street tree, or "model tree," to satisfy three functions: beauty, comfort, and the increase of property value (figure 1.14). For these purposes, the Tree Planting Association advised that trees should be raised in nurseries, should be straight and sound stemmed, and should have a symmetrical crown and compact, fibrous roots. These recommendations reflected a standard that was shared by city foresters, shade tree commissioners, and tree experts in handbooks and pamphlets throughout the country. In the 1920s, Wayne C. Holsworth elaborated further: "From the point of view of design the ideal shade tree is one that: (1) makes a fairly dense mass of leafage early in the spring and holds this foliage late into the autumn; (2) that is straight and symmetrical; (3) that has cleanliness of habit; (4) that is perfectly hardy and not too slow a grower, so that the desired effect may not be too

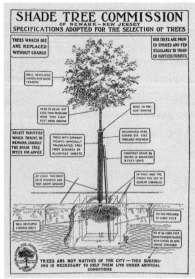 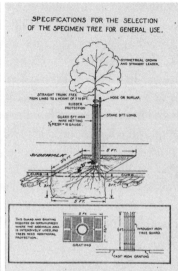 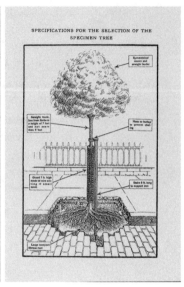

FIGURE 1.14.

The ideal street tree pictured in brochures issued by the Shade Tree Commission of New Jersey (*left*), and the New York City Department of Parks for the Boroughs of Manhattan (*center*) and Brooklyn (*right*). ("With the Vanguard," *American City* 4 [1911]: 148; Department of Parks, *Street Trees for Manhattan and Richmond* [New York: Clarence S. Nathan, n.d.]; J. J. Levison, *Improve Your Street: Plant Trees* [Brooklyn: American Association for the Planting and Preservation of City Trees, 1912])

long delayed; (5) that is long-lived, so that the effect is lasting." Furthermore, the roots of "this ideal shade tree" would not injure sidewalk pavements or interfere with water pipes. Of course, these requirements induced much skepticism at first. In 1895 a similar set of character traits had been presented by the Tree Planting and Fountain Society of Brooklyn, which commented that "one [might] well inquire whether it is possible to get a perfect tree."[31]

The definitions of an ideal street tree and of a street tree standard meant that species with character traits beyond easy human control were considered unsuitable for the urban environment. For example, William F. Fox declared the cottonwood inappropriate for cities because it shed its seed in a downy, cottony tuft that clung to the clothing of passersby, causing annoyance. The related fast-growing Carolina poplar not only was difficult to keep in shape but often turned out to be a nuisance, clogging drain tiles and sewers. Bemoaning its lack of "character or dignity," tree experts criticized the ubiquitous use of a tree species that had such a short life span. Cities throughout the country were affected by "Carolina poplaritis," as one commentator put it. To remedy the situation, Jersey City's city forester A. T. Hastings proposed the cautious removal and replacement of Carolina poplars by inter-planting existing poplar rows with other species that would eventually replace the short-lived poplars. According to Hastings, to prevent wind damage and enhance the tree's aesthetic, another way to tackle the many existing poplars standing along city streets was to trim their crowns, encouraging compact growth.[32]

Street tree advocates in the early twentieth century generally also rejected the male ailanthus despite its hardiness because of its smelly flowers. Although it is no longer planted as a street tree, opinions about the ailanthus varied over time and moved between assessments of most and least ideal street tree. There was no other tree species in New York City like the ailanthus. Originally imported to the United States in the late eighteenth century, in New York City it caused intense discussion, even inspiring an affectionate poem and the title of a famous book—Betty Smith's 1943 *A Tree Grows in Brooklyn*.[33]

In 1841 famed landscape gardener Andrew Jackson Downing gave the ailanthus his blessing as an ideal street tree, and in 1846 his colleague the horticulturist Daniel Jay Browne dubbed the species the "Metropolitan Tree," thus indicating its urban resilience. Among the reasons for this "metropolitan favorite" were its survival in "meagre and barren soil," its short leaf fall period coinciding with the first frost and, most important in Downing's view, its being "perfectly free from insects." In the late 1830s and the 1840s, its unattractiveness to insects had turned the tree of heaven into the number one street tree. In a craze of "ailanthus mania," it had been used to replace horse chestnuts and lindens with their creepy crawlers along many Manhattan streets. But the enthusiasm did not last long, even with Downing himself. In the 1850s, a debate erupted about the suitability of the species, which had recently been planted in great numbers along Manhattan streets, especially in the leafy uptown sections of the island. The discussion, which included fierce outcries against the tree as well as appeasing but definite arguments for it, was incited first by the odor of the male trees' flowers and later, in 1859, by the discovery that its bark contained poisonous oils. Troubled by the "oppressive," "nauseating," and "poisonous" male flower odor, many citizens began to consider the ailanthus a nuisance and public health hazard.[34]

Moreover, in the political climate of the 1850s, increasingly characterized by the nativist and anti-immigrant sentiment of the Know Nothing movement, criticism of the tree soon turned xenophobic. The imported species was regarded as a "filthy and worthless foreigner," a "fragrant stranger" that should be replaced with "hardy, sweet and indigenous" species. Downing himself contributed to this debate in a posthumously published editorial in his journal the *Horticulturalist* with the warrior cry "Down with the Ailanthus!" Caught by the nativist wave in "patriotic objection," Downing now saw in the ailanthus "an usurper in rather bad odor at home which has come over to this land of liberty under the garb of utility to make foul the air, with its pestilent breath, and devour the soil, with its intermeddling roots—a tree that has the fair outside and the treacherous heart of the Asiatics, and that has played us so many tricks, that we find that we have caught a Tartar which it requires something more than a Chinese wall to confine within limits." In contrast, those in favor thought that the ailanthus absorbed as much odor from the city's streets as it emitted "aroma." They considered the tree one of the city's "principal ornaments" and cherished its resilience against insects, its capacity to shield and shelter from the sun and rain, its unassuming

fast growth, its fresh foliage that endured the seasons until the first frost and that created a "very beautiful sight" when it moved in the wind, and its useful and aesthetically pleasing wood. In 1902 the Department of Parks' landscape architect Samuel
Parsons reported that whereas fifty-six trees planted along Fifty-Ninth Street between
Fifth and Ninth Avenues had died due to too shallow soil above the rock subsurface, a
nearby feral ailanthus had split the rock and was growing profusely.[35]

By the time the Department of Parks had finally taken charge of street trees in
the 1910s, fewer ailanthus trees were being planted, but the tree was still on the planting list, and in 1913 the department planted thirty-two ailanthus on Mangin Street, the
second-largest number of one species planted in Manhattan after the Oriental plane,
which by far outnumbered any other species planted. Street tree advocates on the one
hand appreciated the tree of heaven for its lush green foliage, "tropical effect," and
"luxuriant oriental appearance" as well as its hardiness and resilience. But they also
warned of its brittle wood and the tree's straggly, uneven growth that caused a "rather
unsightly appearance in the winter." While the ailanthus was not a natural among the
symmetrical trees, if pruned properly, Laurie D. Cox considered, it was one of the
most suitable species for the dense, congested downtown below Fifty-Ninth Street.
The debate of pros and cons continued throughout the decades, reignited whenever
the lack of street trees was felt most badly. This was the case, for example, in 1934 when
the Department of Parks, alarmed by the uneven distribution of trees and their poor
conditions, conducted the first citywide tree survey, and again in the 1960s. However,
by this time the ailanthus had been struck from the desirable species list and its planting as a street tree was no longer permitted. The species was simply too costly to maintain and manage as an ideal and standardized street tree, the argument went. Undesirable along streets, it was now also objected to in other types of plantings, where it
had become a "weed" because it displaced more desirable tree and shrub species.[36]

The conception of the ideal street tree depended as much on the political context of the times as it did on species characteristics and the respective urban environment. It also depended on the human ability to control and shape nature in the city,
from the scale of the single tree to the scale of the urban forest. Writing about street
trees in 1901, landscape architect Emil T. Mische, who later became park superintendent in Madison, Wisconsin, and then Portland, Oregon, stressed that the city was
"essentially an artificial creation" to which trees and their planting pattern had to adapt.
Some years later, in his 1915 street tree survey and plan for Manhattan, Cox pointed
out that the street tree system was entirely artificial. None of the early city foresters
doubted urban nature's artificiality, and yet they knew it consisted of trees with their
own life force.[37]

# 2

# Street Tree Aesthetics

*Uniformity and Variety along New York City Streets*

T ree experts realized quickly that for street trees to fulfill their various functions they had to be planted, cared for, and managed systematically and in a certain way. Trees' function was not only to provide shade and an amenable microclimate. Their aesthetic effect was equally important. Already in 1871, when systematic treatment was still lacking, the annual report of the New York City Department of Parks pointed out that trees provided an "antithesis" to "the streets and houses" and could "act remedially, by impressions on the mind and suggestions to the imagination." Trees offered contrast to the lifeless walls of buildings in multiple ways. In 1913 landscape architect Elbert Peets explained that street trees could equally enliven an austere street with monotonous architecture or unify and order a street with variable architecture. Trees provided streets with identity; they could frame and thereby highlight architecture or, if the street lacked an architectural viewpoint, the trees themselves could stand in. Planted in rows and in a regular order, street trees rendered the city beautiful, civilized, refined, and attractive. Regular series of trees along North America's city streets were seen as a sign of the country's modernity and progressiveness. A curated and cared-for street tree system could, its supporters believed, prevent both the monotonous and the jumbled appearance of urban streets. In this aesthetic capacity they could provide both visual calm and stimulation, and as a form of nature in the city, street trees were a means of public health that could cure what by 1869 had been diagnosed as neurasthenia, the "exhaustion of the nerves."[1]

The street tree system that Henry R. Francis and Laurie D. Cox had suggested for New York City involved the planting of the same tree species along a straight line throughout one street at equal distances, producing a serial aesthetic. It also determined that trees should be trimmed to the same symmetrical form and size. According to tree expert Peets, a row of trees had "the added beauty which comes from the rhythmic repetition of form and color." William F. Fox, the superintendent of New

York state forests at the time, had pointed to the same effect but in somewhat less elegant prose when he compared trees with lamp posts: "Lamp posts as well as trees, are deemed ornamental by many people; but no one would even think of erecting posts of different heights, size and appearance on the same street."[2]

Within the uniformity that street trees could provide, it was also considered important to have variety, following a common design principle. Monotony was to be avoided at all costs. If regular rows of street trees were understood to provide the order necessary to render dense urban centers beautiful, in urban areas that were less dense and considered less prone to uneven development, some thought that regular street tree planting ran the risk of appearing monotonous. For example, landscape architect Sidney J. Hare, practicing in early twentieth-century Kansas City, argued that regular street tree planting created "spite fences" along streets and made boulevards look like "ten pin alleys." He had urban subdivisions in mind when he suggested that straight streets should be planted with irregular groups of trees. Hare argued that picturesque group planting could frame views, let sunlight penetrate onto the streets, and prevent buildings from being boxed in (figure 2.1). Indeed, as landscape architect Phelps Wyman argued some decades later, a natural effect could best be achieved by irregularly spaced group plantings. According to Wyman, this effect would be supported most effectively by the laying out of sinuous streets in the first place.[3]

But in contrast to beliefs prevalent in Germany, where several early twentieth-

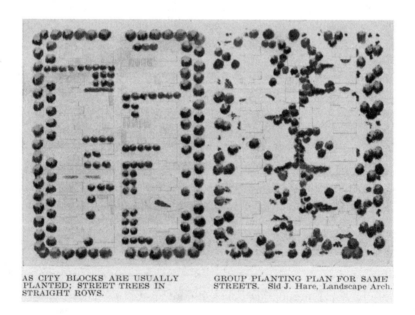

AS CITY BLOCKS ARE USUALLY PLANTED; STREET TREES IN STRAIGHT ROWS.
GROUP PLANTING PLAN FOR SAME STREETS. Sid J. Hare, Landscape Arch.

FIGURE 2.1.
Sid J. Hare's juxtaposition of two possibilities for street tree placement in a suburban neighborhood: the customary straight rows and suggested group planting along straight streets, 1907. (Sid J. Hare, "Avoiding Monotony in City Street Planting," *Park and Cemetery and Landscape Gardening* 17, no. 4 [1907]: 101)

century landscape architects and urban designers opposed regular street tree rows, often giving expression to their anti-urban and anti-modern sentiment and tendencies, Hare's demand for an irregular planting scheme along streets was an exception in the urbanizing areas of the United States. In New York City and the northeastern United States, where dense urban fabrics and the jumble of urban life prevailed, it was enough that the tree—that is, the material of the street tree system itself—came with an inherent variability and that tree species varied throughout the city. With its entangled branches and individual growth patterns, every tree was by nature different. Furthermore, most city foresters in the United States looked at the common practice in European cities of pollarding and stunting street tree crowns with some disdain or skepticism. They proposed it only under exceptional circumstances, as in the case of some of New York City's narrow streets and plazas. Thus, in most situations, trees of the same species and age provided variety even if they were planted in a series at equal distances along a street.[4]

But besides the variety inherent in a single tree, the street tree system at large also required variety. This was necessary for more than aesthetic reasons. As Henry R. Francis noted in 1915, a maximum of different species throughout the city was also a measure of precaution against insect calamities that could otherwise destroy all urban trees at one time. Another consideration that was becoming increasingly relevant in the early twentieth century was air pollution, so the degree to which different species were tolerant of smoke and aerosol was important. Tree experts like William F. Fox, Jacob H. Prost, Elbert Peets, and Wayne C. Holsworth pointed out that areas affected by heavy industry limited species selection as not all trees were resilient enough.[5]

Washington, DC, had set an early example for a comprehensive, unifying, yet variable street tree system in the 1870s. Under the aegis of influential civic leader Alexander Robey "Boss" Shephard, gangs of workers had planted different tree species in different streets (figure 2.2). Similarly, the Tree Planting Association of New York City proposed one species per block (figure 2.3). In case the same species was used across several blocks, Lewis Collins, who served as secretary of Brooklyn's Tree Planting and Fountain Society, suggested planting elms at block corners to diversify the urban canopy within a regular and uniform scheme. Despite this early attention to variety, in the 1930s Laurie D. Cox criticized the lack thereof in many American cities, where systematic street tree planting had led to the adoption of one single design and species that was used along all streets.[6]

But besides the intended variability in planting patterns, street trees—especially those used in temperate climates—also had an intrinsic variability provided by their seasonal and life cycles. City foresters had to plan and design keeping in mind what landscape architect turned city planner Charles W. Eliot II in 1945 called the "fifth dimension": life. Tree growth, decay, and death, potential insect calamities, storms, and other disturbances had to be anticipated as best as possible. Suggestions for how to deal with growth, providing for both immediate effect and continuous green shade in the summer over the years, included planting at half distance with the intention of removing every other tree once they had reached a certain size; in the case of double

FIGURE 2.2.
Different street tree species lining streets in Washington, DC. (C. Lanham, *The Tree
System of Washington* [Washington, DC: CF Judd and Detweiler, 1926])

rows, planting each row with trees of a different age; planting only one side if the street
was narrow; and planting trees on both sides in staggered rows. Another proposal was
to plant fast-growing softwoods near houses and slow-growing hardwoods along the
curb line. Once the "permanent trees" along the curb had grown large enough to fur-
nish shade, the "quick-growing . . . temporary trees" near the houses could be removed.

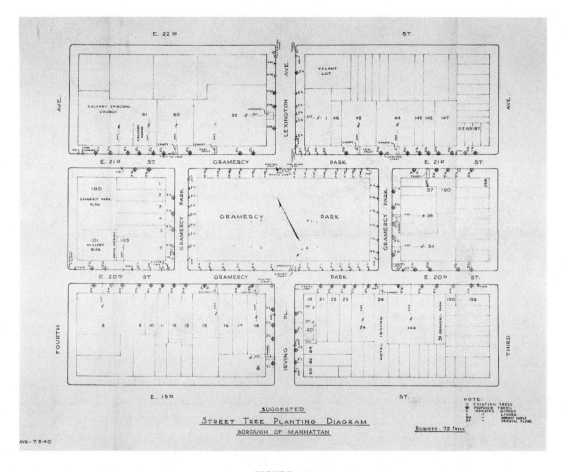

FIGURE 2.3.
The Department of Parks' street tree–planting plan for the blocks surrounding
Gramercy Park in Manhattan showing the alternation between linden, ginkgo,
Norway maple, and Oriental plane trees, 1940. (Department of Parks and Recreation,
General Files, New York City Municipal Archives, Manhattan, 1940, folder 9.
Courtesy NYC Municipal Archives)

If mature trees were not standing too close, young trees could be planted in between, replacing the old trees once they died or were cut down. Similar proposals for the planting of both urban and country roads had been made by German horticultural and landscape authors in the late nineteenth and early twentieth centuries. While aesthetic and climatological reasons for inter-planting along urban and country roads prevailed in late nineteenth-century treatises by the German landscape gardeners Eduard Adolph Petzold and Carl Heicke, timber yield lay at the heart of a most elaborate proposal for a "perpetual planting pattern" along highways and country roads by their French colleague Adolphe Chargueraud (figure 2.4). This arboriculturist's "plantation perpétuelle" proposed a perpetuating pattern of fast- and slow-growing tree species that would provide continuous shade and timber yield along the road. Every

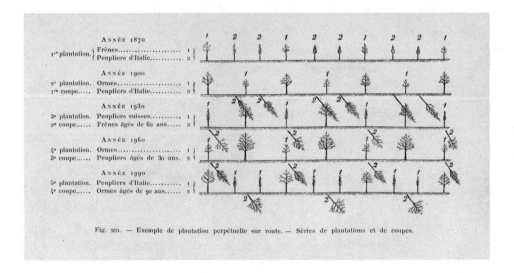

ANNÉE 1870
1ʳᵉ plantation. { Frênes...................... 1 }
                { Peupliers d'Italie........... 2 }

ANNÉE 1900
2ᵉ plantation. Ormes...................... 1 }
1ʳᵉ coupe..... Peupliers d'Italie........... 2 }

ANNÉE 1930
3ᵉ plantation. Peupliers suisses........... 1 }
2ᵉ coupe...... Frênes âgés de 60 ans..... 2 }

ANNÉE 1960
4ᵉ plantation. Ormes...................... 1 }
3ᵉ coupe...... Peupliers âgés de 30 ans. 2 }

ANNÉE 1990
5ᵉ plantation. Peupliers d'Italie........... 1 }
4ᵉ coupe...... Ormes âgés de 90 ans...... 2 }

Fig. 221. — Exemple de plantation perpétuelle sur route. — Séries de plantations et de coupes.

FIGURE 2.4.
Adolphe Chargueraud's perpetual planting plan for country roads in France. (Adolphe
Chargueraud, *Les arbres de la ville de Paris; traité des plantations d'alignement et
d'ornement dans les villes et sur les routes départementales* [Paris: Rothschild, 1896])

30 years throughout a 120-year cycle, there would be a harvest and replacement plantings involving oaks, elms, and Lombardy and Virginian poplars.[7]

Seasonal change in the northeastern United States meant fall color and bare
branches of deciduous street trees in the winter. Although these aspects were hardly
discussed in the technical street tree literature at the time, there were some notable
exceptions. In 1921 Wayne C. Holsworth pointed out that the bark and winter outline
of street trees was important to keep in mind when selecting and pruning them. He
and other authors argued that species like the American elm, pin oak, ginkgo, sweet
gum, white birch, tulip tree, white oak, and sycamore were noted for their even growth
habit and winter beauty. As a young landscape architect who had graduated from the
School of Landscape Architecture at Harvard, aesthetics was not foreign to him. In
addition, Holsworth had been engaged in tree planting in New York City, in tree surgery work in Dallas, in spraying and pruning in Buffalo and Cambridge, and in general forestry in Tennessee.[8]

William F. Fox, too, was interested in the seasonal aesthetics of street trees and
like Holsworth he was writing for a readership in the northeastern states. Fox transferred a selective reading of German forest aesthetics to the American urban context.
*Forstästhetik* (forest aesthetics), an expression coined by German forester Heinrich
von Salisch in the romantic and nationalist spirit of the late nineteenth century in
reaction to Germany's monocultural forests, aimed at turning utilitarian forests into
total works of art for public enjoyment. Von Salisch had argued that creating forest
scenery with attention to effects of color and enabling olfactory and auditory experiences approximated "forest art" while at the same time maintaining the forests' eco-

nomic function. The small and slow-growing German movement of forest aesthetics inspired U.S. foresters like William F. Fox and G. Frederick Schwarz, who began to add to their utilitarian perspective of forestry an affective and romantic view that paid attention to the trees' seasonal life cycles, the shape of their habitus and foliage, and the effects of light and color.

According to Fox, fall color and its phenology—the study of this biological occurrence in the annual cycle—were as important as the shape and outline of the tree canopy. Part of Fox's rationale was based on the fact that different tree species exhibited fall colors over a period of three to four months—almost as long as the green leaf period (plate 3). While fall color therefore played a significant role in the aesthetic appreciation of urban trees, it also turned out to be a means of triumph over European foliage that, as the author explained, never reached the brightness and spectacular effects of trees growing in the U.S. even if the trees were the same species. Although Fox admitted that this—like the change of color itself, which, as he put it, was "as hard to explain or understand as why hair turns gray"—was a mystery, it nevertheless offered an opportunity for a subtle one-upmanship: the natural wonders of the American continent, many believed at the time, superseded European nature in wildness, grandness, and spectacle. Another characteristic that added variety to uniform street tree planting was the shadow cast by the trees. As an aesthetic function of shade, shadow created effects that, like color, were ephemeral and dependent on the seasonal and life cycles of the respective tree species. The shadows cast by trees created rhythmical patterns on the ground and along building facades.[9]

Even if street trees' seasonal and ephemeral aesthetic effects were often mentioned only in passing in the technical literature at the time, these were what caused public commentary, which was frequently romantically inspired. On the occasion of the 1939 elm tree planting along Fifth Avenue (see chapter 3), the *New York Times* observed, "Watchers who study a tree in the shadow of huge buildings struggling so artfully in the Spring to give its every tiny leaf a place in the sun find pleasure in this companionship. Who does not enjoy the change of mood and color that leads from the sharp, lemon-yellow shoots of late March to the placidly rustling Summer foliage and then to the prismatic tapestry of Autumn? Trees in the city link us to Nature. They remind us that bleak times pass and good times return in the old endless cycle. There isn't a street in New York that would not be happier for them."[10]

## Straightening and Shaping Trees

To achieve a serial aesthetic, systematic tree care and management was necessary. Proper treatment begun in the nurseries had to continue once the trees were planted on city streets. There they were exposed to air pollution and gas leaks underground, and they were subject to mechanical injuries by carts, carriages, automobiles, and horses chewing on their bark. Lack of space, light, and direct sunlight were other compromising conditions that affected street trees in certain parts of New York City. To keep the standard and maintain street trees in a beautiful and healthy state despite these challenging urban conditions, trimming, pruning, bracing, and the filling of cavities became necessary practices, carried out by gangs of pruners and climbers as well as "tree doctors" and "tree surgeons." Incidentally, the ordered, regular planting of street trees of the same age at equal distances facilitated tree care and maintenance. Trees planted at regular intervals between the curb and the sidewalk—ideally in tree belts or parking strips, as the grass areas between the curb and sidewalk were called— could be pruned more efficiently and sprayed more easily from vehicles and machines on the street. They could also be inventoried and surveyed more efficiently as well as protected more easily against vandalism.[11]

To maintain the street trees' aesthetic appeal meant maintaining symmetrical crowns because asymmetry signified improper care and disease. Already around the turn of the century, the Massachusetts Forestry Association had advised tree wardens "to assist nature in giving the trees . . . a symmetrical form." Some tree experts discouraged the use of tree species that were less likely to achieve or maintain symmetry in their natural growth, like ailanthus, silver maple, and Carolina poplar. To produce vital street trees of the same symmetrical form and dimension in big numbers, tree experts like Charles Sprague Sargent, William F. Fox, Bernhard E. Fernow, and William Solotaroff suggested the employment of a new optical instrument: the dendroscope (figure 2.5). This handy contrivance had been invented in France in the 1860s by the count Amédée Joseph de Pérusse des Cars. Presented at the 1867 World's Fair in Paris, it captured the attention of foreign foresters.[12]

Holding the dendroscope—a cardboard or wooden sheet with a hole in the shape of a tree in it—at eye level while moving around the tree, its operator could give instructions to the pruner to achieve this desired tree shape. The comte des Cars suggested four different crown shapes and crown/trunk ratios, depending on the tree's age (figure 2.6 and plate 4).

FIGURE 2.5.
Dendroscope, a device for pruning trees. (William Solotaroff, *Shade-Trees in Towns and Cities* [New York: John Wiley and Sons, 1911])

*Fig.* 21. – Manner of using the dendroscope.

FIGURE 2.6.
An illustration of how the dendroscope is used. (A. des Cars, *A Treatise on Pruning Forest and Ornamental Trees*, trans. from the 7th French ed., 3rd ed. [Boston: Massachusetts Society for the Promotion of Agriculture, 1894])

Young tree crowns were to be egg shaped, an elongated oval that should measure two-thirds of the tree's entire height. In contrast, an old tree's crown should have a more rounded outline and could have a height half as tall as the entire tree. By cutting the tree into the shape that was projected onto it through the dendroscope, three-dimensional nature was made to copy the ideal shape on the two-dimensional picture plane. Through the use of the dendroscope, street trees could be "forced" into shape.

Although the dendroscope had been invented to assist in the pruning of forest trees for economic ends so that their vitality and consequently their wood production would be enhanced, the comte des Cars had argued that the tree outlines used in his dendroscope were also the most beautiful. They were the shapes that "nature gives the most perfect and most beautiful trees." Like their French colleagues Jules Nanot and Adolphe Chargueraud, who by the 1890s were promoting the dendroscope's application to the trees of Paris and street trees in general, Sargent, Fox, Fernow, and Solotaroff sought to apply the dendroscope to the context of urban America. It could help to economize arboriculture and standardize tree shape and urban beautification, even though Fernow had some reservations about the actual tree shapes proposed by the French count, and some years later landscape forester Wayne C. Holsworth cautioned that pruning for beauty should not distort the outline typical of the respective tree genus. Regardless, the dendroscope's ideal abstract canopy shapes were presented in all prominent manuals at the time. The Massachusetts Forest Association, the Tree Planting and Fountain Society of Brooklyn, and the Tree Planting Association of New York City all referred to the comte des Cars's work and actively sought to distribute it

FIGURE 2.7.
Celebrating the tree surgeon's work, Dermody Square, Queens, 1930.
(Courtesy NYC Municipal Archives)

through their bulletins and reports. Although this widespread advertisement suggests that the dendroscope could have found wide-ranging application, including in New York City, it remains unclear how many tree pruners were actually using the device. Labor scarcity and the city's hesitancy to invest in the urban tree canopy point to the fact that the dendroscope's impact may have been more indirect. Instead of being an instrument practically employed, it could have become a means to mentally envision an ideal urban forest that was impossible to create due to the circumstances.[13]

In contrast, the filling of cavities with cement (figure 2.7) was a practice that was certainly applied in New York City. It had gained renown in the early twentieth century as best management practice to preserve more or less mature trees. In 1910 the *New York Times* reported that twenty-five thousand trees, of which four thousand were counted in only one year, had been treated in Brooklyn by "surgical operations." Tree experts considered fillings a cosmetic treatment—in some instances the cement was even sculpted to imitate the bark, although the practice was soon considered dysfunctional as it could hinder healing. Fillings were also thought to have a necessary functional and lifesaving purpose that could enhance stability and slow down decay. It was this particular treatment, based on "the science of doctoring trees," that turned tree experts into "tree doctors" and "tree surgeons," and that frequently also led to comments anthropomorphizing trees. Observing some tree work in 1876, one Chicago citizen had commented on the tree doctor's activity, which reminded him "of a surgeon engaged in removing a tumor from the human frame, or a dentist excavating a tooth preparatory to filling." In the first decades of the early twentieth century, similar analogies were frequently drawn to illustrate the importance of employing knowledgeable and trained men for the treatment of trees.[14]

Especially old trees attributed with historic or cultural significance were treated by tree surgeons. Cavity filling was perhaps the most spectacular tree-care practice with the biggest popular appeal. In October 1912, a tulip tree on the bank of the Harlem River that was estimated to be 225 years old, and was thought to stand in the location of a former Native American encampment where Henry Hudson first set foot on Manhattan Island in 1609, was treated by a tree surgeon from the Department of Parks. He cleaned out two large cavities, treated them with copper sulfates, coated them with coal tar, and finally filled them with concrete. The tree's restoration was celebrated in the presence of three hundred children from a nearby public school and other neighbors by park commissioner Charles Stover, who moderated addresses delivered by president of the Tree Planting Association Stephen Smith, director of the Botanical Garden Nathaniel L. Britton, historian James Grant Wilson, and archaeologist Reginald Pelham Bolton. Some eighteen years later, the Bureau of Forestry discovered that the cement fillings had cracked, causing decay of the sapwood. Although they were replaced with new sectional fillings following the latest scientific methods (figure 2.8), the tree was dead by 1938.[15]

In the first decades of the twentieth century, the two competing commercial arboricultural companies offering tree surgery in North America, the Davey Tree Expert Company and F. A. Bartlett Tree Expert Company, either had offices in New York

FIGURE 2.8.
The tulip tree in Inwood Hill Park with new sectional fillings installed at its base, 1930. (New York City Parks Photo Archive, ANR975)

City or were present in the city on a regular basis to consult on tree care and maintenance. John Davey, one of the most well-known tree doctors in the nation, popular-
ized expert tree knowledge and care in his lectures and in his illustrated book *The Tree
Doctor*—sold for $1 so that it would "be within reach of all." Emigrating from his na-
tive England in 1873, Davey had built a business around trees in Kent, Ohio, where
he founded his company in 1880, followed by establishment of the Davey Institute
of Tree Surgery in 1908. Among his early patents was a new "Process of Treating and
Dressing a Bruise or Wound in the Trunk or Live Branch of a Live Tree," which he
filed together with his sons Martin and James Davey in 1908 (figure 2.9). James Davey
advised on the tree rehabilitation work of Central Park in the 1920s, and his company
was also asked by the Women's League for the Protection of Riverside Park to examine
its trees in the memorial grove of Riverside Park when they were suffering from red
spiders and red mites. Francis A. Bartlett, who built a similar company in Stamford,
Connecticut, and was Davey's only competitor until the Asplundh Tree Expert Company was founded in 1928, was present in the city as well. At the 1920 Flower Show,
Bartlett exhibited models of tree surgery and methods of preserving and restoring tree
health.[16]

Despite its long history, reaching back millennia, the filling of tree cavities has
been found to be detrimental to trees and is no longer used. Yet in the late nineteenth
and early twentieth centuries, when Davey's and Bartlett's businesses took off, the filling of cavities with cement, concrete, or "cementitious material" was widely considered an innovative and state-of-the-art method. The practice, like the material itself,
became internationally known and was employed in many major cities with notable
street tree systems.

According to Davey and other tree experts at the time, "The repair of tree cavities [was] very much like the process of filling a tooth," leading some to talk not only
of "tree surgery," but even of "tree dentistry." The edges of cavities were to be cut
smooth and even, and all decomposed matter and debris was to be removed from the
interior. A coating of coal tar was to be applied to the surface of the cavity before it
was filled with cement or with a mixture of cement, broken stone, and brick. In order
to facilitate the process of the healing callus overgrowing the filling, the cement plug
was stopped at a depth just below the bark and shaped to follow the contour of the
tree. Davey's "improved" treatment included a narrow pipe that could drain any efflu-
ent from the interior of the tree and a metal shield that covered and overlapped the
filling. Many other, more elastic materials and methods of filling were used and tested
in the subsequent decades, including magnesite, rubber, polyurethane, and asbestos
fiber. By 1920, the Davey and Bartlett firms were trying to outcompete each other
through the material and method of cavity fillings. Whereas Davey remained a stead-
fast defender of concrete, Bartlett experimented with a new material consisting of
wood fiber and magnesium that was used on the decks of ships and that Philadelphia
arboriculturist Albert Vick had modified for use in tree cavity filling and sold to Bartlett. The Bartlett company hailed the material, fittingly called NuwuD, as a more
flexible alternative to concrete. It appeared to offer an opportunity to beat Davey in

No. 890,968.                           PATENTED JUNE 16, 1908.
J., M. L. & J. A. DAVEY.
PROCESS OF TREATING AND DRESSING A BRUISE OR WOUND IN THE TRUNK
OR LIVE BRANCH OF A LIVE TREE.
APPLICATION FILED JAN. 22, 1908.

2 SHEETS—SHEET 2.

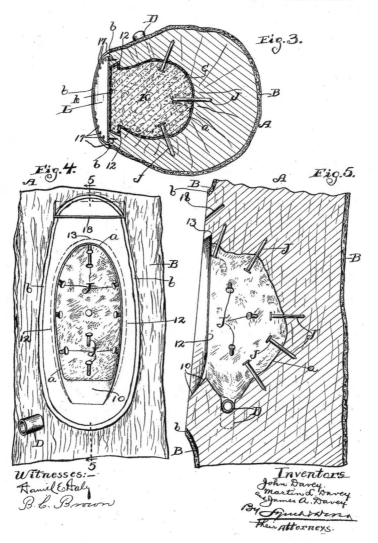

FIGURE 2.9.
Patent filed by John, Martin, and James Davey for the "process of treating and dressing a bruise or wound in the trunk or live branch of a live tree," 1908. (U.S. Patent 890,968, filed 22 January 1908, and issued 16 June 1908)

the lucrative market of tree cavity filling until it was discovered that NuwuD released toxic salts that harmed the trees. Sectional concrete fillings were also thought to prevent cracks, a common occurrence in the monolithic cement plugs. As Davey's son Paul reported in 1939, by which time many tree experts were already questioning the benefit of fillings in the first place, the modern material, when applied in sections, could "bend with the natural movement of the trunk," a vital characteristic given that in long cavities concrete often "made the trunks so rigid that in high winds, the branches blew off in all directions." According to Paul Davey, sectional concrete cavity fillings were the best because of their durability, longevity, flexibility, inexpensiveness, easy installment, and antiseptic effects.[17]

During the first decades of the twentieth century tree experts, regardless of their affiliation, had hailed cement as the material of choice, replacing the simple wooden boards and simple tin and zinc covers that had often been employed previously. Cement was used profusely in the increasingly standardized building industry during this time; arboriculture provided yet another, seemingly unlikely field of employment for this ubiquitous material, which began to be produced in the United States in the 1870s. As George E. Stone, botanist at the Massachusetts Agricultural College, pointed out, the same Portland cement was to be used in both tree work and construction work. Despite its existence for millennia, around the turn of the century, cement began to be considered a modern and progressive material that would transform "nature, . . . ourselves, and our relationships with each other." In trees, concrete became part of nature, yet stood in contrast with the changeable lively qualities embodied by the tree. Their material properties were opposites: wood is warm, concrete is cold; trees and their wood are living and porous; concrete is inert, compact, and nonporous. Whereas trees and wood can burn, concrete is fireproof. The ubiquitous concrete stood, and often still stands, for modernity and progressiveness. It was the quintessential material of the International Style, which opposed ornament. In contrast, the tree is rooted in the ground in a particular location; it stood and stands for the vernacular, nature, and ornament. But through the application of concrete and other maintenance and management practices as outlined above, in the early twentieth century, trees became increasingly standardized and urbanized while cities became naturalized.[18]

## Serial Trees

The interest in introducing systematic street tree management and a serial aesthetic affected more than tree-care methods and practices. It was also reflected in contemporary handbooks that contributed to the systematization and standardization in the nascent tree-planting movement and its scientific management. As part of efforts to professionalize urban arboriculture, comprehensive street tree management went hand in hand with the production of serial images to illustrate the best pruning and management practices. In street tree planting, seriality was not only an aesthetic but also a practice and a representational tool. It was a method both in the actual performance of planting and caring for trees and in its illustration.

Many of the early city foresters and tree experts who produced handbooks and pamphlets used their own photography to illustrate tree species, damage, and the various practices of tree planting and care. Once the halftone method enabled the easy and fast mechanical reproduction of photographs in the 1890s, they proliferated in books, periodicals, and newspapers. John Davey illustrated his 1904 *Tree Doctor* profusely with his own photographs—175 of them. They showed correct and wrong arboricultural practices; tree injuries; portraits of trees in their full splendor as well as suffering from insect attacks, fungi, polluted soil, and other stress factors; and tools and equipment. Some photographs, while being more or less staged, also showed people and appeared as if they were taken from Davey's family album. Children and women were shown assisting the photographer by holding specimens, clipping branches, and pointing out the areas the viewer should be paying attention to. Children were also shown using trees in their play, thus illustrating trees' educational and psychological function. Besides providing scale, the representation of men, women, and children alongside trees also signaled and contributed to *The Tree Doctor*'s popular appeal. The book discussed everyday topics that were of interest and accessible to anyone. Its photographs turned tree care into an area of expertise attainable to anyone with a "tree sense," as Elbert Peets would later posit. For Davey, the photographs functioned as testimonials and documentary evidence. In his eyes, photographs were "something from which you cannot get away." Their immediacy, purported truth value, and relative inexpensiveness turned them into an effective educational tool for the public at large. For educational and advertising purposes, the firm built an image archive that soon also included lantern slides and motion pictures.[19]

Like Davey, William Solotaroff took most of the photographs that he used in his book, the 1911 *Shade-Trees in Towns and Cities*, himself. He had captured the serial images illustrating correct and wrong arboricultural practices during work operations in East Orange, New Jersey, where he served as secretary and superintendent of the Shade Tree Commission. His photographs illustrating tree pruning and step-by-step treatment of cavities comprised a "cinema of stills" that required sequential reading (figure 2.10). In a rough and simplified way, they share character traits with the early motion study photography by Étienne-Jules Marey and Eadweard Muybridge that inspired Thomas Edison to develop one of the first motion picture machines in his industrial research laboratory in West Orange, not far from Solotaroff's area of operations.

The continuity of the individual photographs of each series assembled by Solotaroff portrayed and made understandable the work processes of pruning and treating tree cavities. For example, two series of six exposures each revealed the micromechanics of two different methods of pruning a tree limb. The photographs stopped the flow of motion at the moments that were most relevant to instruct the reader about correct pruning methods. They codified and sought to standardize work methods and their movements, adding a nonverbal language to the book's explanatory texts.

Solotaroff's photographic series transferred and translated to urban arboriculture Frank and Lillian Gilbreth's contemporary time and motion studies of workmen in the construction industry (figure 2.11). The Gilbreths used photographs, film, and

PLATE 28.—FIRST METHOD OF REMOVING BRANCH.

1. The branch.  2. Start with an undercut, saw about half way through.  3. Then saw close to the shoulder.
4. Branch will split horizontally and fall off.  5. Finish cut.  6. Branch is removed.

PLATE 29.—SECOND METHOD OF REMOVING BRANCH.

1. The branch.  2. Start with an undercut, saw half way.  3. Then saw close to the shoulder, about half way.  4. Saw
above undercut, branch will drop off.  5. Remove stub.  6. Work is done.

FIGURE 2.10.
A "cinema of stills" illustrating the best pruning practices for street trees, 1911. (William
Solotaroff, *Shade-Trees in Towns and Cities* [New York: John Wiley and Sons, 1911])

This is the era — *now*.  We have a scientific method of attack, and we have also scientific methods of teaching.

The stereoscopic camera and stereoscope, the motion picture machines, and the stereopticon enable us to observe, record, and teach as one never could in the past.

The following motion study pictures, charts, and diagrams are typical and have been used for teaching journeymen and apprentice bricklayers our standard methods.

### PICK-AND-DIP METHOD — WORKING RIGHT TO LEFT

Fig. 24. — Spreading mortar on exterior face tier.

Fig. 25. — Cutting off mortar before the brick is laid on exterior face tier.

Fig. 26. — Buttering the end of the laid brick on the exterior face tier.

Fig. 27. — Cutting off the mortar after the brick is laid on the exterior face tier.

### PICK-AND-DIP METHOD — WORKING RIGHT TO LEFT (Continued)

Fig. 28. — Throwing mortar on interior face tier.

Fig. 29. — Spreading mortar on interior face tier.

Fig. 30. — Cutting off mortar before brick is laid.

Fig. 31. — Tapping down brick after laying on interior face tier.

Fig. 32. — Cutting off mortar after brick is laid on interior face tier (here working left to right).

Fig. 33. — Handling mortar for two brick at one time on interior face tier.

### STRINGING-MORTAR METHOD

Fig. 34.—Working right to left.—Spreading mortar on exterior face tier.

Fig. 35. — Working left to right. — Buttering end of laid brick on exterior face tier.

Fig. 36.—Working left to right.— Cutting off mortar after brick is laid on exterior face tier.

Fig. 37. — Working right to left.—Spreading mortar on interior face tier.

Fig. 38.—Working left to right.—Cutting off mortar after brick is laid on exterior face tier.

Fig. 39.—Working left to right.— Spreading mortar on the interior face tier.

### STRINGING-MORTAR METHOD (Continued)

Fig. 40.—Working left to right.—Spreading mortar on the interior face tier.

Fig. 41.—Working left to right.— Throwing mortar on the interior face tier.

Fig. 42.—Working right to left.—Buttering right-hand end of brick in hand to be laid on interior face tier.

The "pack-on-the-wall" method is the latest development and is an actual direct result of motion study.  It has again changed the entire method of laying brick by reducing the kind, number, sequence and length of motions.  It reduces the fatigue of the bricklayer and he is therefore able to make more rapid motions.

### FIGURE 2.11.

Frank and Lillian Gilbreth used photographs to illustrate the most efficient methods of bricklaying. (Frank Gilbreth, *Motion Study* [New York: D. Van Nostrand, 1911]) later chronophotography to establish the most efficient and physically compatible work practices in the building industry and beyond. While their original studies aimed at enhancing the productivity and health of workmen in the construction industry, such methods of scientific management, when applied to urban arboriculture, also had to take into account the growth and health of the trees themselves. In contrast to a construction site, which could be totally controlled and laid out in a particular way, the tree, as the arboriculturist's "construction site," had its own life force, changed with time, was irregular, and could never be entirely controlled. It was also subject to exterior forces like storms and hurricanes. While the pruners' motions themselves could therefore hardly be standardized as they had to respond to the respective trees' growth and habit, attempts were made to rationalize and standardize the job, grounding it in scientific tree culture. The photographs were based on belief in a best pruning practice that could be codified, learned, and performed repeatedly.

Besides photographs illustrating tree-care practices, photographs documenting the characteristic form of different tree species and details of tree injuries and pathologies were common in the early manuals and reports of street tree commissions and associations. Inadvertently echoing early human clinical photography, these types of photographs also became an instrument of managerial and administrative power.

## Systematic Street Tree Management

The practices and their representations described above were methods to produce and maintain the standard, or ideal, street tree, and a comprehensive street tree system. The "efficiency craze" instigated by Frederick Winslow Taylor's work on labor productivity in industry had reached street tree planting as part of city planning. In 1921 Wayne C. Holsworth summed up the prevailing notion of city planning at the time: "In this age of advancement the prime thought of the street must be for efficiency." Yet as a landscape architect, he also cautioned, "We cannot ignore the beauty that should surround us in our daily life." The goal, then, was to create beautiful streets as efficiently as possible, and street trees had an important role to play in this effort. City officials believed that the best way to achieve efficiency in street tree planting was to manage and care for it systematically on the basis of a managerial and administrative system, using ever-more-efficient tools and methods. This sometimes happened without reflection or recognition of negative side effects.[20]

Once Mayor Fiorello H. La Guardia had consolidated the borough Departments of Parks into one citywide agency and appointed Robert Moses as park commissioner in 1934, the latter took charge. Realizing that the department staff's authority had often been compromised in public, Moses initiated several changes that enhanced the department's public visibility. The consolidation and changes also affected street tree planting and care. Before Moses's tenure, street trees had been planted and cared for differently in all five boroughs, despite the early twentieth-century standardization efforts by Henry R. Francis and Laurie D. Cox. For example, Manhattan allowed individuals to plant trees at their own cost but only with permission from and under supervision by the Department of Parks. The Bronx offered the same but also provided trees and their planting for a nominal fee to be paid by the adjacent property owner. Despite good intentions, there was no citywide system for street tree inspection, and street tree care was often lacking or lagging. Some laborers were inadequately trained, and it was discovered that some men were unable to climb, inducing them to stunt trees so that they could be pruned from the top of the highest ladder. Furthermore, according to a 1934 Department of Parks press release, trees were pruned or removed only when adjacent property owners complained, leading tree-care employees to be nicknamed "Inspectors of Tree Complaints."[21]

This situation needed reform. To improve street tree planting and care and facilitate a more uniform image, the Department of Parks announced a new policy that turned out to be a centralized regulatory approach for all boroughs: the department would plant trees along parkways and boulevards at no cost to adjacent property owners, would issue permits for street tree planting to individuals and organizations under specified conditions, and would take care of the trees after they had been planted.

Furthermore, the procedures of street tree work were reformed to render them more efficient. In the past, the individual boroughs' tree crews had responded to planting, pruning, and other tree-care requests in order of their receipt, causing work gangs to lose much time traveling between different locations. In contrast, by the late 1930s each borough had been divided into sections consisting of several blocks that had similar numbers of trees. Work crews of five or six men and a foreman would complete all necessary tree pruning, planting, and care in one section before moving on to the next. A truck from the central garage picked up the brush and limbs removed during pruning operations and disposed of them in a landfill. The crews' work was based on instructions from the respective borough's arboriculturist, who in turn depended for his overall management on tree inspectors, who traveled on foot or by bus or trolley car throughout the different sections. While recording the trees' conditions and other necessary field information, they also inspected individual tree-planting requests and tree complaints by New York citizens.[22]

Shortly after Moses had assumed his new office, the department undertook the first comprehensive tree survey since the first attempts by Cox in 1915, taking advantage of funds from the Works Progress Administration (WPA). Moses's plans for a citywide street tree overhaul were largely dependent on these work relief funds, which were pouring into the city as a result of La Guardia's clever politics. Moses knew how to take advantage of this opportunity despite the initial challenges, including the lack of foremen who could supervise the work of the additional laborers along the streets and in the parks. Over $40 million in relief funds were spent on parks and trees in the first year after the departments' consolidation. During 1935, an average of 750 men employed through the Civil Service and the WPA were undertaking the department's forestry work, including the spring planting of four thousand new street trees throughout the five boroughs. Street tree plantings consisting of plane trees, Norway maples, pin oaks, scarlet oaks, and lindens—by this time thought to be the most adaptable species—were considered ideal projects in unemployment relief work in more ways

FIGURE 2.12.
Uniforms for male park workers,
New York City Department of
Parks and Recreation, 1936. (New
York City Parks Photo Archive,
2668, 2650, 2667, 2652, 2655, 2654)

than one. They were a visible outcome that not only enhanced the urban environment but also boosted the nursery industry.[23]

To further promote the department's visibility, a standard uniform—featuring
service stripes and an insignia in the shape of a leaf—was introduced in 1934 (figures
2.12 and 2.13).

Although enforcement of the uniform was difficult at first—many regular laborers as well as WPA and Civil Service workers could not afford to pay for the uniform
that they were required to purchase—its effect was almost immediate. For example,
on Arbor Day in 1935 uniformed supervisors ensured that tree plantings at the public
schools were properly carried out and no trees were damaged or destroyed by schoolchildren. But just as important as the uniform's signification and regulatory effect in
relation to the general public was its effect on morale and corporate identity within
the workforce and the fact that it provided laborers with work clothes adapted to their
tasks at hand. Tree climbers and pruners were equipped with breeches and high boots
to facilitate work in the trees (figure 2.14). Government instructions for the tree workers of the Civilian Conservation Corps issued only a few years later also advised gloves,

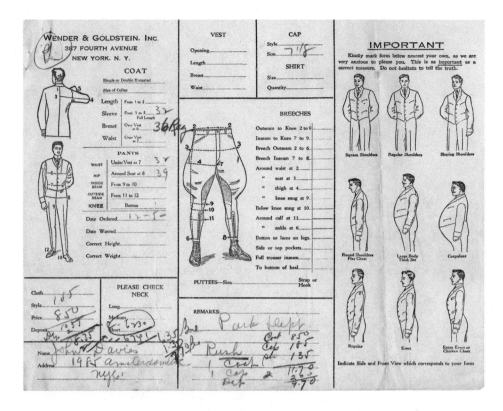

FIGURE 2.13.
Order slip for park worker uniform. (Department of Parks and Recreation,
General Files, New York City Municipal Archives, Administration, 1935, "Uniforms,"
folder 55. Courtesy NYC Municipal Archives)

caps instead of hats, and snug-fitting wool or leather jackets instead of long coats so that
clothing would not obstruct work and labor safety conditions would be improved.[24]

The centralized management of street trees that was finally formalized in 1937
also proved relevant to another contingency at the time: Dutch elm disease. The
disease was introduced in the United States through log shipments and first reported
in the summer of 1930 in Cleveland; in 1933 and the following years, New York City
and the area within its thirty-five-mile radius became the most widely affected area
in the United States. By the summer of 1934, 375 infected elms had been detected in
the city's five boroughs. The Department of Parks issued a call to arms. It explained,
"Fighting the Dutch Elm disease is like fighting a fire," necessitating the felling and
burning of every diseased tree. Although only some of the comprehensive management practices developed in the first decades of the twentieth century were carried
out, the department's consolidation facilitated the treatment of the disease. It could
direct and apply resources more efficiently throughout the individual boroughs and
the city in its entirety, and it could maintain an overview over the spread of the disease
more easily. The disease challenged the incipient street tree management that had

FIGURE 2.14.
Department of Parks workers parading in new uniforms in East River Drive
celebration, August 1935. Tree climbers and pruners are in the front right.
(New York City Parks Photo Archive, 6364)

been implemented in fits and starts in the early decades of the twentieth century. The
need to cut down trees along entire streets to curtail the disease also showed how
important tree experts' early twentieth-century warning calls for species diversity and
aesthetic variety were for the continuity of an urban tree canopy throughout the city.[25]

To effectively treat trees for Dutch elm disease and protect them against insect
pests, technological advancement and scientific research were employed to develop
new efficient ways of tree pruning and care. Tools and clothing were adapted and
developed to further facilitate tree work. For example, the power saw was introduced
more widely in the 1940s, and better spraying equipment was adopted. Around 1900,
such equipment consisted of bucket pumps, barrel hand pumps, and early power
sprayers that required several men, climbing trees to reach the highest branches, to
operate. By 1910, more powerful and precise hydraulic spraying equipment could
reach the tops of tall trees. After Brooklyn's Department of Parks purchased a high-
power sprayer in 1915, its laborers were able to spray twice as many trees as in the year
before (figure 2.15). In the late 1940s and 1950s, when the Department of Parks was
chronically lagging behind in its street tree work schedule, in particular in the bor-

FIGURE 2.15.
High-power spraying in Brooklyn, ca. 1926–27.
(New York City Parks Photo Archive, ANR1355)

oughs of Brooklyn and Queens, upgraded equipment, including new mist blowers that could spray trees more quickly, improved the situation. Before Rachel Carson's 1962 *Silent Spring*—a widely received warning about the dangers of widespread use of DDT and other chemicals, by this time applied by mist blowers—various toxic solutions, including lead arsenate, were used indiscriminately as insecticides in cities and densely inhabited neighborhoods. Men sprayed large trees with giant hoses attached to power sprayers without masks and protective clothing. Although the trees may have been protected, New York City dogs died from arsenic and lead poisoning as the result of this spraying. Citizens were cautioned not to leave pets and children too young to care for themselves unattended in areas that had been sprayed with lead arsenate.[26]

Nevertheless, Malachi Cox, a child living in Staten Island, was reported to have died from a preparation used in tree spraying in the summer of 1959. By this time, the Department of Parks was no longer posting spray warning signs, arguing that the public no longer paid attention to them. The lethal influence of insecticides on wildlife had long been observed; laboratory and environmental studies on the toxicology of DDT and other chemicals began during World War II and confirmed the substances' direct and chronic toxicity for animals and humans. However, ambiguous statements and differing opinions voiced by scientists and physicians on the basis of these studies led to a rather lax handling of these toxic substances by employees and homeowners alike. When one concerned New Yorker addressed the Department of Parks in 1957 about the effects of tree spraying on bird life in Central Park, the department brushed concerns aside, arguing that DDT was used as a contact spray that would not harm young birds and that it was applied only very carefully in the Ramble and the bird sanctuary, the park's favorite nesting areas.[27]

Thus, while the efficient creation of street tree beauty successfully brought nature into the city, it also claimed its casualties, tragically involving tree, animal, and human death.

# 3
# Tree Ladies

*Women, Trees, and Birds in New York City*

T ree planting and protection in the city was a cause taken up by many women who realized trees' potential to civilize life in the city. At the beginning of the twentieth century, street tree planting was a means for them to conquer and control the public urban realm, extending their influence from the domestic into the public urban sphere. In its first annual report in 1897, the Tree Planting Association of New York City highlighted the interest urban tree planting elicited among women. "It was quite proper," Vice President Cornelius B. Mitchell wrote, "that men should take the initiative, start the work and organize an Association . . . , but the best practical results are to be obtained . . . by women. . . . Women are naturally more interested in this question than men are." Therefore, steps would be taken to form a Ladies' Auxiliary Tree Planting Association, a suggestion that was repeated in the following years but came to fruition only around 1908–9, when the Greenwich branch of the Women's Municipal League formed a committee with the purpose of planting trees in the Washington Square district. The Tree Planting Association also appropriated $100,000 for collaboration with women's organizations in 1909. Of its 173 initial members, 28 were women. By 1904, the female percentage of the association had grown: 40 out of 155 members of the Tree Planting Association, and 6 out of 16 members of its Tenement House Shade Tree Committee, founded in 1901, were women.[1]

Ten years later, Mary Cynthia Dickerson became the first woman to hold one of the association's offices. She was an obvious choice as secretary. Having graduated with a BS from the University of Chicago in 1897, Dickerson had held various teaching positions at high schools, the Rhode Island State Normal School, and Stanford University before assuming her job as assistant in the department of the Jesup Collection of North American Woods at the American Museum of Natural History in 1908. She had quickly risen through the ranks at the museum, becoming the editor of its journal in 1910, the curator of the Department of Woods and Forestry in 1911, and the associate curator of herpetology in 1913. Soon after she joined the museum, Dickerson wrote a small popular book, *Trees and Forestry*, that was to accompany a new center of public education on the subject. Unfortunately, many other planned ventures were cut short by Dickerson's death in 1923.[2]

Dickerson, as a highly educated, professional woman, represented a minority at the time. Most women in the Tree Planting Association were members of the city's upper class: the wives of wealthy industrialists and financiers for whom charity and philanthropy were common pursuits. Despite their often wide-ranging involvement in charitable societies, tree planting and protection held a special appeal for these female philanthropists, as they did for early female professionals, social reformers, suffragettes, and animal rights activists.

Already by the middle of the nineteenth century, women in New England had stood at the forefront of village-improvement initiatives largely centered around street tree planting. Women initiated the planting of trees along streets to provide shade and beauty and to stabilize the edges of dirt roads. One of the first village improvement societies, founded by Mary Hopkins in Stockbridge, Massachusetts, became the model for similar societies in other parts of the country (figure 3.1). While New York sanitary engineer George Waring considered this "duty of 'cleaning up' . . . [to] grow more naturally out of the habit of good housekeeping than out of any occupation to which

FIGURE 3.1.
Main Street elms, Stockbridge, Massachusetts, between 1905 and 1915. The planting,
initiated by one of the first village improvement societies, founded by Mary Hopkins,
was emulated widely in northeastern towns. (Detroit Publishing Company
Photograph Collection, Library of Congress)

men are accustomed," the early feminist Mary Ritter Beard stressed women's suitability for this job in loftier terms, arguing, "Men too often cannot see the moral issues at stake in living on treeless streets."[3]

Soon the first-wave feminists took up forest conservation issues, leading to their organization of numerous local, regional, and statewide subcommittees of the General Federation of Women's Clubs for the cause. Women's clubs were instrumental in protecting trees like the sequoias of the Calaveras Grove of Big Trees in California's western Sierra, and in lobbying Congress for the conservation of forests in the northeastern and western United States in more general terms. In 1908 Lydia Adams-Williams, who directed the General Federation of Women's Clubs Forestry Committee, claimed that it was the work of women's clubs that had led to "the almost universal sentiment in favor of preserving forests" in the first place. In Adams-Williams's view, men were the spenders and industrious developers, while women were the conservers and savers who through their "concern for the welfare of the home and the future" and their education of their husbands and children could save the country's resources. This was, she contended with feminist zeal, "peculiarly women's work, as the time is short, and as men are slow in action, even when knowing the facts." Only a few years before, in 1905, the weekly newsmagazine the *Chautauquan* had reported, "Hundreds of women's clubs have forestry or Arbor Day programs, and the forest and civic committees of the federations of clubs enlist the active aid of influential women throughout the land." According to the *Chautauquan*, it was women who were leading the tree-planting movement, in which male-dominated organizations like park commissions and commercial clubs were participants.[4]

Conserving and improving was the women's credo, and the conservationist initiative regarding trees and forests soon extended into the urban realm, and there from private to public spheres. As Mary Ritter Beard noted in the early twentieth century, "Arboriculture for decorative purposes has always been an interest of [women] in their own home plots and now they have extended it to the decoration of their municipal homes." Tree planting was therefore considered a relevant part of women's "municipal housekeeping," which also consisted of the beautification of yards and lobbying for playgrounds, sand gardens, and public urban parks both large and small. Such activities extended "women's roles and values as housekeepers and mothers from the individual home into the larger 'household' of the community" and was a way for them to transcend the domestic sphere and transgress the boundaries between inside and outside while remaining within the bounds of activities that were considered appropriate to their gender.[5]

If in the nineteenth century the luscious tree-lined streets of the residential neighborhoods outside Manhattan's business district were still indicators of the refined domestic female sphere, by the turn of the century initiatives to plant trees in the business district and near the city center were a way for women to quite literally break out of the domestic sphere and claim and mark their place and space in the public realm. In the late nineteenth and early twentieth centuries, parks and tree-lined avenues belonged to the few public urban spaces that middle- and upper-class women

FIGURE 3.2.
A scene along Madison Square, 1901. (Photo by the Byron Company. Geographic File,
box 39, folder: Fifth Avenue 3, image 91237d, New-York Historical Society. Collection
of the New-York Historical Society)

could inhabit while adhering to the social norms of their time (figure 3.2). In these
tree-planted places they were free to engage in what Frederick Law Olmsted had
in 1870 called "gregarious" and "neighborly" activities—socializing and mingling in
crowds—as well as walking, reading, and private conversations. To protect these spaces
against development, therefore, also became an important female concern. Thus,
when Riverside Park was threatened by the encroachment of railway tracks and plans
to build a sea plane station in the early twentieth century, women protested vehemently, criticizing the male business world. In 1929 a member of the Women's League
for the Protection of Riverside Park asked: "Why cannot men think in something besides dollars and cents? Must they even do this on an idle stroll on our beautiful Riverside Park?" And she asked further, "Must this battle to keep from commercializing
any more of our residential shore line always be fought by women, or woman? . . .
Where are the thousands of men who crowd all the walks of Riverside Drive, fill every
available bench to the exclusion of tired women and children strollers; the men who
overflow the shade from saplings called *shade-trees* . . . ?"[6]

Although the role of women in the establishment of a public entity dealing with
street tree planting was not as significant in New York City as it was in Chicago, where
members of the Women's Outdoor League and the Women's Club successfully lobbied for the installment of a city forester, street trees and urban trees in general figured

prominently in New York women's engagement with the city in the early twentieth century.[7]

## Planting the City

In support of urban tree planting, New York women not only participated in organizations like the Tree Planting Association, they also initiated campaigns and founded a variety of organizations themselves. A case in point was the Municipal Art Society's 1916 tree-planting campaign led by Mrs. Hagemon Hall and the initiatives undertaken by Frances Peters, who founded the City Gardens Club of New York City in 1918. The club began its efforts in private yards in the domestic sphere but soon extended them into the public realm through tree planting. Already before 1913 Peters had teamed up with landscape architect Mrs. Frederick Hill to convert the backyard of the headquarters of the Women's Municipal League at 46 East Twenty-Ninth Street into a garden with flowers, vines, a graveled walk, and places to sit. An active suffragette, Peters founded the City Gardens Club to follow this model on a citywide scale, inspiring townhouse and apartment dwellers to transform empty and unattractive urban lots by creating their own backyard gardens, roof gardens, and window boxes. At the Horticultural Society of New York's flower show in March 1921, the club presented a life-sized reconstruction of a typical backyard before and after its improvement (figure 3.3). Around this time, Peters sought to extend the club's influence into the public sphere by leading a project to plant street trees in honor of the men whose lives had been lost in World War I, an initiative that had many statewide counterparts organized by the nation's women's clubs. To support such campaigns, in the early 1920s the club formed a tree-planting committee, and in 1923 it protested against the destruction of street trees through increasing traffic and the widening of Park Avenue.[8]

After her death in 1924, Peters was remembered in particular for her tree-planting initiatives. Fittingly, in 1926 the Women's League for the Protection of Riverside Park, which she had served as secretary, planted a tree in her honor (representing Rhode Island) in its Memorial Grove of State Trees. The City Gardens Club founder was remembered on this occasion for her "unselfish devotion to the cause of saving the City trees and of interesting property owners in transforming their back yards into beautiful restful spots." A later eulogy by the league stressed her vision of tree-lined streets: "Frances Peters dreamed of a city beautiful, with tree lined streets: houses with flower boxes at the windows; attractive back yards and roof gardens. She dreamed and worked—worked until her strength failed, but 'Thoughts are things,' and today many who never heard of Frances Peters are enjoying beauty of flowers, oases of green and shade, in the desert of brick and stone, which constitute so much of our city."[9]

Although founded in 1916 with the explicit objective of protecting Riverside Park against encroachment by the port and terminal facilities that were part of the New York Central Railroad Company's West Side Improvement Plan, the Women's League for the Protection of Riverside Park shared more general goals with the City Gardens Club, the New York Bird and Tree Club, and the Women's League for Animals. Tree

FIGURE 3.3.
The City Gardens Club of New York City exhibition at the International Flower Show,
held in March 1921 at the Grand Central Palace, New York. (Photo by Frances
Benjamin Johnston. Frances Benjamin Johnston Collection, Library of Congress)

planting and conservation played a more or less direct role in the programs of all these
women's organizations founded around issues of nature conservation and education.
As in the case of Frances Peters, who served as both president of the City Gardens
Club and secretary of the Women's League for the Protection of Riverside Park, it was
not uncommon for members and officers of one organization to also be involved in
another. For example, Helen Culver Kerr served as president of the Women's League
for the Protection of Riverside Park and chaired the Park Spaces Committee of the
City Gardens Club, and Ellin Prince Speyer was president of the Women's League
for Animals and a member of the Tenement Shade Tree Committee. These personal
connections created a strong network among the various women's organizations and
with other associations dealing with tree planting. News could be passed on more
easily and quickly, and stronger support could be garnered for certain campaign issues
if necessary. Organizations co-hosted events as well. For example, in January 1927, the
New York Bird and Tree Club joined with the Women's League for the Protection of
Riverside Park in a special meeting featuring three speakers giving lantern slide lec-
tures: S. Harmsted Chubb, associate curator at the Museum of Natural History ("Wild
Birds of New York City"), tree expert James A. G. Davey ("Some Diseases of City
Trees"), and Department of Parks landscape architect Jules Burgevin ("What the Park
Department of Manhattan Is Doing to Save the Trees").[10]

## Trees and Birds

The talks showed the close connection of interests in tree and bird conservation at the time. With increasing urbanization and pristine wilderness disappearing fast, one of the big contemporary concerns of all these organizations was children's nature education and the provision and preservation of nature in the city. In the late nineteenth century, concerns about the conservation of forests and trees began to be connected to concerns about birds and their protection. In the last two decades before the turn of the century, increased deforestation, sports hunting, and hunting for the millinery trade (which used large quantities of feathers for hat decoration) had resulted in insect calamities plaguing agricultural and forest crops, leading farmers to employ increasing amounts of insecticides. Foresters, farmers, and the millinery trade had paid little attention to the relationships between vegetation, birds, and insects. In 1914 the noted conservationist William T. Hornaday deplored that the "habits and interrelations" of "our wild creatures" had lacked scholarly attention in favor of their classification.[11]

The problems that conservationists had observed outside of the fast-growing cities were also affecting urbanizing areas by the late nineteenth century. Once street trees were planted more systematically and parks were established, insect calamities occurred. No sooner had the Department of Parks proudly reported at the end of 1900 that the trees in Manhattan were "remarkably free from insects," and that the city's parks in general had been "remarkably free from the swarms of insects that have at times infested the neighboring country," than another major insect calamity hit the city. In 1901 and 1902 an unusual number of tussock moths ravaged many trees in the parks and along avenues, requiring the Department of Parks to employ additional workmen in the winter to destroy the cocoons as soon as they appeared on the tree trunks. Extra work was also required to keep the tent caterpillars in check. As a port city with one of the largest import businesses in the country, New York City was a harbor for insects from overseas. Especially trees lining streets and connecting parks could act as further vectors in the spread of diseases and insects. The leopard moth was a case in point. Accidentally introduced in the years before 1879 from Europe, this species was by the 1890s giving Department of Parks entomologist Edmund B. Southwick a hard time. It spread from tree to tree through streets and into parks. In 1894 he called it "one of the worst pests with which we have ever had to deal," and the situation had hardly improved by the turn of the century.[12]

Southwick had first been employed by the Department of Parks in 1883 to keep insects and fungi at bay in Central Park and other city parks. In the last summers and winters of the nineteenth century, additional gangs of workers scraping cocoons and egg masses from trunks and branches with heavy steel wire brushes became a regular sight throughout the city, as did the spraying of trees. Increasing amounts of chemicals and solutions were applied to trees to control insects. In 1908 the insecticides included a solution of whale oil soap, considered a cure against scale insects; a kerosene emulsion used to spray maples and elms against elm bark beetles and the leaf louse, scale insects, aphids, and *Pleurococcus*; and a mixture of water and arsenate of lead thought to be a successful treatment against the tussock moth. Besides the damage they caused to trees, caterpillars and other insect larvae were also a nuisance because they fell onto and covered the garments of people passing underneath the trees that the insects were feeding on.[13]

Southwick, conservationists, the Tree Planting Association, and concerned citizens also noted the decreasing number of birds, especially songbirds, in the city, a reason for the increase of insects. In the 1890s, the Brooklyn Tree Planting and Fountain Society bemoaned the absence of native birds from cities, offering instructions for the construction of a "bird fountain" to attract them and to further bird and tree protection. The hope was that among the birds attracted would be robins, one of the few bird species known at the time to feed upon the larvae of the tussock moth.[14]

Entomologists speculated that one of the reasons the wood leopard moth was spreading more slowly outside of New York City and other urbanizing areas was the relative lack of birds in the city compared with the countryside. Ornithologist W. L. McAtee explained that birds were "health officers" for trees. In order to control insects like the leopard moth and the tussock moth and to preserve urban trees in a more natural way, it was clear that birds had to be attracted to the city and protected. While indirectly affected by the lack of birds, street trees were a means to attract them to the city in the first place. Without street trees, birds would disappear entirely, the Tree Planting Association argued.[15]

Following the example of Arbor Day, which had been institutionalized throughout the United States beginning in 1872, around the turn of the century, teachers in New York City called for the establishment of Bird Day. The first Bird Day had been celebrated in Oil City, Pennsylvania, in 1894. Initiated by Oil City's superintendent of schools, Charles A. Babcock, the celebration was adopted by schools in several cities of eight states in the ensuing years. Educators considered Bird Day, like Arbor Day, an opportunity for the schools to add nature education to the curriculum. Intent on the holiday's establishment, T. S. Palmer of the Biological Survey in the Department of Agriculture suggested that Arbor Day and Bird Day could be combined if schools thought the additional holiday was too much.[16]

In 1901 in New York City, where Arbor Day and Bird Day had been commemorated together, educators wanted Bird Day to be celebrated separately so that it would provide an additional educational opportunity in the name of nature. Educators argued that the celebration would instill in the younger generation respect for wildlife, particularly bird life, and thus foster its protection in the decades to come. Yet Bird Day was an education not only for children but also for the many fashion-conscious women who were wearing birds and feathers on their hats. In an 1894 letter of support of Bird Day, Arbor Day's founder J. Sterling Morton pointed out that two of the most destructive and relentless bird enemies were "our women and our boys," the latter of whom were collecting eggs and killing birds.[17]

Living in the fashion capital of the United States and in the center of the millinery trade that dominated three full blocks on lower Broadway, many New York women at the time were still wearing the "garniture of death," as Charles A. Babcock called women's hats that were extravagantly decorated with birds and feathers. On two afternoon walks in the 1880s along Fourteenth Street—the most frequented shopping street in the city—ornithologist Frank M. Chapman undertook an unofficial count of birds on women's hats. On one afternoon alone, he found that 77 percent of all hats exhibited feathers. While most birds perched on the hats were mutilated beyond recognition, Chapman counted forty different species. A decade later, it was estimated that milliners in the United States in 1896 used 10 million birds. Many more birds were exported to Europe. During the winter season, the New York fashion reporter of *Harper's Bazaar* observed that "in spite of . . . the suppression of wearing feathers, they are . . . seen on every kind of hat. To be sure, the osprey and egret are not so much in evidence, but that there should be an owl or an ostrich left with a single feather apiece hardly seems possible."[18]

By the turn of the century, it had become clear that trees and birds belonged together, and women appointed themselves as their protectors. Women had successfully founded the Audubon Society, the first American association to protect birds, and they also organized clubs and societies that had the protection of both birds and trees as their explicit goal. The recognition of the mutual dependency of trees and birds in the city led to more than the celebration of Arbor Day and Bird Day. In 1913 Anna Maxwell Jones, the first president of the Woman's Association of Saratoga, New York, and a member of the General Federation of Women's Clubs Art Committee, founded the New York Bird and Tree Club. Mineralogist George F. Kunz served as president, with Jones as one of many vice presidents along with Mina Miller Edison, Mary Harrison McKee, and Caroline Goldsmith Childs. Jones also chaired the Tree Planting Committee. The club's local activities included various conservation and protection activities as they related to urban development, but its initiative to replant French fruit orchards particularly illustrated the role that trees in general played in female emancipation. At a time when American male professional expertise was providing reconstruction plans for French villages and cities after World War I, American women initiated and organized the fund-raising for the replanting of ten thousand to twelve thousand fruit trees that had been cut down by the German military in France. In this way they provided their own contribution to the reconstruction efforts in Allied European territory. Besides aiding postwar reconstruction, the tree plantings were also conceived of as patriotic acts, as the trees were to memorialize fallen American soldiers.[19]

By the turn of the century, the bird craze in women's fashion was replaced by a craze for bird protection. In the first three decades of the twentieth century, birds captivated female nature activists, schoolchildren, and conservationists. The 1918 Migratory Bird Treaty Act between the U.S. and Great Britain (for Canada) for the protection of migratory birds further fueled the attention paid to bird conservation. Bird sanctuaries were established, and birdhouses were built and distributed along streets and parkways as well as in parks. When in 1919 the City Gardens Club issued planting suggestions for a backyard, it also proposed adding a birdbath that could be made out of a large wooden chopping bowl. In 1925 the Women's League for the Protection of

Riverside Park inaugurated a bird sanctuary, including a birdbath adjacent to its Memorial Grove of State Trees, then in the process of being planted. Pairing nature conservation with children's nature education, the league enlisted Boy Scouts to prevent vandalism by daily patrols, to keep the birdbath clean and filled with water, and to record the birds.[20]

But already in 1915, following the establishment of committees on bird protection in the Association of Park Superintendents, a bird sanctuary had been created in Brooklyn's Prospect Park by planting trees and shrubs attractive to birds and putting up birdhouses and boxes. In tune with similar initiatives in other cities in the 1910s and 1920s, Queens Park commissioner Albert C. Benninger promoted the establishment of such areas throughout the city. He was, of course, particularly concerned about his own borough, which was seeing the effects of suburbanization. Located below the flight path of many migratory birds, sanctuaries in Queens would be able to serve these bird populations especially and would provide the necessary oases for bird life in a heavily urbanizing area of the city. Three bird sanctuaries were established in Central Park at Harlem Meer and near the lakes at the height of Seventy-Second and Fifty-Ninth Streets. They were maintained in cooperation with the city's Audubon Society, which in 1935 planned five more sanctuaries in the city's larger parks.[21]

Besides the planting of Russian mulberries, good feeding trees for most bird species, the construction and distribution of birdhouses was a common activity at the time, especially among schoolchildren. Promoted in scientific and commercial journals as well as in the popular press, this was part of the nature study movement that had begun in the late nineteenth century. As an educational reform movement, nature study aligned with the preservation and conservation movements and was also largely promoted by women. In initiating birdhouse contests among public schools, New York's female conservation activists found a way to directly apply nature study education to their conservation and preservation efforts in the city. In 1912 boys and girls from Public School No. 46 in the Bronx built birdhouses as a reaction to their observation that insecticides were being applied to street trees and that few birds could be seen in their part of the city. In 1920 the Girl Scouts of Manhattan Troop 50 presented birdhouses to the Department of Parks to be distributed in Central Park, and beginning in 1927, under the auspices of the Women's League for the Protection of Riverside Park, boys and girls from West Side public schools built birdhouses that the Department of Parks hung in the trees along Riverside Drive. In April 1928, the pupils' birdhouses were installed on the occasion of the first Bird Day celebrated in New York City, initiated by the Women's League for the Protection of Riverside Park (figure 3.4). The women's initiative to hold a birdhouse contest as well as a contest for posters that lobbied for cleanliness and proper behavior in parks met with an enthusiastic response by the Board of Education and public school principals, teachers, and pupils—so much so that the contests and Bird Day were repeated in the following years, and the women requested that the Department of Parks continue to hang the pupils' birdhouses in the trees of Riverside and Inwood Parks. In 1929 the department installed 100 birdhouses in street, avenue, and parkway trees, followed by 120 in 1930. It also

FIGURE 3.4.
Birdhouse contest organized by the Women's League for the Protection of Riverside
Park, 1929. (Women's League for the Protection of Riverside Park [MS 139],
box 9, folder 8, image 92638d, New-York Historical Society.
Collection of the New-York Historical Society)

arranged for the schoolchildren's work to be exhibited in the Education Hall of the American Museum of Natural History in March 1930 and 1931. The women were careful in their efforts, making sure that the birdhouse initiative did not undermine its purpose by inquiring at the USDA whether birdhouses, with their crevices, ran the risk of encouraging the breeding of the gypsy moth, a question that was answered by entomologist and moth expert Alfred Francis Burgess in the negative.[22]

Many elaborate birdhouse designs resulted from these initiatives. Often the children sought to copy the design traits of human houses they knew. One Girl Scout produced a miniature Queen Anne cottage. A schoolboy copied Edgar Allan Poe's house in Fordham, and another boy was inspired by a garage he had seen in an automobile magazine. Of course, some well-meaning but spoilsport naturalists criticized this practice, which often did not fulfill the individual bird species' needs and seemed to waste time and energy on representational effects rather than on the birdhouses' function. Yet the children were, probably quite unconsciously, following a contemporary vogue among birdhouse producers. Companies founded for the purpose, like the pioneering Jacobs Bird House Company, built birdhouses, especially the ones for purple martins, in different architectural styles. The Jacobs Bird House Company advertised its houses in its own *American Bird House Journal*. J. Warren Jacobs, who had developed his first martin house in the late nineteenth century, began his birdhouse business in 1908, gradually expanding his product range to include "single-room nesting boxes," an "Audubon Bird Lunch-counter," the "Jacobs Feeding Booth," and

FIGURE 3.5.
Jacobs birdhouses were found
to add "beauty and charm,"
especially if positioned in the
center of an elaborate flower
display or at regular intervals along
a vista. (*American Bird-House
Journal* 5, no. 1 [1921]. From the
collections of the Ernst Mayr
Library, Museum of Comparative
Zoology, Harvard)

the "Government Sparrow Trap." His martin houses came in different styles and sizes, including one thought to resemble the Parthenon. Model 4, rumored to be favored by Henry Ford for his garden, came as a miniature palace boasting seventy-eight "rooms," whereas model 5, which resembled a mansion, was the preferred birdhouse on many estates of the various Rockefeller family members. Model 1 had twenty "rooms" and could be ordered with or without balconies. The journal also gave examples of birdhouse site design (figure 3.5).[23]

Popular scientific literature on birds and their protection, including the construction of birdhouses, birdbaths, and feeding stations, abounded at the time. Furthermore, in the first three decades of the twentieth century, the American Museum of Natural History's public lecture program always included lectures on birds, as did the educational materials the museum provided the public schools after it had begun to collaborate with them on a nature study program in the late nineteenth century. A lantern slide lecture set, "The Birds of Our Parks," which addressed how birds could be attracted through nesting boxes and feeding stations, became a staple offered by the museum in the 1910s and 1920s. Among its public lectures for schoolchildren was biology educator Paul B. Mann's talk "Birds and Their Relation to Man," given on the occasion of the first Children's Fair at the museum in 1928. Associate curator and director of the New York Bird and Tree Club G. Clyde Fisher regularly delivered the lecture "Birds in Their Relation to Field, Forest and Garden," and the famous author, naturalist, and founder of the Meriden Bird Club Ernest Harold Baynes contributed "Wild Bird Guests."[24]

## Meriden Bird Club

Many tree-planting and bird-protection activities that unfolded in New York City at the time can be attributed to the charismatic Baynes and his Meriden Bird Club, founded in January 1911. A small village located northeast of the artist and writer colony Cornish in New Hampshire, Meriden was home to many upper-class New Yorkers in the summer months. Many became members of Baynes's club. Not only did the Meriden Bird Club establish one of the first designed bird sanctuaries in the country, it also disseminated knowledge and ideas about birdhouses, feeding stations, and other ways to encourage and protect bird life. These included killing what were considered the primary bird enemies: English sparrows, red squirrels, and cats. The club went so far as to encourage its cat-loving members to cruelly kill their cats by sending them to "the lethal chamber," as one member unabashedly put it in 1912. With these methods, the club sought to turn the entire village into a bird sanctuary, or "bird village," as it was often referred to. Birdhouses lined Meriden's central street, and inhabitants competed with one another in the number of birdbaths, bird feeders, and berry-yielding shrubs set up in their gardens. The actual sanctuary, however, was a converted old farm that had been acquired with funds donated to the club by Helen Woodruff Smith. The club hired Baynes's friend Frederic H. Kennard, a bird-loving landscape architect from Boston, for the design. Kennard's work for Meriden was one of the first landscape designs that explicitly sought to produce a habitat and refuge for wildlife (plate 5 and figure 3.6). It provided open lawn space and ample edge conditions with numerous kinds of berry- and nut-bearing shrubs and trees, and it also included a field for the growing of bird-feeding crops like millet, buckwheat, and sunflowers.[25]

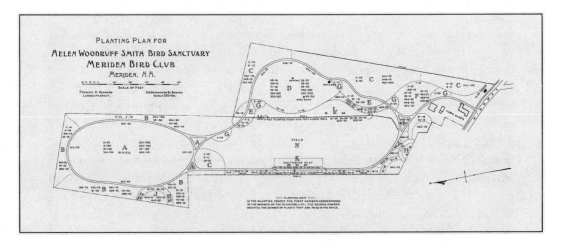

FIGURE 3.6.
Frederic H. Kennard's planting plan for the Meriden Bird Sanctuary, Meriden,
New Hampshire, 1911. (Courtesy of the Frances Loeb Library.
Harvard University Graduate School of Design)

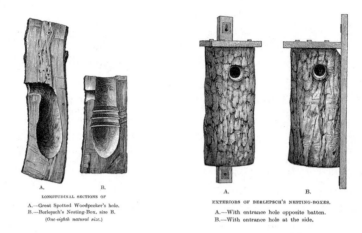

A.                              B.
LONGITUDINAL SECTIONS OF
A.—Great Spotted Woodpecker's hole.
B.—Berlepsch's Nesting-Box, size B.
*(One-eighth natural size.)*

A.                              B.
EXTERIORS OF BERLEPSCH'S NESTING-BOXES.
A.—With entrance hole opposite batten.
B.—With entrance hole at the side.

FIGURE 3.7.
Illustrations showing the von Berlepsch nesting box and its natural model, the
woodpecker's hole. (Martin Hiesemann, *How to Attract and Protect Wild Birds*
[London: Witherby, 1908])

Kennard was a member of the early Nuttall Ornithological Club and the American Ornithologists' Union, and his interest in birds had led him to experiment with bird-feeding stations and birdhouses on his own property in Newton Centre in Massachusetts. But Kennard was not only a man of science. As a landscape architect, he also cared about aesthetics. The rustic appearance of the von Berlepsch nesting box, first produced on the estate of Hans von Berlepsch in Thuringia, Germany, held a special appeal for him. Consisting of a hollowed-out log with a lid and a circular hole on the side, the von Berlepsch nesting box blended inconspicuously and seamlessly with its environs (figure 3.7). It resembled natural bird homes more closely than any other nesting box designs. Kennard's attention had been drawn to the German nesting box through the 1908 English translation of Martin Hiesemann's book *How to Attract and Protect the Wild Birds*, recommended to him by Baynes. Besides the aesthetic appeal the box held for its human patrons, birds in Europe had taken to it. The German baron's lands and forests had been liberated from insect plagues as a result of the nesting boxes attracting birds. Largely due to Kennard's and Baynes's promotion, the von Berlepsch nesting box became an import hit among American bird lovers before the Meriden-based Audubon Bird House Company was finally able to produce successful copies. However, the American fascination with von Berlepsch inventions did not stop here. Kennard adapted the von Berlepsch food station to American needs, and at the Meriden Bird Sanctuary von Berlepsch bird-food recipes were copied and his "food trees" were erected. These trees had been felled, then erected again and covered with a food blend consisting of suet mixed with ground bread and meat, crushed hemp, sunflower seeds, oats, dried elderberries, maw (animal stomach), and ants' eggs.[26]

Expert voices on the von Berlepsch nesting boxes varied, however. While Baynes and Kennard could not envision anything better for species like hairy and downy woodpeckers, screech owls, bluebirds, flickers, and chickadees, and the boxes at their homes and in the Meriden Bird Sanctuary as well as in other locations proved hugely successful, Massachusetts state ornithologist Edward H. Forbush was skeptical. After three years of trial in the United States, he claimed in 1915, birds in Massachusetts still preferred natural hollow limbs or rectangular boxes to the inventions of Hans von Berlepsch. But perhaps Forbush was arguing on the basis of his expert ego and the fact that he himself had already in 1905 proposed boxes made out of hollow limbs and out of bark, in the latter case following a suggestion of his ornithologist colleague William Brewster.[27]

Despite varying opinions about some of the best methods to attract birds and the professional vanity at play in the discussion, it was precisely the Meriden Bird Club's popular appeal that quickly turned it into a model for similar clubs throughout the entire country. One of the vehicles of dissemination was the bird masque *Sanctuary*, written by poet and dramatist Percy MacKaye and performed on the occasion of the 1913 official inauguration of Meriden's bird sanctuary. The audience included the First Lady and President Woodrow Wilson, whose daughters were part of the initial cast. *Sanctuary: A Bird Masque* was a call for bird protection and a condemnation of plume hunters and the millinery trade, and the prominent audience drew heightened attention to these concerns. The masque was so successful that the New York City Civic Forum invited the Meriden Bird Club to perform it in the Hotel Astor ballroom in February 1914. When it subsequently toured for twenty weeks through 120 towns in the Southeast and Midwest, the performance provided Baynes with the perfect occasion to further the cause of bird and wildlife protection.[28]

The movements for forest and wildlife conservation outside of cities, which, especially in the case of forests and birds, were led by women, also reached urbanizing areas like New York City. These movements were linked through the personal experiences of their leaders, who often belonged to the social elite and could observe the transformation of the larger landscape during their summer retreats to the countryside. Attention to nature conservation outside of cities led early environmentalists, social and health reformers, philanthropists, and some members of the social elite to develop and engage in tree and bird protection inside the cities as well. Fewer birds due to urbanization and hunting for a surging urban fashion industry meant more frequent insect menaces both inside and outside the urban centers. At a time when women were increasingly seeking to influence the public realm and urban sphere, they brought these concerns regarding the larger landscape of the United States to the attention of city officials and business leaders.

## Female Designers, Planters, and Pruners

In the nineteenth and early twentieth centuries, many New York women played leading roles in organizing tree planting and mobilizing conservation efforts and many more participated. Their impact was significant, albeit indirect. Only a few women were in positions in which they could determine or make decisions about the physical shape of the city. But in the early decades of the twentieth century, women increasingly moved from voluntary organizational and activist work into professional positions and public posts in which they could actively shape the city forest and its new urban environment through the planning and design of tree plantations. Trees belonged to a class of materials that pioneering female landscape architects, who often came to the field through an interest in horticulture and plants, were very familiar with. Street tree planting seemed to be a logical extension of this interest into the public urban realm.

Female landscape architects like Mrs. Frederick Hill and Marian Cruger Coffin had earlier provided design work in private practice for the City Gardens Club, but a milestone was reached in 1936, when Clara Stimson Coffey, who had trained and collaborated with Coffin and was, like her, a dedicated plant specialist, became chief of tree plantings for the New York City Department of Parks. Under park commissioner Robert Moses, she was in charge of planting designs for the Hutchinson River and Belt Parkways. She continued in her position until 1942, when she returned fully to private practice, both on her own and in collaboration with other landscape architects and architects.[29]

Individual women did not confine themselves to the design and planning departments. They also sought to enter the male-dominated sphere of tree work. When the Department of Parks sought to add to its inadequate workforce in 1939, six women were among the eight hundred applicants for thirty jobs as tree climbers and pruners in the forestry service of the department. Climbers and pruners belonged to the labor class of municipal employees, like bakers, porters, and seamstresses. Only persons between twenty-one and thirty-two years of age were eligible to apply, and the requirements included "extraordinary physical ability" and "three years experience over the age of 18 in a tree nursery or other position where the care and the protection of trees were among the chief duties." An alternate requirement was "a full four year course in agriculture in a recognized institution followed by a degree in forestry or agriculture." Candidates had to pass physical and written examinations, strength and agility tests, and practical tests that included tree identification, elementary tree surgery, tying knots and hitches, and shinnying up a forty-foot tree with ease (figure 3.8). Most women who worked for the Department of Parks at the time were employed as playground directors and supervisors or as matrons in comfort stations. The requirements for these jobs were very different. They demanded less specific skills, less physical strength and agility, and no movements or clothing that would at the time have been considered inappropriate female attire.[30]

Yet already at the beginning of the twentieth century, when John Davey had produced his popular *Tree Doctor* with instructions for the planting, pruning, and care of trees, he had addressed not only merchants, farmers, mechanics, laborers, men, and boys, but also women and girls. Although only three of his sons and not his daughter entered his business, Davey had illustrated to his readers that "ladies" were perfectly capable of pruning the tops of small young trees so that they would grow a "more *bushy*" head. Nevertheless, in the 1930s it was still unheard of for a woman to be em-

FIGURE 3.8.
Practical test for the job of climber
and pruner. (*Official May Civil
Service Bulletin*, City of New York,
Civil Service Commission [May
1940], World-Telegram photo)

ployed as a climber or pruner. This was reason enough for feminist lawyer and judge Dorothy Kenyon to publicly rejoice in the six female applicants to the pruner and climber positions at the 1939 conference of the Women's Institute of Professional Relations, where she vehemently argued for female equality and women's rights. Some years later, when Davey's company was lacking manpower due to the draft during World War II, the company hired two young women who had finished two years of college course work in horticulture.[31]

In general, however, the female presence in tree work and related fields of science was low. When the Eastern Shade Tree Conference was held at the New York Botanical Garden in December 1938 to assess the damage of the hurricane that had ravaged the Northeast in September, only 14 of the 208 registered attendants were women. Between 1937 and 1947, there were between 3 and 5 female members or associates at the National Shade Tree Conferences, which in those years counted between 392 and 752 members and participants working in the commercial shade tree business, in city forestry and forestry, or as scientists. Regular female participants in these years were plant pathologist Cynthia Westcott; conservation chairwoman of the Plainfield Garden Club Mrs. Garrett Smith from New Jersey; USDA forest and shade tree pathologist Alma May Waterman from Connecticut; Margaret E. Watt from the School of Horticulture for Women at Amber in Pennsylvania; Stephanie T. Adams from Pasadena; and Elizabeth H. Burnside from Iowa, a librarian and the daughter of William Burnside, president of the Hawkeye Lumber and Timber Companies.[32]

## The Business with Street Trees

Urban business districts, the marketplaces of the Taylorized city, were determined and controlled by men. For decades they had therefore been considered a male domain, as opposed to the residential areas that were the sphere of women. Around the turn of the century, this distinction gradually began to weaken. The introduction of street trees throughout entire urban areas and especially in the business districts contributed and gave expression to this development. Street trees, like park spaces, were considered elements and symbols of refinement, civility, and culture. City officials and businessmen alike recognized that street trees could also elevate the market value of an entire city, not to mention individual properties, and they performed this function without their owners having to spend a great deal of expenditure and effort. Vigorous and healthy trees with a full, lush canopy were "thrifty" in more ways than one. In New York City, where space was at a premium and air space was discovered as real estate, street trees were an ideal means of supplying the city with beauty and promoting public health. The Tree Planting Association explained, "Instead of spreading its foliage upon the ground which would require a large surface area, the tree lifts its acres of vegetation on a small trunk and spreads it in the air."[33]

No campaign for street tree planting and no bulletin issued on urban trees ever missed the economic argument. In some publications it was even listed first among the many arguments for a shaded city. In the capitalist city, tree advocates knew not only their trees' value but also the value of playing this financial card. It sometimes appeared to be the only way to persuade inner-city business interests. Economics could legitimize the planting of trees and convince the last naysayer. With the economic argument the tree-planting promoters also found a way to rationalize their request. Arguing for tree planting along economic lines was rational, scientific, and progressive—everything the male-dominated business world aspired to be.

Yet despite the trees' economic function, business interests and the fast-paced urban development in late nineteenth-century New York City had soon led to the neglect of the trees planted along the city streets in the business district. Trees could be found along residential streets in the wealthier parts of town, but there were hardly any in the city center or in the city's tenement areas. The *New York Times* commented on this division: "The commercial spirit which dominated the influential classes, diminishes civic pride. The city is regarded as a place of business only, and its attractiveness as a place of residence does not appeal to them as of special personal interest. The suburban home with its acres of lawn, its shaded walks, its groves of rare trees, and its conservatories of flowering plants, absorbs all the thoughts and energies of the leisure hours of the 'merchant prince.'"[34]

The typical downtown New York street, then, was treeless, as a photograph of a barren city street published in the 1901–2 report of the Tree Planting Association showed (figure 3.9). This characterization also applied to Fifth Avenue, the city's wealthy showcase street. While it was lined with upscale retailing, expensive residences, and important religious and secular institutions, street trees were conspicu-

A Typical New York Street.
22

FIGURE 3.9.
A treeless street was considered typical for large parts of New York City at the turn
of the century. (*Annual Report of the Tree Planting Association of New York City,
1901–1902* [New York: Tree Planting Association, 1902]. General Research Division,
The New York Public Library, Astor, Lenox and Tilden Foundations)

ously absent except along the avenue's southernmost part between Washington Square
and Thirteenth Street. The Tree Planting Association wanted to see Fifth Avenue
develop into the shaded street it had once been, a tree-lined access road to Central
Park. The association addressed adjacent property owners in circulars encouraging
them to plant street trees in front of their properties. But by 1910 the situation had not
changed much. Parts of the avenue had begun to be widened, causing tree removal,
and instead of tree plantings, traffic and business enterprises had taken over even
more space along the avenue. The *New York Times* counted the trees on the five-mile
stretch between Washington Square and 110th Street with alarming results: there were
forty-six trees on the east side and thirty-seven on the west. Along the two and a quarter
miles between Fourteenth and Fifty-Ninth Streets, where most stores and businesses
were located, there were only six trees on the east and seven on the west side. In the
following years the Department of Parks removed many more trees than were planted,
and in 1914 Henry R. Francis listed only four trees along the avenue between Fortieth
and Eighty-Sixth Streets, excluding the street trees along the side of Central Park. The
declining number of trees on Fifth Avenue reflected a larger trend in Manhattan as
a whole. In 1913 the Department of Parks reported 402 street tree removals and only
273 tree plantings. Subway construction had caused many trees to die. This was, for

example, the case on Broadway between Fifty-Ninth and Seventy-Second Streets, where only 11 of 208 trees survived.[35]

Artist, architect, and Arts and Crafts proponent Charles R. Lamb commented in 1912 that it seemed that economic forces precluded any artistic building activity and "picturesqueness" in the middle of Manhattan. Lamb served the Tree Planting Association in various offices and was on a mission to improve the city's aesthetic appearance. Once Fifth Avenue had been widened above Forty-Seventh Street in 1911, he presented improvement plans to the Fifth Avenue Association, whose own activities at the time sought to protect retail businesses and control land use by lobbying for a seventy-five-foot height restriction for buildings along the avenue and the banishment of the beggars and loiterers who were affecting the southern parts of the avenue in the garment district. Lamb's proposals included "isles of safety" in the middle of the avenue to be planted with slender trees like Lombardy poplar, and cement planting boxes lining the sides of the avenue between the curb and sidewalk. Besides shrubs, small pruned street trees were to be planted at regular intervals in these boxes, which would be up to eight feet tall (figures 3.10 and 3.11). Indeed, it was the "Pruned Formal Type" of planting that tree expert Laurie D. Cox would recommend for Fifth Avenue below Fifty-Ninth Street some years later. Pruning would adapt trees to the space restrictions

FIGURE 3.10.
Charles R. Lamb's plan for tree planting on "isles of safety" and trees in movable boxes along Fifth Avenue, ca. 1911. (Tree Planting Association of New York City, *Reports of Officers, Charter, By-laws, Committees,* 1912 [New York: Tree Planting Association, 1912]. General Research Division, The New York Public Library, Astor, Lenox and Tilden Foundations)

DETAIL SHOWING TREE BOXES
CLIPPED FOLIAGE, FLOWERS AND CARRIAGE BLOCKS ALONG CURB

FIGURE 3.11.
Perspective sketch by Charles R. Lamb showing his proposed plan for tree planting
in movable boxes along Fifth Avenue, ca. 1911. ("Fighting to Beautify Fifth Avenue
with Trees," *New York Times*, 30 April 1911, SM3)

of the dense and congested business district, add "dignity to the architecture of the
street," and "remove the prejudice of shop owners to trees in front of their places of
business." This type of planting appeared especially appropriate and valuable in New
York City. Despite these good intentions, however, it would be more than thirty years
until the first trees were growing in the areas Lamb had sketched in bird's-eye views
and other perspectives illustrating his visions of improvement.[36]

In the first decades of the twentieth century, treeless streets were the more common sight in cities' business districts. As horticulturalist Furman Lloyd Mulford
pointed out, trees were removed "because they excluded a little daylight, or made a
store less prominent, or were somewhat in the way of using the sidewalk for merchan-
dise." Writing about street trees in 1913, art critic Ada Rainey commented that commercial considerations had come before thinking "of our cities as places in which we
live," but that life and business depended upon one another. What was the use, she
asked, "of extending our vast commercial activities if it narrows our lives, makes them
barren and insensitive to the feeling of nature and its uplifting influence"? However,
Rainey did see a sign of an awakening toward beauty and the intrinsic value of street
trees in a Supreme Court decision in New York City. The Court's Appellate Division
had sustained the judgment of a lower court that had fixed the value of street trees
destroyed by a construction company at $500 each. Besides the mature trees' timber,
the judgment recognized their aesthetic and shade value, which could not be replaced for many years. Commercial business, Rainey noted, was beginning to consider beautiful tree-lined streets a good investment that brought in greater monetary
returns. She observed that "real-estate and other business men have taken to planting
trees in front of their property and are becoming interested in shade trees in cities."[37]

*At the top is Palisades Avenue, Englewood, N. J., as it appears today. Below is the same picture with trees cleverly painted upon it to show how it would look if the proposal to plant trees as a war memorial were carried out.*

FIGURE 3.12.
Two scenes of a commercial street in Englewood, New Jersey, in 1945: its current condition without trees and how it would look if trees were planted. ("Trees in Business Streets: Esthetic or Practical or Both," *American City* 60, no. 7 [1945]: 90)

But opinions varied and continued to vary in the succeeding decades. Most tree advocates, regardless of whether they were city foresters, landscape architects, or the tree-minded public, argued for trees in business streets. Their arguments were often bolstered by visual rhetoric that continued to be used throughout the twentieth century: photographs and photo collages showed a business street or street scene, both with and without trees—the former was considered a significant improvement (figure 3.12). "A well shaded business front gives a cheerful and pleasing appearance which is appreciated by all," Chicago's city forester Jacob H. Prost pointed out. Referring to European business streets, William F. Fox noted that foliage not only provided a pleasing picture but could also hide the marks of trade. But this was precisely what many business owners in American cities feared, besides the tree care and maintenance they would be responsible for. When in 1956 the New York City Department of

Parks wanted to plant trees on Duffy Square in the heart of the entertainment district near Times Square, a fight broke out between property owners and park commissioner Robert Moses. Adjacent property owners feared that the trees would curtail real estate values as they would obscure the flashing electronic advertising signs that were already characterizing the area. Walter J. Salmon of the Atlantic Leasing Company was particularly adamant as one of the signs that he had been renting to the Coca-Cola and Pepsi-Cola companies in different years and that would potentially be obscured by the new trees had an annual rental value "in the high five figures." Moses, on the other hand, saw in the signs "garish monstrosities" that attracted the wrong types of business and had turned the area into a "cheap, disgraceful honky-tonk district." But as Laurie D. Cox had observed in 1932, "The average American business man instinctively feels that business streets and trees do not mix."[38]

## The Fifth Avenue Tree-Planting Controversy

Although not the average businessman, Philip LeBoutillier, president of Best & Company, felt that trees did not belong along Fifth Avenue and certainly not in front of his new flagship department store on the corner of Fifty-First Street (figure 3.13). In 1944 the store moved to Midtown from previous locations on Sixth and Fifth Avenues along

FIGURE 3.13.
Wurts Bros., *Best & Company, final view*, 23 May 1947. No trees occupy the space between the curbs and Philip LeBoutillier's flagship store on Fifth Avenue. (Museum of the City of New York. X2010.7.1.9223)

the "Ladies' Mile." Specializing in women's clothing and children's wear, Best & Company was one of the most popular department stores among New York women. LeBoutillier's position on trees, however, quickly became very unpopular among women and men alike.

The Fifth Avenue tree-planting controversy that raged in 1945 provoked articles, radio contributions, and even a poetry battle. The dispute was incited by a letter to LeBoutillier by Iphigene Ochs Sulzberger, philanthropist, *New York Times* heiress and trustee, and president of the Park Association of New York City, which included a Tree Committee. Founded in 1928 to promote the city's parks and acting in a supervisory capacity to the Department of Parks, the nonprofit Park Association had begun to promote street tree planting in the 1930s. Sulzberger approached LeBoutillier asking him to consider planting trees in front of his new store. LeBoutillier answered brusquely in the negative, reaffirming and elaborating on his reaction to the planting of the Rockefeller elms on Fifth Avenue some years earlier.[39]

In March 1939, the planting of eight mature, fifty-foot-tall English elms along the Fifth Avenue frontage of the Rockefeller Center between Forty-Ninth and Fifty-First Streets, sponsored by the Rockefeller Center Corporation, had been celebrated as a spectacle. The event was reported across the continent. The *Seattle Daily Times* exclaimed that "men, women, children, and dogs were excited" when they found "a tree—a real one, twenty-five feet tall—spread over the sidewalk of Fifth Avenue." The choice of elms was considered ideal for the Fifth Avenue location because their high branches enabled buses to pass unhindered and shoppers to peer into unobstructed store windows. The elm plantings were thought to set an example and inspire emulation among adjacent property owners (figure 3.14). The Department of Parks was dependent on private sponsors because it lacked funds and more than 21,000 park and street trees had been destroyed during a fierce hurricane on 21 September 1938. Moreover, the department's forestry workforce was struggling to keep up with the thousands of citizen complaints and the regular care and maintenance of some 977,750 street trees throughout the five boroughs. The department was therefore intent on the Rockefeller tree plantings, not least because the trees could also improve the city's image during the 1939 World's Fair year. The plantings drew crowds over a period of eight nights. Although the Fifth Avenue Association remained skeptical and did not favor trees, other business owners in the vicinity, like Alfred L. Aiken, president of the New York Life Insurance Company on Madison Square, supported the idea, and the Rockefeller elms inspired requests for more trees along both Fifth and Sixth Avenues. Most of these were, however, thwarted by underground obstacles like the subway station on Sixth Avenue and underground sidewalk vaults in front of Saks Fifth Avenue.[40]

Inspired by the Rockefeller Center tree plantings, radio host Major Edward Bowes gifted four forty-foot elms and eight thirty-foot Schwedler's maples to St. Patrick's Cathedral opposite Rockefeller Center. The elms, which were planted in front of the cathedral along Fifth Avenue in October 1939, were thought to complement the Rockefeller Center's work along this stretch of the avenue (figure 3.15). In the wake

FIGURE 3.14.
Fifth Avenue's first elms, ca. 1940. (Photo by Edward Ratcliff.
Museum of the City of New York. X2010.11.4689)

of the Rockefeller elm tree plantings and the World's Fair beautification efforts, the number of private property owners who requested and sponsored street tree plantings surged from 139 in 1939 to 278 in 1940. Inspired by these plantings, in 1942 the Live in Manhattan Committee led a drive to plant ten thousand new trees in the borough, an initiative backed by the Department of Parks (figure 3.16).[41]

Whereas these property owners looked upon the elms favorably and supported the Department of Parks in its efforts to plant more street trees, LeBoutillier remained opposed and defensive. His arguments in 1945 were simple: trees did not belong in the city; urban conditions were unnatural and therefore unsuitable; trees attracted birds that soiled sidewalks; and they interfered with the beauty of some of the avenue's important buildings such as the New York Public Library, St. Nicholas, St. Patrick's Cathedral, the Pulitzer Fountain, and the Plaza Hotel.[42]

In an effort to bring LeBoutillier around, the office of borough president Edgar J. Nathan Jr. conducted a survey among seventeen prominent architects, landscape architects, and artists. The professionals were architectural etcher John Taylor Arms; architectural artist Hugh Ferris; architects William Adams Delano, Harvey Wiley Corbett, Otto R. Eggers, Eric Gugler, Arthur Holden, Ely Jacques Kahn, James W. O'Connor, Perry Coke Smith, Ralph Walker, and Edgar Williams; landscape architects Charles Downing Lay, A. F. Brinckerhoff, Alfred Geiffert Jr., and Ellen Shipman; and Department of Parks landscape architect Gilmore D. Clarke. In contrast to Le-

FIGURE 3.15.
St. Patrick's Cathedral, 1946. (Photo by Irving Underhill.
Museum of the City of New York. X2010.29.375)

Boutillier, all seventeen thought that trees enhanced architecture, and with the exception of one abstention they affirmed that trees would add to the avenue's beauty.
A. F. Brinckerhoff pointed to the "humanizing effect" of street trees, which, with their
welcome shade in the summer and their "tracery of shadows in winter, . . . contribute
to the amenities of life." His landscape architecture colleague Gilmore Clarke considered trees an amenity on residential and business streets and an indicator "that we are
becoming just a little bit more civilized."[43]

     LeBoutillier's steadfast, stubborn, and recalcitrant reaction inspired some hu-
morous attacks as well. He himself had made the dispute public by sending copies of
his and Sulzberger's letters to the *New York Times*, the *Herald Tribune*, and the *New*

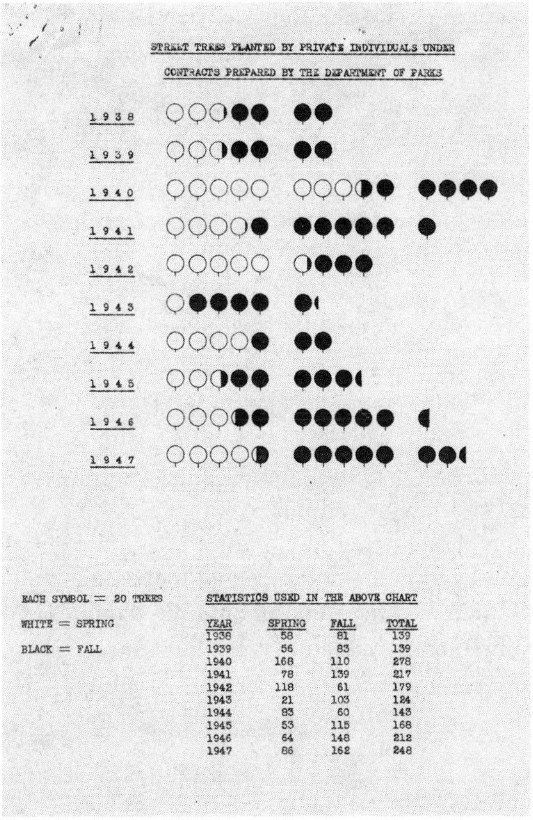

STREET TREES PLANTED BY PRIVATE INDIVIDUALS UNDER

CONTRACTS PREPARED BY THE DEPARTMENT OF PARKS

1938  ○○○●● ●●

1939  ○○○●● ●●

1940  ○○○○○ ○○○◐● ●●●●

1941  ○○○○● ●●●●● ●

1942  ○○○○○ ○●●●

1943  ○●●●●● ●◖

1944  ○○○○● ●●

1945  ○○◐●● ●●●◖

1946  ○○○●● ●●●●● ◖

1947  ○○○○◐ ●●●●● ●●◖

EACH SYMBOL = 20 TREES

WHITE = SPRING

BLACK = FALL

STATISTICS USED IN THE ABOVE CHART

| YEAR | SPRING | FALL | TOTAL |
|------|--------|------|-------|
| 1938 | 58 | 81 | 139 |
| 1939 | 56 | 83 | 139 |
| 1940 | 168 | 110 | 278 |
| 1941 | 78 | 139 | 217 |
| 1942 | 118 | 61 | 179 |
| 1943 | 21 | 103 | 124 |
| 1944 | 83 | 60 | 143 |
| 1945 | 53 | 115 | 168 |
| 1946 | 64 | 148 | 212 |
| 1947 | 86 | 162 | 248 |

FIGURE 3.16.
Diagram illustrating private sponsorship of street trees in Manhattan, 1938–47. (Department of Parks and Recreation, General Files, New York City Municipal Archives, Manhattan, 1948, folder 19. Courtesy NYC Municipal Archives)

*Yorker.* Reactions in the latter magazine followed quickly. Poet Frank Sullivan took up the controversy in "Friendly Advice to a Recalcitrant Mercer":

. . . . . . . . . . . .
Why, forward looking business leaders
All speak well of elms and cedars;
Every man these days foresees
A city thickly sown with trees.

. . . . . . . . . . . .
I see that block forlorn and bare,
Sizzling in the summer air

. . . . . . . . . . . .
Be warned, Phil, be less adamant!
Before it is too late, recant!
Get back to nature, man, be free!
Disport yourself, go climb a tree;

. . . . . . . . . . . .
Why, you may be the first employer
To shoot the works and plant a sequoia![44]

But LeBoutillier was not so easy to defeat. He took up the challenge and responded in style:

Little boy Sullivan
Go blow your horn.

. . . . . . . .

For everything rates
In its proper place
The corn on the cob
The nose on the face
And things out of order
Like fish out of water
Arouse my deep pity
Like trees in the city.[45]

. . . . . . . .

LeBoutillier's poetic verve did not stop here. He wrote another poem commenting on Chief Magistrate Henry H. Curran's criticism of his anti-tree stance voiced at
the dedication of two tree-planting ceremonies in Greenwich Village:

When Curran dedicates a tree
He has both eyes on publicity.

. . . . . . . . .

Without the slightest provocation
He makes the smallest situation
A public one—

. . . . . .

So when he dedicated a tree,
He takes a crack at Phil LeB.

. . . . . . . . . .

Though there is no remote connection
'Twixt Avenue trees and the quiet section
He dedicates—he does confuse
The issue so's to make the news.

. . . . . . . . . .

So Curran stands 'neath a village tree
A small man, pompous as can be,
He hardly even sees the tree,
He has both eyes on publicity.[46]

As amusing as these publicity-reeking poems were, LeBoutillier knew they would
not win the battle. He commissioned the Connecticut Tree Protection Association to
review the controversy and assess the site's suitability for tree planting. Perhaps not
surprisingly, the association's Edward A. Connell and Albert W. Meserve sided with

LeBoutillier, arguing that the soil conditions were unsuitable and that indiscriminate tree planting in unfavorable urban locations would be unsuccessful and endanger public support for trees in general. For them, trees on Fifth Avenue between Thirtieth and Fifty-Ninth Streets were untenable unless planting strips existed.[47]

Unfortunately for street tree proponents, history would play out in LeBoutillier's favor, confirming his argument that the site was ill adapted for the tree species chosen. When the third cathedral elm was planted, rock was discovered only six inches below the topsoil, requiring an extensive drilling operation to create a planting pit. Although most Rockefeller elms seemed to begin to thrive in their new environment—only one had to be replaced in spring 1940—in 1948 all had to be cut down as they had supposedly succumbed to seasonal climatic changes and gases from motor traffic. Yet the stress on the trees had already been apparent in August 1940. As Department of Parks director of horticulture David Schweizer had reported then, their leaves had prematurely yellowed and fallen, the result of the heat reflection from shop windows and buildings and of soot clogging their stomata, the pores on the leaves' surface that control gas exchange. In 1948 they were replaced with American elms because these native trees were considered stronger despite the Dutch elm disease that had already affected trees along the entire East Coast. The Department of Parks had suggested London planes or honey locusts instead, and Rockefeller Center's horticulturalist and men in charge had warned of Dutch elm disease favoring pin oaks. But both Nelson A. and John D. Rockefeller Jr. preferred the more majestic elms. Three elms soon died of the disease and in 1951 were replaced again, this time with fifty-foot-tall honey locust trees. In 2017 ten honey locust trees along Fifth Avenue's west side in front of Rockefeller Center and two honey locust trees in front of St. Patrick's Cathedral were the only mature, large trees lining the avenue in Midtown between Forty-Second and Fifty-Eighth Streets. Although the early twentieth-century visions were not completely realized, private initiatives have had a more lasting effect than skeptics may have imagined.[48]

Many tree plantings in New York City were initiated, organized, or even designed, planned, and undertaken by women. Women embraced street trees as means and symbols of empowerment, emancipation, and even resistance. Street tree planting and care were ways to "conquer," reimagine, and construct relationships with the city. The initiatives were women's attempts to realize what Mary Beard in 1915 described as the "city beautiful,—beautiful not with a lot of expensive cut stone, formidable fences or marble columns, but beautiful with natural parks, with avenues lined with fine trees, and with front yards covered with verdure and blossoms." Street tree plantings contributed to the creation of a controlled, refined, feminized, and civilized urban environment that stood in opposition to the "wild," morally detestable public urban realm characterized by prostitution and crime that was described by many nineteenth-century European and American writers. Creating a naturalized city beautiful was a means to bridge the private and public spheres and transgress the binary of male-coded architecture and female-coded nature. The trees themselves were both living nature and architectural, structural elements in the modern public urban landscape and therefore provided an ideal material for this transgression. Street trees have been for women a material and tangible way of shaping everyday life in cities, from the scale of the neighborhood to that of the entire city. In this way, they have been material and markers of women's activism and design, and of the "other nature" of cities that many women have embraced, regardless of whether they were design professionals or not. Women also used street trees to foster environmental and sustainability citizenship — the sense of responsibility that leads to actions on behalf of the environment and, in the case of sustainability citizenship, goes beyond these to include human rights, democracy, and equality. While street tree planting was a way for women to promote an environmental as well as a social ethics, since the Progressive Era it has also been a mechanism and emancipatory political act that has allowed women to gain, stand, and quite literally mark their own ground in the public urban realm. This became particularly relevant again during the civil rights movement and second-wave feminism.[49]

# 4
# Planting Civil Rights
## Street Tree Plant-ins in New York City

I n the 1973 children's book *What Are We Going to Do, Michael?* ten-year-old Michael and his neighbor Mrs. Jacobson help to save an eighty-year-old southern magnolia tree that is in danger of being cut down to make way for an urban renewal project (figure 4.1). Nellie Burchardt's story is based on actual events that occurred in Brooklyn's Bedford-Stuyvesant neighborhood in the late 1960s and early 1970s. Yet Burchardt's portrayal of her fictional young hero Michael and his friend Mrs. Jacobson belies parts of the true story. In the children's book, the two protagonists appear as white residents of a rundown racially diverse neighborhood. In reality, Mrs. Jacobson was Hattie Carthan, an African American woman in her seventies. By 1970, her neighborhood had become the second-largest African American community in the United States and "a code word . . . for America's unresolved urban and race problems." Whether the author's choice to change the race of her protagonists was inspired by hopes for higher sales numbers, idealist educational and egalitarian aspirations to cultivate white children's awareness of nature in the city and empathy toward its ethnically and racially diverse citizenry, or even unabashed racism, the story itself as well as the changes made to its principal characters reflect the social concerns and anxieties of the time.[1]

But changing the protagonists' race—while perhaps making the story more accessible to the anticipated majority of readers—also covered up one of the most important facts about it. By rallying for the protection of the southern magnolia, successfully saving it and the three historic brownstone buildings behind it, and founding the Neighborhood Tree Corps for the planting and maintenance of street trees, African American citizens of Bedford-Stuyvesant turned trees into a means of empowerment and emancipation within the civil rights movement (figure 4.2). While the planting, maintenance, and conservation of trees became a grassroots initiative of Bedford-Stuyvesant's African American citizens to assert their rights to city spaces in general, the tree-planting and conservation activities provided in particular the most vulnerable

97

FIGURE 4.1.
Illustration by Dick Kramer in
Nellie Burchardt's *What Are We
Going to Do, Michael?* (All rights
reserved. Reprinted by permission
of Franklin Watts, an imprint of
Scholastic Library Publishing, Inc.)

and powerless groups—women and children—with a way to make themselves heard
and seen. Tree planting and "plant-ins" became their tool of community building as
well as a civil right that could be used against ghettoization. Thus, the events leading
up to the magnolia tree's landmark designation in 1970, the implementation of the
Neighborhood Tree Corps in 1971, and finally the foundation of the Magnolia Tree
Earth Center in 1973 as a nonprofit educational institution illustrate grassroots initiatives for the planting, care, and protection of trees that stood squarely within the civil
rights, environmental, and women's movements of the time.[2]

Hattie Carthan, described in a 1979 *Daily News* article as "the matriarch of
[Bedford-Stuyvesant's] greening movement" and in many other newspaper articles of
the time as Bedford-Stuyvesant's "tree lady," assumed her leadership position sponta-
neously in 1964, the first year of the civil rights riots in Bedford-Stuyvesant. The riots
were a response to the shooting of a fifteen-year-old African American student by a
police officer. Ensuing protest rallies led to violent fights between the police and predominantly African American citizens and to the destruction and looting of stores
in Harlem and Bedford-Stuyvesant. They exacerbated a situation Carthan had long
been observing: many streets had turned from safe, tree-lined havens into rat-infested
"slums" littered with trash (figure 4.3). As in other parts of New York City, in Bedford-
Stuyvesant it had become unsafe to go outside. "Airmail"—the residents' description
of garbage sailing out of open windows—often failed to reach the trash bins that land-

FIGURE 4.2.
Magnolia tree at 697 Lafayette Avenue, February 1970.
(Photo by Irvin I. Herzberg. Brooklyn Public Library—Brooklyn Collection)

lords and the Department of Sanitation provided in too few numbers, and littered the streets.[3]

To alleviate the situation, in 1964 Carthan, together with seven neighbors, formed the Vernon Avenue T & T Block Association. By gearing the first block association's activities toward the planting of trees, Carthan responded to the lack of trees in Bedford-Stuyvesant. A social survey of the Stuyvesant Heights section conducted in 1964–65 by the Community Education Program in Pratt Institute's Planning Department found that many families considered street improvements and the provision of "benches, tables, and, above all, trees" a pressing need. Carthan's block association's modest beginnings consisted of a fund-raising event that brought in money to buy four trees. In 1966 a visit by Mayor John V. Lindsay to the association's fund-raising

FIGURE 4.3.
A street in Bedford-Stuyvesant, 1963. (Photo © Bob Adelman)

barbecue finally brought the necessary attention and publicity. Carthan's invitation to the mayor and her successful lobbying led to the implementation of the New York City Department of Parks tree-matching program: for every four trees a block association planted, it would receive six additional ones from the city. The tree-matching program would in the following years be copied in other parts of the city, for example, in Manhattan's West Village. As a result of her success, Carthan soon became chairwoman of the Bedford-Stuyvesant Beautification Committee. One of its principal objectives was the formation of more block associations that would plant trees under the tree-matching program. By 1970, almost one hundred block associations had been formed, and more than fifteen hundred ginkgo, sycamore, and honey locust trees, species that were resilient and adaptive to urban environments, had been planted in Bedford-Stuyvesant.[4]

Carthan, her collaborators, and the block associations were unwittingly seeking to restore a condition that had existed around the turn of the century, when many Brooklyn streets had been shaded by trees, numbering around seventy thousand in total (figures 4.4 and 4.5).

Observers in late nineteenth-century Brooklyn noted, "As you ride past intersecting streets in any or all directions, you look down a long, beautiful vista." Due to the initiatives of the Tree Planting and Fountain Society of Brooklyn, the city had by 1895 earned the reputation of being the second city after Washington, DC, when it came to numbers of street trees. Since its foundation by a group of businessmen, financiers,

FIGURE 4.4.
Norway maple trees along a
street in Brooklyn, ca. 1908. (New
York City Parks Photo Archive,
ANR545)

FIGURE 4.5.
Street tree planting on Lafayette
Avenue, looking east from the
corner of Adelphi Street, ca. 1915.
(New York City Parks Photo
Archive, ANR863)

and philanthropists in 1882, the Tree Planting and Fountain Society had encouraged neighborhood cooperation for the purpose of establishing "a regular system of shade-tree planting on the avenues and streets of our city, thereby adding to its attractions as a place of residence, the pleasure, comfort, and health of the people." Besides the protection and planting of trees, the society's objectives included the erection of drinking fountains. It argued that in these efforts of "street ornamentation and cleanliness," united action and an "organized association of the residents and property owners on a block, or in a section comprising several blocks, . . . would enable . . . the best results with the least expenditure." At a time when Brooklyn had not yet been incorporated and New York City had not yet passed its 1902 street tree law, the society maintained further that neighborhood associations were the ideal entities to supervise and report to the city authorities any necessary actions with regard to street trees and civic embellishment. The society even envisioned neighborhood associations to be in charge of planting, pruning, and caring for trees, securing a desirable uniform tree management. Already before Carthan's initiatives, therefore, neighborhood groups had been collaborating with the Department of Parks to plant trees on their blocks. For example, in the early 1940s one such cooperation led to the planting of several hundred trees in the fifteen-block area between Dekalb and Greene Avenues, and Cumberland and Washington Streets southwest of Carthan's block.[5]

## A "Treeroots" Initiative

With the help of a grant from the New York State Council on the Arts in 1971, Carthan and her committee members were able to turn their tree-planting initiatives into opportunities for the nature education of neighborhood youth. The Neighborhood Tree Corps consisted of up to thirty children between nine and sixteen years old who attended classes on tree care and gardening, cared for the community's street trees, and in return received a modest stipend (figure 4.6). To augment the support of the tree corps and the neighborhood trees, the children were equipped with buckets and cans to water the trees rather than with water hoses, which could easily have been hooked up to the fire hydrants (figures 4.7 and 4.8). Knocking on doors to ask for water for the trees, it was thought, would create a larger awareness of the street trees and their needs and draw the neighborhood together on the issue of street tree planting.[6]

In 1975 Joan Edwards, the first executive director of the newly founded Magnolia Tree Earth Center, enabled an extension of the Neighborhood Tree Corps project. In collaboration with the Department of Parks, the tree corps members, college students, neighborhood youth corps workers, and adult volunteers conducted a pilot tree survey of seven hundred blocks. The initiative provided summer jobs and an educational experience for the participants; moreover, the data collection was to be processed by a computer and used to improve future tree planting and care.[7]

As a key tool of neighborhood transformation, tree planting was used for children's nature education, community and identity building, and to rejuvenate the neighborhood and improve its living conditions. Fittingly, the founding motto of the

FIGURE 4.6.
Members of the Neighborhood
Tree Corps loosening the soil of a
street tree pit in Bedford-Stuyvesant
in the early 1970s. (Reprinted from
Joan Edwards, *Caring for Trees on
City Streets* [New York: Scribner,
1975])

FIGURE 4.7.
Members of the Neighborhood
Tree Corps setting out with
buckets to water Bedford-
Stuyvesant's street trees in the
early 1970s. (Reprinted from Joan
Edwards, *Caring for Trees on City
Streets* [New York: Scribner, 1975])

Magnolia Tree Earth Center was "Save a Tree, Save a Neighborhood!" In contrast to earlier initiatives by groups such as the Brooklyn chapter of the Congress of Racial Equality (CORE), which addressed discriminatory municipal policies, for example, irregular garbage collection, Carthan's project was more tangible and direct, and it was initiated by a citizen who actually lived in the spaces that were in need of attention and transformation. Tree planting was a creative and constructive activity that could— its supporters certainly hoped—replace the neighborhood's more common hostility and alienation. It had immediate aesthetic and spatial impact and could be carried out by the citizens directly. While based upon intrinsic values that were implicit in the planting activities themselves—neighborhood improvement, nature education, community building, and mutual social support—the tree plantings were also representational. One of the things distinguishing the tree-planting activities from other civil rights initiatives, therefore, was not only their grassroots do-it-yourself and self-help character but their aesthetic. Trees made a difference in a visual and spatial sense as well as in the more ephemeral atmospheric sense that includes their ability to change the microclimate. Trees could fill the visual and phenomenological gap between top-down policies and the actual experience in the neighborhood. Not only was tree planting an act of community building that could turn the neighborhood into a healthier and more sociable place, it left a tangible result and made the neighborhood *look* and *feel* different. It could give the neighborhood a new identity while at the same time recognizing that the street had been and could be a community social space.[8]

It was this aesthetic and representational aspect of the tree planting campaigns that complemented—and in some cases influenced—federal programs like the Model Cities Program of the time, which was directed toward rehabilitating the inner city and modernizing its fabric. Indeed, Carthan's campaign to save the southern magnolia meant that the initial design for a new Model Cities development in the area had to be altered so that the tree and its three sheltering brownstone buildings could ultimately be preserved. Before the preservationists won the entire battle in the early 1970s, a first, intermediary success was the Housing and Development Administration's agreement to erect a thirty-five-foot-high and forty-foot-long wall that was to replace the brownstones to shelter the tree from wind and protect it from the new urban development. Ultimately, this agreement proved insufficient, and in 1976 the New York City Housing Development Agency was finally persuaded to sell the three brownstones to the Magnolia Tree Earth Center. In the battle for the southern magnolia's protection, which among other things became a political issue in the local 1970 Democratic primary election for assemblyman, the unique physical character traits of the local neighborhood stood against the unifying and standardized designs of the Model Cities program. Through the Neighborhood Tree Corps, the grassroots magnolia tree preservation campaign became quite literally a "treeroots initiative" that carried its values into the adjacent neighborhood.[9]

## Planting Civil Rights

The time when Hattie Carthan began her project, which ultimately led to vacant-lot gardening and other greening projects, was rife with similar activities throughout the city. The activities in Bedford-Stuyvesant inspired emulation, although not all plantings took place in areas as desolate and poor as parts of that neighborhood. For example, in the fall of 1964 on Manhattan's Upper West Side, the 88th Street Block Association, supported by the City's Salute to the Season's initiative for beautification, planted twenty-six trees for which it had raised the unprecedented sum of $3,000 in less than three weeks. The tree planting ceremony on West Eighty-Eighth Street, attended by Mayor Wagner and other city officials and civic leaders, celebrated "the largest number of trees ever donated by private interests in the City," as a press information fact sheet noted.[10]

In Bedford-Stuyvesant itself, five blocks near Lafayette Avenue between Marcy and Nostrand Avenues were part of a pilot window-box program subsidized by the Park Association of New York City, which collaborated with block association leaders for the first time in 1965 (figure 4.9). The program, which consisted of the installation of window boxes filled with red geraniums and ivy, was modeled on an initiative begun in Philadelphia in the early 1950s by Louise Bush-Brown, who directed the Pennsylvania School of Horticulture for Women from 1924 until 1952. Although transferred first to Bedford-Stuyvesant, the project was planned to be emulated also in other rundown neighborhoods of New York City. Similar programs were established in other cities, including Boston, Detroit, Chicago, Buffalo, Pittsburgh, Washington, DC, and Cleveland. Its declared objectives were to use "windowboxes, trees, and other methods of street beautification . . . 1. To make the streets of deprived neighborhoods more beautiful, giving residents a sense of pride and pleasure in their block. 2. To give

PARK ASSOCIATION OF NEW YORK CITY, INC.
15 Gramercy Park, New York, N.Y. 10003 OR 4-8822

# WINDOWBOXES CAN DO WONDERS FOR A CITY BLOCK!

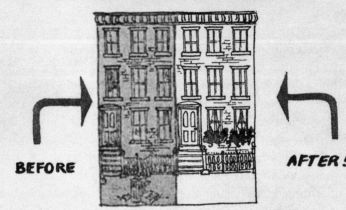

**BEFORE** → ← **AFTER!**

HAPPY BLOOM DAY!

Your windowbox has geraniums and ivy.

The ivy is planted in the windowbox in separate pots. They are removable so that you may bring them into your house in the fall.

The geraniums are planted directly in the soil in your windowbox. They cannot be planted in their pots and grow well. If you wish, you may remove them separately in the fall and replant them in pots. But geraniums really should be new and fresh each year.

KEEP YOUR WINDOWBOX WATERED.

For replanting: Use your present windowbox and soil.
Just use a big spoon -- soften up the soil and make room for the pots of ivy. Plant them as they are -- IN THEIR POTS!

Dig holes for your geraniums -- take them OUT OF THEIR POTS. Be gentle so as to leave their soil around them. Transfer them to your windowbox. Now, pat the extra soil from your windowbox around them all.

Water very well -- keep your windowbox watered.

NOW YOUR WINDOWBOXES MAY BE LEFT OUTSIDE ALL YEAR. JUST REMOVE THE IVY IN THE FALL -- IN THE POTS THEY ARE PLANTED IN. YOUR HOUSE WILL LOOK PRETTY WITH IVY PLANTS INDOORS.

When next spring comes, you will have ivy all ready to put into your windowboxes again.

Your Park Association appreciates the way you are improving your block. We'll continue to help you every way we can.

THANK YOU!

**BRING BEAUTY TO YOUR BLOCK — JOIN THE WINDOWBOX CAMPAIGN NOW!!**

FIGURE 4.9.
Leaflet issued by the Park Association of New York City on the occasion
of the window-box pilot program carried out in Bedford-Stuyvesant in 1965.
(Parks Council Records, Drawings and Archives Department, box 11, folder 25.
Avery Architectural & Fine Arts Library, Columbia University)

children and adults an opportunity to handle, care for, and study growing plants, and awaken an appreciation of natural beauty. 3. To use the windowbox program as a tool to create viable block leadership which can spark block residents to other self-help activities." Besides the window boxes, trees were considered an important component in turning blocks that showed signs of deterioration into "garden blocks" that were cared for. From the beginning, the Park Association aimed at involving the Department of Parks to sponsor and plant trees on the blocks participating in the window-box program. Since its inception, the association had promoted street tree planting, and tree plantings were also used to celebrate the program's accomplishments.[11]

The Park Association was convinced that "natural beauty" had a role to play in the improvement of deteriorating areas. But although the window boxes were appreciated by black neighborhood leaders for fostering pride and neighborliness, critical voices saw them as a cosmetic treatment incapable of causing real change. Their otherworldliness and ineffectiveness was commented upon cynically by one resident: "Man, if you want those boxes to relate to what's happening, fill 'em with grass [marijuana]." Indeed, the first studies of the program undertaken in 1966 by political science professor Charles S. Asher, Donald Neuwirth, Jeff Offerman, and "Mrs. Sedlin" suggested that window boxes and good intentions were not enough to resolve social and physical problems in the long term. The scientists also pointed out that the first blocks targeted by the program were already in relatively good condition before the program began, with homes belonging to a stable middle class. Although the Park Association sought to collaborate with groups such as MEND (Massive Economic Neighborhood Development), ACTION (American Council to Improve Our Neighborhoods), and HARYOU (Harlem Youth Unlimited), it also realized that its activities as an outside white organization in African American neighborhoods could cause significant tension. Somewhat paradoxically, its mandate therefore became to keep the window-box program "simple, politics-free, and controllable."[12]

In contrast to Carthan's bottom-up initiative, organized by her and other women who succeeded in obtaining governmental support, most other activities, therefore, had a top-down structure. They were led either by government officials or by the volunteers of philanthropic organizations like the Park Association, and they were sponsored by the city, federal grants, and particularly private donor money. For example, the New York City Housing Authority began sponsoring a summer garden contest among its six hundred thousand public-housing tenants in 1963. Based on the experience of a similar program in Chicago, the Tenant Flower Garden Competition contributed to a reduction of vandalism and "airmail." It grew consistently in popularity in the first two decades, with 105 garden design entries at the beginning and 283 in 1971, and by the end of the 1970s reaching over 1,000 group entries in 250 different housing projects throughout the city. Involving more than 10,000 tenant gardeners by then, the program had become one of the largest of its kind in the United States (figure 4.10).[13]

In addition to the elaborate flower garden displays, which created a more pleasurable and amenable environment, adding, as the housing authority stated, "color, beauty and a pervading sense of warmth to the surroundings," a vegetable garden

FIGURE 4.10.
The garden at Tilden Houses in Brownsville, Brooklyn, ca. 1970. (The La Guardia and
Wagner Archives, La Guardia Community College/The City University of New York.
The New York City Housing Authority. 02.015.15051)

category was added to the competition in the 1970s. Beginning in 1974 during Mayor
Abraham D. Beame's term, when the financial crisis hit the city so hard that it almost
had to declare bankruptcy, by 1977 the vegetable garden competition already yielded
362 entries. Its winners went by names such as The Busy Bees of 305, Jolly Green
Giant, Organic Oasis, Getting It Together, and Martin Luther King, Jr. Garden. Besides its social and financial benefits, the successful annual competition also helped
to countervail the image of the housing authority's cold and sterile housing developments. But to do so, strong management and control were deemed necessary. In 1969
Ira S. Robbins, who sat on the board of the New York City Housing Authority, cited
the competition as proof that even initiatives from the top could lead to cooperation,
which he considered "the antithesis of paternalism." He thought they could flow "over
into behavior which respects other facets of physical maintenance." According to him,
the competition could "even stimulate involvement in larger aspects of community
betterment." Centrally organized and managed by the housing authority, all applicants received $25 for materials, a budget that was sponsored by a $25,000 grant from
the Vincent Astor Foundation. In the competition's first round, gardens that lagged
behind in their planting schedule, lacked maintenance, or showed signs of vandalism
were banned from further participation. A second round, which entailed the screening

of all remaining gardens, brought the number down to the finalists, who competed for first prizes in their borough and for one overall prize in their category.[14]

On a national level, tree planting initiatives soon figured prominently in the First Lady's beautification campaign, which followed the May 1965 White House Conference on Natural Beauty. Among the initiatives for the beautification of the national capital in 1967, Lady Bird Johnson stressed the planting of new street trees as well as the establishment of benches and trash cans. Street trees soon became the theme of beautification efforts in Washington, DC, and although money from private donors was allotted only to street trees in the wealthier areas of the city, the First Lady made sure that government funds were found to plant street trees in the poorer parts of the district as well. In particular, money was appropriated for the planting of trees in the Shaw neighborhood, home to many African Americans, which had suffered severely during the riots following the assassination of Martin Luther King Jr. on 4 April 1968. The International Shade Tree Conference honored Lady Bird Johnson for her efforts with a Special Merit Award.[15]

In New York City in 1967, after he had witnessed Carthan's campaign, Mayor Lindsay appropriated $1.2 million for tree planting and removal. A drought in 1965 had killed many street trees and tree care had been neglected, a situation that earned much criticism from concerned citizens. Park commissioner August Heckscher, who assumed office in the spring of 1967, was leading the Department of Parks in a new direction, emphasizing the decentralization of facilities and programs, including community- and neighborhood-oriented street tree planting. For him, "street trees [were] miniature vest pocket parks." Although Heckscher had pushed for a much larger appropriation of $1.9 million for street tree planting and $6 million for their care, the city had never before set aside so much money for street trees, and it was also seeking to attract private donor money. To entice neighborhood involvement and educate citizens about street tree planting, in 1969 the Department of Parks organized a display of street tree planting at the first Bryant Park Flower Show, held in October that year. With their initiatives, Mayor Lindsay and his new park commissioner were also jumping on the bandwagon set into motion under mayoral predecessor Robert F. Wagner on the occasion of the 1964–65 World's Fair. Mayor Wagner had granted an appropriation of $675,000 for the planting of around 5,000 new street trees along the city's major thoroughfares in 1963 and again in 1965. The tree-planting program had been initiated by philanthropist Mary Lasker and political consultant Anna Rosenberg (Hoffman). Lasker, who had already in the 1950s become an influential financial backer of street trees in Manhattan, sponsored the project and consequently also had a large say in where the trees were planted. A 1960 street tree survey had revealed, perhaps not surprisingly, that Manhattan had the fewest street trees of all five boroughs. In the city's central borough, only 15,200 trees were counted, compared with 48,000 trees in the Bronx, 117,000 in Brooklyn, 293,000 in Queens, and 76,000 in Richmond (today Staten Island). The World's Fair tree planting program addressed Manhattan's lack of street trees by planting 3,982 of the 5,000 new trees in this borough.[16]

Although these planting efforts in the 1960s targeted all boroughs, Richmond—not a favorite of Lasker—and the Bronx were allotted only 187 and 131 new street trees respectively, and street tree planting in Harlem remained contentious and complicated. East Harlem's inspired community leader the Reverend William Henry Kirk, who served as executive director of Union Settlement from 1949 until 1971, had begun a street tree–planting program with interested citizens in 1960. Kirk, who on that occasion lacked the Department of Parks' support for more than regular pruning and spraying activities of street trees, sought a few years later to ensure that his part of the city would not lose out on the World's Fair street tree–planting project. After all, East Harlem provided transportation links not only to the Bronx but also to the Queensboro Bridge and the Midtown Tunnel, routes that would lead visitors to the fairgrounds. Park commissioner Newbold Morris conceded that the World's Fair street tree–planting project would indeed include street trees on First Avenue between East Fifty-Ninth and East 124th Streets and on Second Avenue from East Forty-Second to East 128th Streets, streets that Kirk feared had been neglected in the initial plans.[17]

But in the early 1970s, Anthony V. Bouza, a police officer then commanding Harlem's police force, was still appalled by the "nakedness of the streets." A 1960s survey in Harlem had indeed revealed (as was the case in Bedford-Stuyvesant) that residents' most common complaint about their neighborhood block, after the larger and more general concerns about housing quality, drug addiction, and crime, was the lack of trees and grass. A first attempt to plant new street trees in summer 1965 as part of Project Uplift, an experimental anti-poverty program launched in Harlem to release the tensions that had led to the riots the summer before, had largely failed. Last-minute planning and failure to recognize that summer was not a good planting time meant that of the 1,500 trees originally planned only 371 were finally planted by the end of November. Realizing the potential of trees in impoverished neighborhoods, Officer Bouza began his own tree-planting campaign, mobilizing Harlem's neighborhood children as well as the Department of Parks, which had abandoned the area out of fear. Bouza's initiative was based on his firsthand observations of the lack of trees in Harlem. Yet it also supported a common and much older belief in the moral, educational, and psychological function of nature.[18]

Both the Tree Planting Society of New York City and the Tree Planting and Fountain Society of Brooklyn had built on this belief around the turn of the century when they had sought to establish connections with the Brooklyn and New York City police forces. According to Brooklyn's Tree Planting and Fountain Society, arboriculture supported "morality—the foundation of noble manhood and good citizenship," and it had a positive effect on citizens' emotional balance and stability. What, therefore, was more logical than to enlist police officers—the city's upholders of moral standards in the public urban realm—in the protection of street trees and to turn the police force into street tree advocates? While they could prevent tree vandalism and initiate charges against offenders, if trained appropriately, police officers could also report different types of tree damage. The Tree Planting and Fountain Society stressed that besides the public benefit, an officer attuned to civic embellishment would undergo personal growth and development and be better prepared to "perform his duties satisfactorily, and better fit . . . for the higher positions he may be called upon to fill." Thus undergoing lessons in nature appreciation and conservation themselves, the police officers supported tree-planting and preservation campaigns in multiple ways. The Tree Planting Association of New York City also cooperated with the police force. In this case, patrolmen were instructed to inform the association about any arrests for tree injuries so that the organization's counsel Charles Thaddeus Terry could attend court cases and make sure that those responsible for malicious injuries would be prosecuted appropriately. The association also saw in this cooperation an opportunity to educate the public about proper tree care.[19]

## Environmental and Social Justice

In contrast to these more or less successful personal initiatives and governmental programs involving street tree planting, the most comprehensive midcentury effort to plant and more evenly distribute street trees throughout the city failed to garner the support of park commissioner Robert Moses and the city council. In 1958 Charles Abrams, chairman of the State Commission against Discrimination, proposed what would become known as the DiCarlo bill. Named after councilman and Bronx Democrat Joseph C. DiCarlo, who introduced it in the city council in June 1959, the bill would require property owners in so-called tree areas—designated either by the City Planning Commission, the Department of Parks, or the office of the respective borough president—to plant street trees in front of their properties. One year after they were planted, the Department of Parks would take over tree care and maintenance. Owners who failed to comply with this new tree ordinance would be charged for their tree and its planting carried out by the Department of Parks.[20]

For Abrams, the bill was an obvious solution to the city's lack of adequate attention to "the natural elements which must be a part of any program of sound urban growth." The bill would finally provide a way for the city government to enforce street tree planting rather than leaving the matter entirely to private initiative. With his proposed amendment to the New York City Charter and the designation of "tree areas" on the city map, Abrams sought to "help preserve and maintain [the respective area's] economic or residential soundness, and . . . help prevent its deterioration into slum or other form of blighted area." For him, street trees increased property values and contributed positively to a neighborhood's appearance, fostering neighborhood identification and pride and thus serving as a means to revitalize neglected areas. Abrams hoped that a uniform, concerted planting of trees on entire blocks would "spark the beginning of a city-wide effort to instill greater community pride and cooperation in general neighborhood up-grading, better street sanitation, compliance with housing laws and improvements of buildings."[21]

For Robert Moses, on the other hand, the bill was impractical, threatening, and unconstitutional. Its consequential increase of street tree planting was a liability. In 1958 his department could hardly keep up with the care of the existing street trees.

The sum of $600,000 was needed to deal with the current backlog, and an additional $250,000 annually was necessary for external contracts that would enable coping with the present workload. The bill also appeared to Moses as an attempt to curtail the power of his department, as it was at first unclear which department would be in charge of designating tree areas. He also considered the proposed bill a zoning law "not related to the normal use of zoning" that interfered with the rights of private property owners. Moses sought to bolster this argument through legal council from attorney Mike Meyer Scheps. But Moses's objections were also personal; in his view, Abrams was "not only a fanatic but an ambitious and malicious one," and his proposal was "ridiculous."[22]

A personal tension had developed between the two men in previous years, when Abrams had been part of the group of neighborhood activists, intellectuals, and professionals, including William H. Whyte, Jane Jacobs, and Lewis Mumford, who successfully fought against the commissioner's plans to build a highway through Washington Square Park and today's Soho. Abrams, who was fighting racial discrimination and seeking to preserve urban diversity, stood against Moses, who was often prejudiced and showed little concern for the poor, the disenfranchised, and ethnic and racial minorities.[23]

But whereas Abrams had stood on the winners' side in the case of Washington Square Park, the DiCarlo bill failed to gain traction and ultimately came to nothing. Abrams had sought to garner support by founding the Temporary Committee for Tree Planting in the spring of 1959. Supporting committee members included Mary Lasker as honorary chairwoman; lawyer and former assistant to President Eisenhower Maxwell M. Raab; Judge Millard Midonick; Katherine W. Strauss, the president of the Women's City Club; Stanley Lowell, former deputy mayor; Goodhue Livingston, member of the City Planning Commission; George Hallett, executive director of the Citizens Union; William North Jayme, who served as secretary of the Municipal Art Society; political advisor Raymond Rubinow; city planner, economist, and administrator Morton Schussheim; and landscape architect Nancy Grasby, among others. Once the bill had been handed off to the city council's Committee on Streets and Thoroughfares, Abrams and his committee proposed amendments that addressed the bill's constitutionality and the reservations of some of the tree committee members. For example, Katherine W. Strauss feared that not all owners would be able to afford tree plantings, thereby preventing an even distribution of trees. Robert Randall was apprehensive about empowering the borough president to designate tree areas. The bill, Randall noted, could easily favor those with access to the political system who did not need trees and harm those without access who wanted trees. One of the amendments to the bill that Abrams proposed was that the borough president should be able to designate a tree area and charge the cost of tree planting to the owners only if at least 25 percent of them agreed to plant trees. Another potential amendment stipulated that a tree area could be designated only if 50 percent of the owners did not dissent. In addition, Abrams pointed out that owners could charge the cost of the tree as an expense on their income tax return or, when the tree was turned over to the city, as a deductible contribution. Although Abrams was able to bring around Manhattan borough president Hulan Jack on the first amendment, many council members opposed not only the bill but trees in general: trees bulged out of sidewalks and blocked windows. Given these reactions, Abrams realized that the bill was "premature politically" and would require "a long educative campaign."[24]

While the protection of private property rights was ultimately decisive, public opinion offered further reasons for the bill's defeat. In a November 1959 letter to the *New York Times* Jesse Kashner used an argument common in residents' street tree complaints of the 1940s to query the DiCarlo bill: the obstruction of streetlights. Kashner argued that more and not less light was needed to prevent the "murder, muggings, hold-ups and all types of crime" that were "rampant to the point where people fear to go out at night." In the eyes of Kashner and others, street trees were a threat to public safety. Rose Lane, who was soon recognized as a "chronic kicker" by the employees in the Department of Parks forced to answer her repeated letters, complained in 1940 that trees on Brooklyn's Avenue I between East Thirty-Fourth and Thirty-Fifth Streets rendered the area dark "as a closet." She herself had been threatened by a man attempting to hold her up on the avenue that was "pitch black" because of street trees, and she colorfully warned, "A body could be lying on the sidewalk or dirt-strip and a passerby walk on them unnoticed. Every once in a while, to my knowledge, a person walks into another person because of this awful darkness. On a rainy night, people have to stop and feel around with their feet (the sidewalk is not in very good condition) as they can't see where they are going."[25]

In a period of increasing crime and mounting inner-city tension, street trees were variously seen as protection, a means of releasing pressure, and a threat. Sociologists and environmental psychologists soon began to explore these different reactions. In the 1960s and 1970s, governmental tree planting and urban greening efforts were sustained by what sociologists had identified by that time as the needs of distressed urban areas: stimulation that could break the monotony of everyday life; a sense of community and beneficent territoriality that could grow out of spontaneous activities; and democratic participatory control over the environment. Implicit in the sociologists' recommendations was an argument sustained by sociologists and psychologists Kenneth Clark and Isidor Chein, who both worked to see their theories applied to federal policies and court cases like *Brown v. Board of Education.* They argued that environmental conditions and their aesthetic expressions played a role in determining an individual's behavior. One line of argument went that environmental degradation reinforced poverty and the humiliation that African Americans encountered in all realms of life. Clark maintained that a "clean and decent and even in some way beautiful" environment would promote a stronger sense of self because it reflected what its inhabitants were worth. Although Clark was basing his argument on the immediate indoor living environment of the house or apartment, his theory could be applied to the larger outdoor living environment as well. In fact, sociologists had realized that it was life on the streets that stood at the center of community life in most "ghettos." The recognition that citizens in poor neighborhoods treated streets not only as pathways but as places that functioned as an extension of the indoors led social psychologists Marc Fried and Peggy Gleicher in 1961 to talk about the residents' *"territorial* space," which, with its permeable boundaries between indoors and out, contrasted with the clearly bounded and often enclosed "selective space" of the middle class.[26]

"Territoriality" was a concept at the forefront of sociological and psychological studies and debate at the time. First conceptualized by geographers of the Chicago School in the 1920s, it regained relevance in the 1950s and 1960s when the problems abounding in urban centers were related to the observation that urban dwellers were increasingly losing "territorial identification with areas immediately surrounding their homes." When the New York City Housing Authority's first Tenant Flower Garden Competition manual announced in 1963 that "the flower gardens you make will be your gardens, just as though they were growing in the yard of a small house," it implicitly relayed one of its objectives: facilitating the tenants' territorial identification and giving them the opportunity to create and shape what sociologists at the time called a "home territory."[27]

Following Jane Jacobs's interest in sidewalk life and her call to action in her 1961 book *The Death and Life of Great American Cities*, the design of streets and sidewalks was also of more general concern among a range of civic and professional organizations at the time. They were reacting to abounding transportation conflicts and increased motor traffic that proved hazardous and an obstacle to pedestrians, and that caused streets to be widened and sidewalks narrowed. Consequently and in protest, in 1964 the New York Chapter of the American Institute of Architects, the New York Metropolitan Chapter of the American Institute of Planners, the Park Association of New York City, the Metropolitan Committee on Planning, the Citizens Union, Citizens Housing, the Planning Council, the Municipal Art Society, and the Committee to Keep New York Habitable sponsored a conference called "The Sidewalks of New York." Talks by architects Simon Breines and Norval White as well as landscape architect Robert Zion's lecture "Designing for the Pedestrian" drew attention to how sidewalks could be turned into attractive pedestrian walkways. Some years later, in 1969, August Heckscher of the Department of Parks and Doris Freedman of the Department of Cultural Affairs promoted turning entire streets into temporary neighborhood festival spaces. A mobile festival truck designed for this purpose created new ground for neighborhood alliances and pride, stressing the idea that the street was the common arena or theater in inner-city neighborhoods.[28]

Building on this more general discussion and on scholarship in the social sciences and behavioral psychology, in the early 1970s, environmental activist and policy analyst Charles E. Little criticized the common devaluation of city streets and the general lack of recognition that they were a commons for inner-city dwellers and "habitation grounds of the urban neighborhood, just as much as the yards of suburbia are habitation grounds." Little had founded the Open Space Action Committee (later renamed Open Space Action Institute, then Open Space Institute) for the preservation of open space in the Tri-State New York Metropolitan Area in 1963. The initiative soon included a "field program" to enable disadvantaged children and teenagers to

**Open space and the inner city**

FIGURE 4.11.
Title image of a special report on open space and the inner city from a 1968 issue of
*Open Space Action*, a publication of the Open Space Action Institute. (Copyright
Open Space Action Institute. Parks Council Records, Drawings and Archives
Department, box 9, folder 7. Avery Architectural & Fine Arts Library,
Columbia University/New Yorkers for Parks)

spend time in nature outside of the city. However, Little also engaged in efforts to
improve the situation inside the city (figure 4.11). He was critical of the double standard
of open space that characterized attitudes in particular toward African Americans living in impoverished urban areas. Demanding environmental and social justice, Little
provocatively and ironically argued that "policies seem to be based on the assumption that poor people, especially blacks and Latins, do not even like greenery, except
maybe watermelon rinds and plantain leaves. The logic of our policy seems to rest on
this syllogism: inner cities have no greenery; poor people live in inner cities; therefore

parks, open space, and wilderness are not necessary for them." In contrast, an egalitarian open-space policy, he argued, could not only improve the quality of the landscape, it could also contribute "in the struggle to achieve a decent, human, and dignified way of life for those Americans who have thus far been denied it."[29]

Besides policies affecting social and environmental activism, some also believed that design practice could contribute to reducing crime. The concept of the street as territorial space, for example, informed architect Oscar Newman's 1970s ideas about how architectural practices could help prevent crime. His design guidelines for "defensible space" in urban renewal areas physically tied together interior living space with the space of the street, leading the residents of the adjacent buildings to have—as Jane Jacobs had said—their eyes on the street. Residents would regard the street as an extension of their own dwelling and therefore as being under their sphere of influence and their responsibility. Newman established design guidelines in a project sponsored by the Department of Justice and the Department of Housing and Development after he had undertaken extensive research into existing designs for public housing in various U.S. cities, in particular New York City. In general, his guidelines suggested replacing the high-rise buildings of midcentury public housing standing in shared open space with low-rise buildings located in open space that was differentiated into semi-public and semi-private territories. But Newman's guidelines focused on architectural and site design; they were ambiguous with regard to trees. On the one hand, trees were an amenity that could reinforce the residents' claim to territory and increase their identification with and care of the environment. On the other hand, trees and shrubs could also heighten the sense of insecurity—for example, when they blocked views and hindered sight on the street. Although detailed examples were lacking in the guidelines, sketches of prototypical site plans suggested layouts in which street trees would not cover facades and ordered plantings would not diminish oversight.[30]

Less preoccupied with direct applicability to specific questions of racial discrimination, civil rights, and criminal activity but following social and environmental psychologists like Clark and Chein, social scientists and sociologists in the 1970s and 1980s continued to explore the relationship between human behavior and environmental perception. Their theories formulated what activists like Hattie Carthan were unconsciously acting upon. The goal was an environment supportive of human well-being, one that lacked distraction in the form of auditory and visual noise like "obtrusive advertising, clutter, and confusion." To create such an environment, those involved would need to be equipped with information that would enable them to make good choices, and they would need to be afforded the opportunity to participate in local control of their environment.[31]

Since the 1970s, studies by social scientists and urban forestry researchers have confirmed some of the general ideas of these theories. For example, a set of studies by U.S. Forest Service researchers has established that trees, in particular large street trees, can deter crime because they can be interpreted as signaling a well-cared for and socially controlled neighborhood; because they are located far enough away from single family houses to prevent visual obstruction; and because they make public spaces more desirable and thus encourage their use. Further studies have shown that people tend to associate street tree plantings with lower crime rates. In keeping with these findings, as early as the 1970s a study had attested that citizens of five midwestern cities found street tree plantings aesthetically desirable in urban settings and that they preferred trees taller than twenty-five feet. In the 1980s, studies of Detroit residents revealed that they preferred street tree plantings in residential areas compared to other types of tree plantings in parks and other parts of the city. The studies ascertained that African American residents in particular preferred ordered or manicured tree environments, in contrast to white residents, who favored "wilder" and more forested scenes. While many researchers were hesitant to draw any conclusions with regard to ethnic and racial differences, arguing instead that some natural respite from the urban environment, heavily characterized by stone, brick, and concrete, was generally considered desirable, the Detroit studies suggested that the effects of "human influence, neatness and openness" in urban nature were "more vital to some groups than to others." In fact, based on their own experience, African American residents often associated more densely vegetated and wooded areas as well as "weedy" grounds with fears of physical danger. In the 1980s, more studies revealed that natural elements like individual trees rather than large open spaces and parks contributed substantially to people's life satisfaction and well-being, confirming that "nature that is most immediately available does, in fact, make the most difference." But as numerous studies by then could attest more generally, people derived "psychological and physiological benefits from trees in a wide range of situations and through both active contacts with trees (e.g. tree planting . . .) and passive exposures."[32]

While such studies again confirmed assumptions that have been voiced throughout human history, it has taken longer for researchers to address the inequities in the distribution of this healthy nature, although they were noted decades before. When Joan Welch approached the question of the relationship between tree canopy cover, land use, and socioeconomics in her 1991 dissertation, she had few precedents to build on. Investigating the Boston neighborhoods Roxbury and North Dorchester, since the 1950s home to many African Americans, she revealed a close relationship between canopy cover, race, income, and poverty rates. The public urban canopy structure was most intact and densest in areas with greater residential stability and good-quality housing. These neighborhoods were inhabited by residents who earned a moderate to high income and had access to the political system. This meant and still often means today that poor neighborhoods, frequently inhabited by citizens of various ethnic and racial minorities, have fewer street trees. Although urban trees have always influenced human activities and the distribution of wealth, the study also confirmed the role of socioeconomic and political factors in the uneven distribution of nature in the city. Welch therefore concluded that an even and equal distribution of urban trees could ultimately also contribute to the improvement of the social milieu.[33]

With regard to community building as well, studies have confirmed the positive impact of planting urban trees. That community participation in urban tree planting could also have a beneficial impact on the trees and reduce tree vandalism was confirmed by projects such as the Oakland Tree Task Force (OTTF). Formed in the 1970s to make tree planting a cooperative venture stressing citizen participation, OTTF successfully planted new trees and led many citizens to take over the care of individual trees. This stood in contrast to the unsuccessful planting of two thousand trees in the context of Oakland's Model Cities program, of which only a few were still standing two years later. Thus the participatory project appeared as a more successful paradigm than the planting projects undertaken independently by the city's Department of Parks. For sociologist Richard G. Ames, tree planting facilitated inhabitants' identification with their particular neighborhood, or territory, allowing them to exercise "mastery and control over their environment, part of the core values of success in American life."[34]

In the 1990s more studies in Detroit showed that tree planting and vacant-lot gardening could enhance a community as a whole as well as people's individual sense of identity and lead to residents' involvement in other projects throughout the neighborhood. Using MillionTreeNYC as their case study in the 2010s, Dana R. Fisher, Erika S. Svendsen, and James J. T. Connolly explored the relationship between the planting of trees as an expression of environmental stewardship and civic engagement more generally. Their study revealed that planting trees has often been a starting point for citizens to also become engaged in other civic realms. Respondents wanted to improve life in New York City, educate the next generations, and strengthen their communities. Although the tree-planting volunteers the researchers interviewed in the 2010s did not enjoy the decision-making authority that Hattie Carthan exercised in the 1960s and 1970s, they expressed a sense of joy and pride when reflecting on their tree-planting experiences.[35]

## Street Tree Activism

In the United States, street tree activism, urban tree stewardship, and civic engagement have contributed to creating the foundations of urban forestry. Early Arbor Day planting activities in the nineteenth century were used for both the nature education of children and the planting of city streets. In 1896 the Tree Planting and Fountain Society of Brooklyn argued in charitable and philanthropic spirit that as far as possible the poor should be given the same excellent living environment as the wealthy. In New York City, tree-planting activities that were specifically geared toward the most disenfranchised began only some years later, during the Progressive Era, with the activities of the 1902 Tenement Shade Tree Committee (figure 4.12).[36]

As a subgroup of the Tree Planting Association founded five years earlier in 1897, the Tenement Shade Tree Committee followed the association's mission to promote urban tree planting and protection but applied these measures to the city's tenement districts (figure 4.13). As was the case with its mother organization, the committee's members included many illustrious men who were either working for or were related to charity and social reform initiatives, like Lawrence Veiller, the Reverend David H. Greer, and Archibald Alexander Hill, secretary of the Tenement-House Committee of the Charity Organization Society and later secretary of the Metropolitan Parks Associ-

FIGURE 4.12.
Model tree planting along the model tenements on West Sixty-Eighth and Sixty-Ninth
Streets. (*Annual Report of the Tree Planting Association of New York City, 1903*
[New York: Tree Planting Association, 1904]. General Research Division,
The New York Public Library, Astor, Lenox and Tilden Foundations)

ation; men who engaged in philanthropy, like Abram S. Hewitt, co-founder with his
father of the Cooper Union; as well as industrialists and financiers like John D. Crim-
mins and Robert Fulton Cutting. Among the committee's sixteen members in 1903–4
were six women, including the philanthropists Edith Carpenter Macy and Ellin Prince
Speyer.

As shown in the previous chapters, most of the inner city's streets lacked trees at
the time, but the situation was particularly tenuous in the densely populated and
disease-stricken tenement districts. There in the 1870s noted physician and later president of the Tree Planting Association Stephen Smith had observed during the hot
summer months the largest number of deaths from what was then considered "foul
air" thought to cause "diarrheal diseases." Besides turning Tompkins Square on the
Lower East Side of Manhattan into an umbrageous park, Smith therefore suggested
that the city's avenues be lined with trees, thereby infiltrating the residential districts
and in particular the tenements with their "cooling and health-giving influence."[37]

The trees' healthy and positive influence was also noted by Jacob Riis, who famously captured the conditions of the tenement districts in his photographs and books
and became a member of the Tenement Shade Tree Committee. One of the few
photographs in which Riis focused his camera on a tree—there were indeed few of
them in the tenements—was of the last trees remaining in the mulberry bend (figure

FIGURE 4.13.
Street tree planting along East Thirty-Ninth Street initiated by the Tenement Shade
Tree Committee in the early twentieth century. (*Annual Report of the Tree Planting
Association of New York City, 1903* [New York: Tree Planting Association, 1904].
General Research Division, The New York Public Library, Astor, Lenox
and Tilden Foundations)

4.14), one of the most infamous tenement areas in the city, which through Riis's efforts
in the 1890s was demolished and replaced with Mulberry Bend Park (today Columbus
Park). Rather than focusing on the trees' public health function, Riis emphasized
their edifying and educational role in the livelihoods of the poor and their children.
Incited by a *New York Times* editorial that had criticized the Tenement Shade Tree
Committee's activities on the grounds that it was following the wrong planting practices and thus wasting money on the poor, Riis voiced an early concern for citizen
engagement: "Let us have the trees, and the nearer the homes of the poor the better;
as they grow good citizenship will grow with them." To encourage citizen engagement
and responsive tree care, and to prevent vandalism, each of the trees planted by the
Tenement Shade Tree Committee bore a small enameled sign stating: "This tree is a
gift to all children. Be its friend" (figure 4.15). Besides improving the urban climate
with street tree plantings, one of the committee's objectives was to cultivate "a love of
nature among street children."[38]

Early proof of the empathy and engagement fostered through the tree plantings
directed by Archibald Hill came in observation of the tenement children's behavior.
Visiting the fifteen trees planted along an East Side block between First and Second
Avenues, a committee member was reported to have "found a small boy digging vig-
orously around the roots of a thriving Carolina poplar, with a broken fork." Although

FIGURE 4.14.
One of the few photographs in which Jacob A. Riis focused his camera on a tree, 1895.
(Museum of the City of New York. 90.13.1.275)

the boy "was not making much impression on the soil, . . . his chubby face was beaming with a sense of importance and public service rendered. 'This is the Tree what I takes care of,' he explained. 'It's the biggest of 'em all, but it wouldn't be 'less I did this to it every day.'" As reported in the weekly newsmagazine the *Chautauquan*, "Of the fifteen trees planted in one block, where over 600 children made their homes, only two had died in the year, and neither of these from willful injury or neglect. Every

FIGURE 4.15.
To protect trees from vandalism, the Tenement Shade Tree Committee fastened small enameled signs to the trees it planted in the tenement districts. (*Annual Report of the Tree Planting Association of New York City, 1905* [New York: Tree Planting Association, 1904]. General Research Division, The New York Public Library, Astor, Lenox and Tilden Foundations)

child in the neighborhood could give the individual history of any one of the trees, and considered each almost in the light of a personal friend."[39]

Throughout the early decades of the twentieth century, educators and social reformers attributed increasing importance to children's nature appreciation and education, which also came to be included in school curricula. As shown in chapter 3, education in the fundamentals of tree and bird life also became a staple in the extracurricular activities of various children's clubs. But progress in nature education was slow and uneven. In the 1940s, magistrate J. Roland Sala blamed the lack of nature education for contributing to tree vandalism by two juveniles. Two high-school boys from Bedford-Stuyvesant were charged with damaging a tree on Marcy Avenue located nine blocks south of the southern magnolia tree. In his attempt to further the boys' appreciation for nature, the magistrate imposed a unique sentence: he ordered them to copy Joyce Kilmer's poem "Trees" twenty times and recite the poem in court some days later. Since its publication in 1914, "Trees" had become popular verse recited at many Arbor Day celebrations and memorial tree dedications throughout New York City and the entire nation. Its recital now served to relieve two delinquent juveniles from tree vandalism charges.[40]

Regardless of whether tree-planting initiatives were begun as grassroots activism by those immediately affected, as in the case of Hattie Carthan in 1960s Bedford-Stuyvesant; as a presidential initiative in these same years, as in the case of Lady Bird Johnson's tree-planting campaign in Washington, DC; or by more or less patronizing public health and social reformers, as in the case of the Tenement Shade Tree Committee in the early twentieth century, in all instances the activities ultimately came to be based on a hybrid of bottom-up and top-down, private and public initiatives that could support each other. As physical elements that were rooted in the ground yet changed throughout the seasons and their life cycle, street trees indeed occupied a ground in between these different spheres that offered itself as a new space for the negotiation of power. Street trees occupied a liminal zone that was often less obviously public than urban parks and playgrounds. They also often affected private property through their aesthetics and microclimatic functions, which could lead to the increase of adjacent property prices. From the beginning, in many U.S. cities private persons were encouraged to plant their own street trees on public ground within a regulatory framework established by the respective city government. The above initiatives show that landscape, especially in times of crisis, has been a means and method of both public and private activism.

PART TWO
# Berlin

# 5
# Burning Trees
### *Street Trees in Wartime and Early Cold War Berlin*

In the years after the fall of the Berlin wall, the number of buildings whose facades still bore witness to the 1945 battle of Berlin decreased fast. Today, due to the city's rebuilding efforts after reunification and the increasing gentrification of its central districts, only a few facades still have bullet holes. Renovation and restoration have turned these remaining holes that once appeared as ugly patterns on the buildings' surfaces into much-famed spots of interest that act as tangible reminders of and evocative testimony to Germany's troubled past. Tourist groups flock past the buildings, and their guides describe the scenes of war-ravaged Berlin. But buildings—residential, commercial, and industrial; and hospitals, schools, and churches—were not the only physical aspects of Berlin that were torn apart by the bombing raids and that suffered from sharpshooters' gun shells in the last European battle of World War II. The city's street and park trees were also affected. For example, artillery fire heavily damaged the majestic plane tree planted in Margaretenstraße (today standing along Potsdamer Straße) in the middle of the nineteenth century and the two plane trees on Schinkelplatz that had been listed as natural monuments only in 1939, an act intended to protect them against willful destruction. Many trees were scorched, and their bark was loosened and torn as a result of high air pressure from explosions. Trunks were splintered and damaged by bombs and bullets, especially between 1943 and the end of the war (figure 5.1). Although exhibiting resilience and recovery like all living organisms, trees were also subject to the war's lasting effects. Many succumbed during the war's aftermath. They lost limbs, becoming unsafe and unseemly, and their injuries made them more susceptible to fungi and insects, often causing their premature death. As a result of urban trees' destruction, Berlin's postwar planners had to deal with both desertification and predictions of an elevated water table by one and a half to two meters.[1]

Nevertheless, during World War II and the ensuing Cold War, Berlin's street trees played various roles in the city's defense. They provided protection and resources,

FIGURE 5.1.
Trunk damaged by bomb splinters,
1950s. (K. H.-P., "Kontrolle von
Baumschäden unter besonderer
Berücksichtigung der Verkehrs-
sicherheit und Haftpflicht,"
*Das Gartenamt* 3, no. 2 [1954]:
22–24)

alleviated the extremes of hot and dusty summers, and both incited and mitigated the extremes of an increasingly tense political climate.

While real and elaborately constructed artificial trees played an important role in military camouflage, in the last days of the battle of Berlin, the city's street trees provided another, more literal means of defense. Trees lining Ritterfelddamm at the corner of Potsdamer Chaussee on Berlin's northwesternmost periphery were equipped with explosives so that they could be used as anti-tank traps of last resort. The explosives were fixed to the trunks by German army pioneers after the trunks had been indented on the street side. Setting off the explosives would have caused the trees to fall across the road, and in this way could have slowed down the Allies' access to the nearby airport at Gatow. This late and desperate act of defense by the Germans was rediscovered only in the summer of 1955, when one of the trees equipped with explosives was hit badly by a storm and had to be cut down. Many other trees along the roadside had been outfitted with munitions, by this time partly overgrown. As a result, the forgotten explosives had gone unnoticed, and the trees had survived. By 1955, certain methods of tree surgery had become best management practices both in the United States and Germany, and efforts were undertaken to remove any remaining explosives in the trees along Ritterfelddamm to increase the trees' chances for a long life. The police stood by, blocking the road, as workers treated the trees (figure 5.2).[2]

By the last months of the war, Berlin's street trees had become a means of defense and a resource of last resort in more ways than one. They provided much-sought-after firewood, and in March 1945 the Berlin district department of the Reich's Committee for the Utilization of Unfermented Fruit (Reichsausschuss für gärungslose

FIGURE 5.2.
Workers removing explosives from trees along Ritterfelddamm, July 1955.
(Photo by Gert Schütz. Landesarchiv Berlin, F Rep. 290, Nr. 0041794)

Früchteverwertung) aspired to enlist all mountain ash standing along the city's streets as vitamin providers for Berlin's population. Since 1934 research into the sources and industrial production of vitamins had been part of Nazi biopolitics, which sought to create a healthy, resilient, and strong people ready to fight and win a war. Research into the vitamin content of berries of native plants like sea buckthorn, dog rose, and mountain ash was begun in the mid-1930s. Due to the high vitamin C and ascorbic acid content in mountain ash berries, the species' upright growth, and its high and loose yet beautifully shaped canopy, by 1944 the Reich's Committee for the Utilization of Unfermented Fruit considered the mountain ash an ideal roadside tree (plate 6). The use of mountain ash berries had already been promoted during World War I by, among others, Berlin nursery owner Hellmut Späth. By the time of World War II, the Späth nursery was cultivating mountain ash varieties that it had introduced from Russia around the turn of the century and the nonbitter variety *Sorbus aucuparia* L. var. *moravica* from Austria. These cultivars as well as their bitter cousins were planted increasingly along country roads as well. Scientists were advocating mountain ash as an ideal fruit tree that survived harsh winters and could therefore provide the country and its military with lifesaving vitamins. With the advancement of the Allied forces on German territory, by the beginning of 1945 it had become clear that Berlin itself would need to provide the resources necessary for its citizens' survival. In response to the request by the Reich's Committee for the Utilization of Unfermented Fruit, Charlottenburg's Department of Gardens reported around 680 trees of the genus *Sorbus*—the name deriving from the Latin for *eating* and *edible*—standing along the district's streets, including mountain ash, Swedish whitebeam, and the whitebeam variety *lutescens*.

As a result of its attractive spring blossom and orange autumn fruit, its tolerance with regard to soil quality and drought, and its relatively small canopy, mountain ash had been used especially in narrow streets since the late nineteenth century. Charlotten-burg's Department of Gardens estimated that each tree could provide one and a half kilograms of fruit. Skeptical of the endeavor, however, and weary of the additional work involved, it stressed the berries' bitterness and the department's inability to provide the labor necessary to collect them. The Allies defeated Germany and Berlin only seven weeks later but, of course, the necessity to provide food and lifesaving vitamins was as urgent as ever. A year later, mountain ash was among the species subject to a new decree mandating the protection of all wild woody fruit species in Berlin. The decree sought to secure lifesaving resources and solve a vital conflict: many fruit-bearing trees and shrubs were being used as firewood. Instead, the planning and building director Hans Scharoun demanded that the district Departments of Gardens should protect and propagate fruit-bearing trees and shrubs and use them in the city's afforestation projects, in particular its planned shelterbelts.[3]

Street trees provided another vital resource in the war and immediate postwar years: firewood. Already scarce during wartime, the wood supply for heating and cooking soon became one of the most contentious issues in postwar Berlin. The city's street trees figured prominently in the debate and battle about the city's energy supply, again becoming a vehicle of defense—this time against the cold in the winters and against the dust clouds of rubble and ruined landscape in the summers. And they became a fulcrum of contention in some of the first ideological battles of the Cold War that were carried out between the Communist regime of Soviet Russia and the Western Allies during the Berlin Blockade from 24 June 1948 to 12 May 1949.

Besides their incendiary effect in the Cold War, street trees were also used in the early 1950s, after the Berlin Blockade, to consolidate the border between West and East: in West Berlin near S-Bahnhof Düppel-Kleinmachnow and Königsweg, and in Kohlhasenbrück, the police ordered Zehlendorf's district foresters to paint tree trunks marking the border between the American and Soviet sectors. Girdled with white paint, they were to warn and prevent unsuspecting West Berliners from accidentally crossing the border into the Soviet zone. In East Berlin, the Volkspolizei felled three trees rooted in West Berlin and used their wood as posts for the fences they were erecting between the Soviet and French sectors in Frohnau.[4]

Later, once construction of the Berlin wall had begun in August 1961, trees on St. Hedwig Cemetery in the Soviet sector were pruned to prevent escapes across the border, and East Berlin street trees standing on border territory were felled. This occurred, for example, along Heidelberger Straße, one of the narrowest border areas between East and West Berlin that in 1962 became the location of two of the earliest escape tunnels. Street trees had to be felled not only to make room for the various elements of the border system but also to facilitate clear oversight and control (figure 5.3).[5]

Besides these acts of border consolidation, however, there was also an exceptional incident on Heidelberger Straße much later when street trees provided the East Berlin Volkspolizei with both the reason for a temporary breach of the wall and

FIGURE 5.3.
East Berlin border police cutting down poplars on Heidelberger Straße, 20 April 1963.
(Photo by Gert Schütz. Landesarchiv Berlin, F Rep. 290 [09] Nr. 0088907)

an opportunity to reenact territorial control. In the winter of 1978, East Berlin police cut a section of the wall in order to subsequently fell several rotten and unsound poplars standing along Neukölln's western side of Heidelberger Straße. Located on East Berlin territory but on the West Berlin side of the wall, the trees had been endangering West Berlin pedestrians. Seeking the help of the American military government, the West Berlin Senate had tried in vain to receive permission to fell the trees. East Germany wanted to take care of its tree matter on its own even if it meant cutting a hole in the wall. Initiated by West Berlin and carried out by East Berlin, the cutting down of the poplars perhaps left the inhabitants of Heidelberger Straße living in a safer place, but it also left them staring at a gray wall.[6]

The effects of the Berlin border system on West Berlin trees were also felt when— as reported by *Der Tagesspiegel* in June 1966—pesticides used by the East German border patrol to treat the carefully curated grass on the border strip drifted across and caused the premature coloring and loss of leaves of black locust trees that were standing near the border in West Berlin. On the other hand, the border and the use of pesticides on borderland could not prevent insects from traveling between East and West Berlin and attacking trees. In 1971, a year after the caterpillars of the browntail moth had denuded entire trees in the East Berlin districts of Friedrichshain, Prenzlauer Berg, and Lichtenberg as well as in the neighboring West Berlin district of Kreuzberg, the insects were causing similar trouble in the other districts of West Berlin. Caterpillars were entering homes, falling onto pedestrians, and causing rashes to such an extent that special eradication task forces were established on both sides of the wall, albeit without direct coordination and communication.[7]

## The Berlin Blockade and the Battle over Trees

In the immediate postwar years, street trees, especially the famous linden trees of Unter den Linden, the boulevard first planted in the mid-seventeenth century under the Great Elector Frederick William to connect his city palace with the nearby hunting grounds (Tiergarten), quickly became symbols of rebirth and recovery for Berlin's citizens. As paintings by Werner Heldt, who chronicled Berlin's cityscape in the years before and after World War II, attest, street trees had inherent emotive value (plate 7). Bringing color and liveliness into urban streets, trees indicated the seasons, and their forms, which changed with their life cycles and the seasons, contrasted with the often gray and lifeless building facades. In the early twentieth century, Berlin architect and art nouveau promoter August Endell had highlighted these qualities that could render cities like Berlin beautiful in a lyrical essay on "spring trees." Like Heldt and Endell, many citizens appreciated the trees' aesthetic qualities. They were emotionally attached to street trees and other urban trees that in the war and the postwar years became victims of the initiatives to secure the city's wood supplies.[8]

Already at the end of July 1945, not quite three months after Berlin's capitulation on 2 May, guidelines for the replacement of street trees were established. In August, newspapers announced that the trees of Unter den Linden would need to be replanted, and by 1950, Berliners were again able to wander below a new crop of young silver linden planted in four rows between Staatsbibliothek and Pariser Platz. Street trees were, of course, an integral component of the open-space plans for the new Berlin that were drawn up by a team led by Reinhold Lingner. Lingner had been appointed director of the newly established Department of Open-Space Planning and Horticulture (Hauptamt für Grünplanung und Gartenbau) in 1945 under the Communist-led Berlin Magistrate formed immediately after the war.[9]

The general idea of the planners' reconstruction plans was to rebuild Berlin as a "city landscape" firmly grounded in the ecology of its natural environment. Landscape planning units were to be formed on the basis of Berlin's geophysical character traits like the Spree and Havel valleys and the Barnim and Teltow plateaus. The city was to have a green core, and green open space was to expand into the surrounding residential districts, connecting the core to adjacent agricultural and forested areas. Street trees fulfilled an important function in this scheme. But despite this intent, street tree replanting and afforestation in general were slow as nurseries with an appropriate stock of trees were hard to find.[10]

During the war, tree work and the work of the district Departments of Gardens in general had already been a balancing act. The increasing lack of tree workers due to the war effort meant that the pruning of street trees could no longer be carried out.

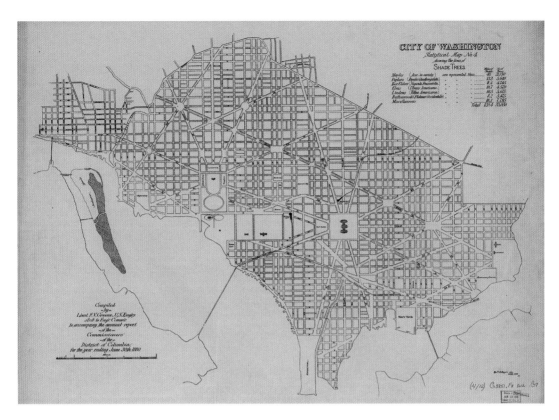

Plate 1. F. V. Greene's map showing the lines of shade trees in Washington, DC, accompanying the annual report of the Commissioners of the District of Columbia, 1880. (Library of Congress)

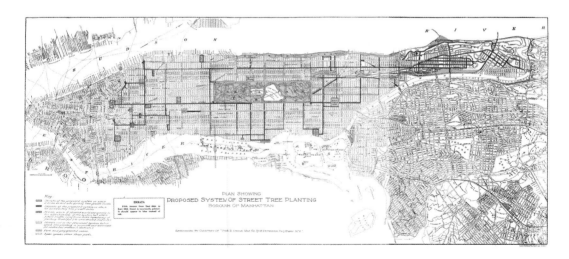

Plate 2. Laurie D. Cox's proposed system of street tree planting, Manhattan, 1915. (Laurie Davidson Cox, *A Street Tree System for New York City, Borough of Manhattan: Report to Honorable Cabot Ward, Commissioner of Parks, Boroughs of Manhattan and Richmond,* New York City [Syracuse: Syracuse University, 1916])

HARD MAPLE
⅓ NATURAL SIZE

RED OAK
⅓ NATURAL SIZE

Plate 3. William E. Bruchhauser's portraits of the fall foliage of trees used along city streets and highways. (William F. Fox, *Tree Planting on Streets and Highways* [Albany: J. B. Lyon, 1903])

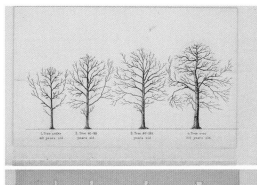

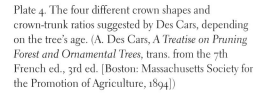

Plate 4. The four different crown shapes and crown-trunk ratios suggested by Des Cars, depending on the tree's age. (A. Des Cars, *A Treatise on Pruning Forest and Ornamental Trees*, trans. from the 7th French ed., 3rd ed. [Boston: Massachusetts Society for the Promotion of Agriculture, 1894])

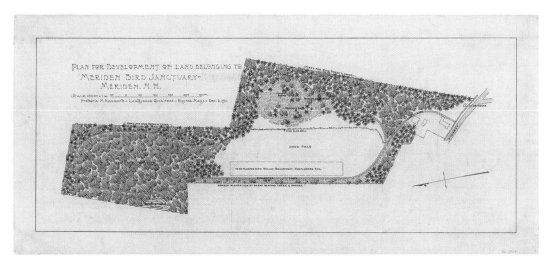

Plate 5. Frederic H. Kennard's design for the Meriden Bird Sanctuary, Meriden, New Hampshire, 1911. (Courtesy of the Frances Loeb Library. Harvard University Graduate School of Design)

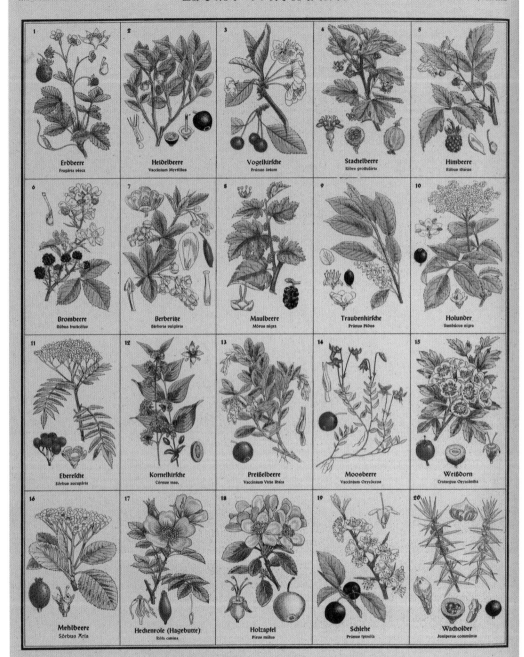

Plate 6. Page from Bernd Hörmann's *Plant Atlas* illustrating edible wild berries.
Mountain ash appears in the third row on the left. Illustrations like these were to
encourage the country's autarky by educating people about native resources.
(Bernd Hörmann, *Pflanzen-Atlas I* [Munich: Franz, 1940])

Plate 7. Werner Heldt, *Berliner Vorstadtstraße, Herbstag, 1948* (Berlin Suburban Street, Autumn Day, 1948).
(© 2016 Artists Rights Society [ARS], New York/VG Bild-Kunst, Bonn)

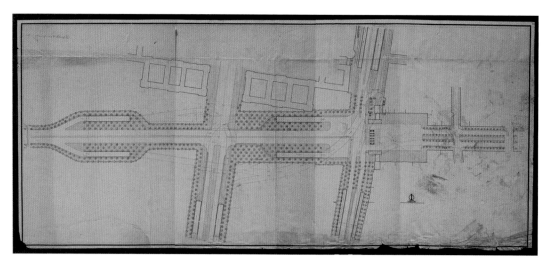

Plate 8. Tree-planting plan for Albert Speer's new east-west boulevard running through the Brandenburg Gate, 1938–40. Between 1937 and 1939, Charlottenburger Chaussee (today Straße des 17. Juni) had been widened from 15.45 to 50 meters, requiring the planting of new street trees. (Landesarchiv Berlin, A Pr Br Rep. 107 [Karten], Nr. 137, T 1–4)

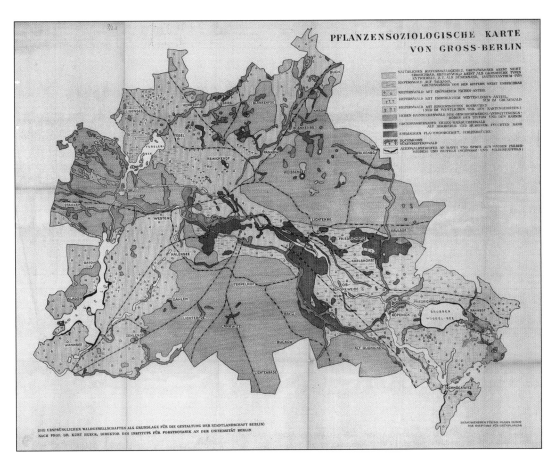

Plate 9. Kurt Hueck's phytosociological map of Berlin, 1948.
(IRS Erkner, Nachlass Johann Greiner, C10, 57, P001)

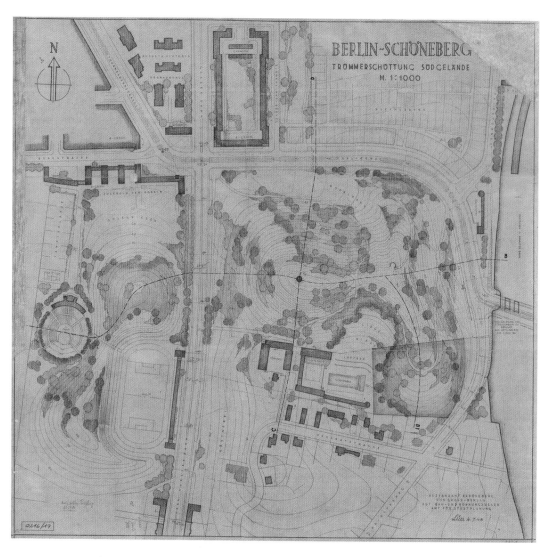

Plate 10. Plan for the Insulaner, a park to be created on rubble, Schöneberg, 1948.
(Landesarchiv Berlin, Insulaner, F Rep. 270 A, 4748)

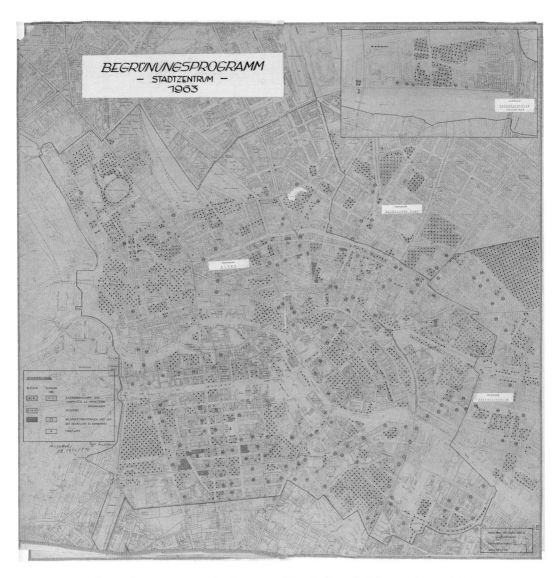

Plate 11. Greening program for the center of East Berlin, 1963. The map shows existing
(black hachure on green background) and planned (numbered, green) areas of the
Wild Grass Program and existing (green with filled black circles) and planned
(green with unfilled circles) areas that were to become permanent parks and gardens.
(Landesarchiv Berlin, C Rep. 110-04 [Karten], Nr. 112, T 1 and 2)

Plate 12. Promotional poster for "Beautify Our
Cities and Municipalities—Participate!"
(Bundesarchiv Koblenz, PlakY7 247)

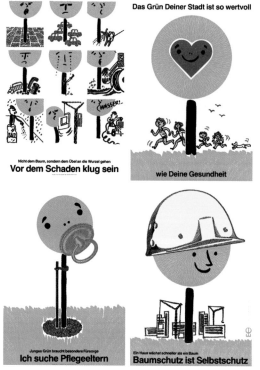

Plate 13. Hajo Schüler's series of posters for the
planting, protection, and care of street trees, produced
for the Erfurt Department of Gardens, 1980s.
(Studienarchiv Umweltgeschichte, Neubrandenburg)

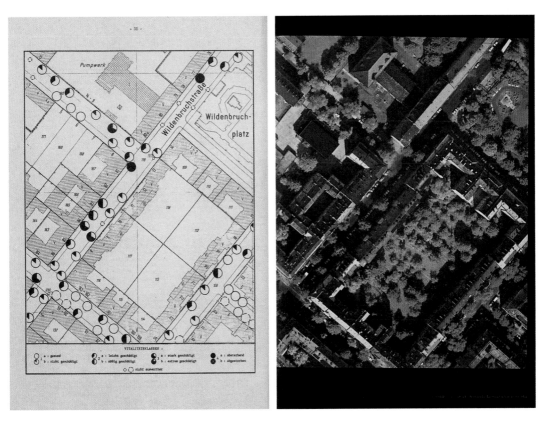

Plate 14. Street tree assessment using aerial infrared photography, Neukölln.
(Michael Fietz, "Lebensbedingungen von Strassenbäumen in Berlin-Neukölln," *Berliner geowissenschaftliche Abhandlungen*, series C3, vol. 3 [1984]: 53–64)

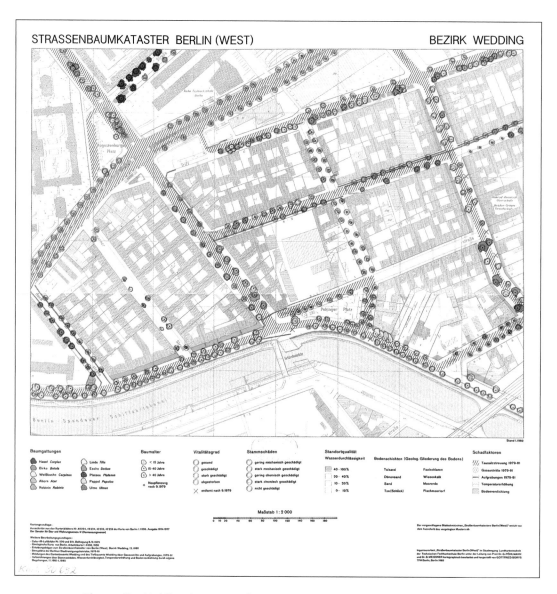

Plate 15. Gottfried Borys's prototypical street tree map for Berlin, including tree species, tree age and vitality, trunk damage, soil type and permeability, and damage causes. (Gottfried Borys, "Karten zum Strassenbaumkataster—Berlin (West)—1:2000," *Berliner geowissenschaftliche Abhandlungen*, series C3, vol. 3 [1984]: 101–20)

Plate 16. Green open-space plan for Berlin, 1957.
(Fritz Witte, "Zur Grünplanung in Berlin," *Garten und Landschaft* 67, no. 10 [1957]: 267–69)

Plate 17. Norway maples of second-rate quality along Dunckerstraße, Berlin, ca. 1986.
(Photo by Wolfgang Krause)

Plate 18. Concrete planting containers along Hufelandstraße, Berlin, ca. 1986. The linden
tree in the distance shows the typical damage caused by underground gas leaks. The tree
in the foreground was planted before the decision to replace trees with planted containers.
(Photo by Wolfgang Krause)

Plate 19 (*left*). Stills of *Street Tree* TV spot produced by the New York City Parks Council. The moving images are accompanied by music and words: "Hey, have you ever seen a tree water itself?/Run away from a dog?/Jump over a motorcar?/Call for help?/No, it needs your help." (Avery Architectural & Fine Arts Library, Columbia University/ New Yorkers for Parks)

Plate 20 (*above*). Whereas in the early twentieth century street trees were used as advertising pillars, today advertising columns are used to promote street trees. Here, the large poster promotes the private-public City Tree Campaign, Samariterstraße, Berlin, 1 January 2017. (Photo by Sonja Dümpelmann)

As a result, street trees shaded apartments and houses, leading their inhabitants to use artificial light even during the summer and at a time when energy saving was a necessity.[11]

The municipal tree nurseries had to deal with another vital conflict. Given the food shortage, they had to give more and more land over to the cultivation of vegetables. But at the same time they were advised to care for the existing stock of future street trees and make provisions for general building inspector Albert Speer's megalomaniac plans for the reconception of Berlin as the future world capital Germania. Around eight thousand trees would be needed to line Speer's new monumental boulevards (plate 8). The first trees for this purpose had already been cultivated in 1938. But in 1940 garden director Joseph Pertl found an additional source. Around thirteen thousand trees, predominantly linden trees with a ten- to twelve-centimeter caliber, were to be transported by train from a confiscated Polish nursery in Kunzendorf to Berlin's municipal nurseries. Most of the tree transfer was completed by May 1941, but the tree care in Berlin's nurseries had to be curtailed in 1943 once Joseph Goebbels had declared "total war," and it ultimately ceased a year later. Nevertheless, the plans for urban planting continued, and in 1944 landscape architect Hermann Mattern submitted to Albert Speer a list of suitable tree and shrub species for the Reich's capital, including for the use along streets.[12]

In the postwar years, it took a while before extensive tree-replanting programs could begin again. As in the war years, in the early postwar years, priority had to be given to the growing of vegetables. Soon all open spaces that appeared somewhat suitable for the purpose were turned into farmland. Despite continuing care and sympathy for tree-planting efforts and newly planted trees (figure 5.4), therefore, the necessities and resource scarcity of daily life often gained the upper hand, also leading to street trees nearly being buried under rubble as well as to frequent tree theft and vandalism (figure 5.5). The status of street trees was so perilous that the British military government warned Berlin citizens to cherish and respect plant life. Any offense against plants would be prosecuted.[13]

Berlin citizens were put to the test for the first time after the war during the cold winter of 1945–46. The low temperatures led people to use every type of wood they could get their hands on, including street trees, the wooden latches of park benches, and the wooden blocks still used as paving for many streets (figures 5.6 and 5.7). In the two years between 1945 and 1947, the Department of Gardens reported the destruction of around one thousand street trees and nine hundred garden benches. Many trees on private property, protected under Berlin's 1922 tree ordinance, were also felled. Newspapers warned of the safety hazards when searching for wood in the ruins and rubble, and several Berliners perished during their desperate searches. The number of trees in Tiergarten was reduced from two hundred thousand to only seven hundred (figure 5.8), and the military governments allowed each family to buy and cut a tree in Grunewald to use as firewood. Once before, at the end of World War I, Berliners had flocked to Grunewald and nearby forests to gather wood because of a coal shortage in the city and the more general wartime economic crisis. By 1946, 85 percent of the city's wooded

FIGURE 5.4.
"I am thirsty": note attached to newly planted horse chestnut tree, Berliner Straße, Charlottenburg, 1948. (Landesarchiv Berlin, F Rep. 290 [09] Nr. 0286822)

FIGURE 5.5.
Trees under rubble in Friedenstraße, 1952.
(Bundesarchiv Koblenz, Bild 183-13249-0018/photo: Martin)

FIGURE 5.6.
Felling street trees for firewood in front of the ruined Sportpalast Hall, February 1945.
(Photo by Wolfram Kahle. The Granger Collection, New York)

area had been deforested, including a third of Grunewald. Although in January 1947 the city council had sought to prohibit all unplanned cutting of trees and to initiate the reforestation and reestablishment of a sustainable forest rotation, political events in the following years demanded a temporary interruption of these plans.[14]

Once the Soviet military government had closed the transportation corridors to the Western sectors to force them under Soviet control on 1 April 1948, Berlin's energy

FIGURE 5.7.
Boy collecting wooden paving blocks to be used as firewood, Charlottenburger Ufer, 1945. (bpk Bildagentur/Art Resource, NY, 30015238)

FIGURE 5.8.
Two men in what remains of Tiergarten cut up a tree stump to fulfill the urgent need for
firewood, 1946. (bpk Bildagentur/Friedrich Seidenstücker/Art Resource, NY, 30003101)

supply became one of the biggest challenges the Allies—in particular the French,
British, and American military governments—and the West Berliners had to tackle.
When winter was approaching, it was the fear of insufficient coal and wood supplies
rather than food shortages or politics that induced some West Berliners to register for
supplies in East Berlin. The counter blockade by the Western Allies prevented East
Berliners from being beneficiaries of coal supplies from the Ruhr region. But supplies
from Poland became a substitute, and in August 1948, all East Berliners—and but few
West Berlin households—received lignite briquettes. The situation in the Western sectors worsened so that on 6 October 1948, the military governments of the three Western
sectors under leadership of the British military governor General Sir Brian Hubert
Robertson ordered the cutting of 350,000 cubic meters of wood by 31 January 1949.
While 200,000 cubic meters could be gained from Berlin's forests, according to the
Western Allies, every second tree in streets, parks, and gardens was to account for the
remaining 150,000 cubic meters (figure 5.9). The military governments' drastic measures also allowed the felling of trees on private property as long as the owner was
compensated by obtaining 10 percent of the wood, or at least one cubic meter.[15]

The British order had immense emotional repercussions, appalling many Berliners in the Eastern and Western sectors alike. They didn't like the idea that trees had
to be felled in a phase of reconstruction. Many Berliners would rather have been cold
than witness another devastation of their forests and urban trees. There also remained

FIGURE 5.9.
Woman collecting kindling wood near felled street tree, June 1948. (Photo: akg-images/Henry Ries, 136890)

an open question about how useful the wood would be in the first place. Before its use in the winter, it would not have had sufficient time to dry out, and the tree species concerned produced wood of relatively little heat value. At the city council meeting of 21 October 1948, delegates from the Socialist Unity Party (SED) failed to appear out of protest. But delegates from the Social Democratic Party (SPD), the Christian Democratic Union (CDU), and the Liberal Democratic Party (LDP) discussed the British order, promoting and unanimously voting for an amendment to it. They stressed the climatological and scientific role and importance of urban trees, and of street trees in particular. LDP delegate Winfried von Wedel-Parlow also emphasized the trees' emotional role. The trees, like the Berliners themselves, had survived the war and were symbols of resilience. Von Wedel-Parlow reminded the councilors that the trees had also suffered the bombings and were "living beings that as our company during the war carried their hardships together with us and who in every spring through their green vest gave us hope for better times."[16]

The Berliners' appalled reaction and discussion of the British order in the city council led the Magistrate to ask for a deforestation in stages and for a revision of numbers: instead of 350,000, only a maximum amount of 120,000 cubic meters should be felled by the end of November. The maximum amount considered justifiable by the Magistrate was an overall 130,000 to 140,000 cubic meters. The city government pointed to the fact that the 150,000 cubic meters of wood that the Western military governments wanted to see harvested in the Western sectors' streets, public parks, and private gardens did not even exist in these locations. Instead, trees with a diameter of at least twenty centimeters were estimated to sum up to approximately 57,000 cubic meters, of which merely 17,500 could be harvested. The Magistrate argued that Berlin's forests had already suffered significant deforestation during the war and immediate postwar years, so that the large-scale clearings ordered by the Western military governments would risk desertification and serious wind damage. And as CDU council Wilhelm Dumstrey argued, street trees were important because they stabilized the dirt of the streets.[17]

The Western military governments finally agreed to the Magistrate's requested revisions, commanding the felling of 120,000 cubic meters of wood instead of the initial 350,000. While they would be making every effort to fly sufficient coal into Berlin, the Western Allies asked the Magistrate to prepare a phased plan for the potential felling of an additional 150,000 cubic meters throughout the winter. It was clear by then that many Berliners would suffer cold despite the many public heating halls that would be established as in the winters before. By the end of November 1948, the Magistrate was preparing the distribution of one box of firewood per household. Those West Berliners who had been able to obtain tree stumps or other wood over the summer were asked to hand in their firewood-ration cards so that their boxes of firewood could be passed along to the needy. The situation of West Berlin's energy supply became so uncertain that at the end of December, the Magistrate indeed ordered more cutting of trees, this time 70,000 cubic meters of wood.[18]

The felling of street trees was one of the most political and contentious issues during the early months of the Berlin Blockade; the East Berlin newspapers—increasingly controlled, censored, and ultimately coordinated by the Soviet military government and the SED—used the issue to further incite the burgeoning Cold War. The Communist media began reporting on the felling of every second street tree in early September when the Western military governments asked the district Departments of Gardens to calculate how much wood could be gained this way. By mid-September, the SED party organ *Neues Deutschland* reported that the district Spandau had decided to cut numerous street trees, against the vote of the SED. The *Berliner Zeitung*, first published by the Red Army and then turned into the official voice of the Berlin Magistrate, followed by accusing the Spandau district government of making an irrational and shortsighted decision that would not result in the provision of sufficient heat and would instead deprive the population of urban green in the years to come. Why did Spandau not exchange industrial goods for coal briquettes from East Berlin? the *Berliner Zeitung* asked, seemingly incredulously.[19]

More sensational was the reportage of *Neues Deutschland*, which described the Magistrate's decision to cut down street trees as a "hysteria" and "catastrophe" that testified to "the incapacity of the much-praised 'bluff airlift' [*Bluff-Brücke*] to supply the Western sectors with sufficient heating material in the winter." There was, so the argument went, enough heating material in East Berlin for West Berliners to use if they only wanted to, and that would save the street trees. In fact, if one could even speak of a blockade, *Neues Deutschland* proclaimed, it was the Western sectors, not the Eastern sector, that had established one. The newspaper appealed to West Berliners to protest "the destruction of the wimpy rest of the tree population" and get heating material from the Eastern sector by registering for food-rationing cards in East Berlin and demanding that the Magistrate transport it to the West.[20]

Photographs published in *Neues Deutschland* in October illustrated the felling of street trees on West Berlin's showcase boulevard Kurfürstendamm. The purported alarming situation was further demonstrated by photographs juxtaposing the felling of street trees in West Berlin with the distribution of briquettes in East Berlin. Other photographs published in *Neues Deutschland* showed the transfer of coal for Berlin from rail onto barges at Königs Wusterhausen. No Berliner had to suffer the cold, the newspaper argued, "if the Western military governments would end their blockade against the citizens of the Western sectors and if the Magistrate would follow its duties and supply the West with products from the East." What the West considered a fight for liberty was, the Communist press argued, "a disgraceful tree crime" and a "barbaric act." As the *Berliner Zeitung* summed up in one of its October articles chronicling West Berlin's deforestation, everybody who muttered about a Berlin Blockade and considered the airlift necessary was guilty of deforesting the city.[21]

Once the firewood shortage in West Berlin had reached an alarming state by November 1948, the East Berlin press capitalized on this situation, satirically illustrating the Western sectors' desperation in a cartoon (figure 5.10). Standing in front of a ruined building along a street strewn with street lamps, one man is commenting in Berlin slang to another man carrying a large saw: "I told you the street did not have any trees no more, but you had to keep sawing in darkness!" By January 1949, parts of West Berlin's Grunewald appeared as a "wasteland" and another bare "battlefield," as reported in the *Berliner Zeitung*.[22]

Not surprisingly, reporting in the Berlin newspapers on this issue reflected the East/West divide. East Berlin's *Berliner Zeitung, Neue Zeit,* and *Neues Deutschland* figuratively turned Berlin's street trees into matches with which to inflame the Cold War. The destruction of street trees and their potential use as firewood figured prominently in these newspapers throughout October and November 1948, including articles with unchecked or manipulated facts as well as letters by readers who were up in arms about the tree crime. *Der Tagesspiegel* in West Berlin, on the other hand, chose to report on the issue rather lightly and in passing, although it made sure that the blockade by the Soviet occupiers was blamed for the Allied order to fell trees. For example, the newspaper reported only briefly on Spandau's winter preparation in mid-September, and the felling of street trees went unmentioned. *Der Tagesspiegel* also corrected reports by East Berlin newspapers that announced the clearing of the botanical garden in Dahlem. Reporting on Berlin's trees in the Western sector appeared calmer, less sensational, and it was perhaps more evenhanded.[23]

One short report in *Der Tagesspiegel* on 10 October discussed the pros and cons of deforestation in and around cities. On the one hand, the newspaper argued in support of the Western Allies' order that the felling of every second street tree would be

".. . .ick habet doch gleich jesaacht, die Straße hat ja keene
Bäume mehr, aba du mußtest ja in der Dunkelheit weita säjen!"

FIGURE 5.10.
"I told you the street did not have any trees no more, but you had to keep sawing in
darkness!" At the height of the Berlin Blockade, this satirical cartoon in the *Berliner
Zeitung* ridiculed the use of street trees as firewood in the city's Western sectors, 1948.
(*Berliner Zeitung*, 7 November 1948, 4)

beneficial in suburban districts in particular, as it would in many cases facilitate the
access of light and air to what through excessive tree growth had become dark apartments. Referencing botanist Ernst Tiegs, the director of the Institute for Water, Soil,
and Air Hygiene (Institut für Wasser-, Boden-, und Lufthygiene) in the Berlin locality
of Dahlem, who had become an indispensable authority for the Allies due to his expertise in public health, pest management, and water pollution control, *Der Tagesspiegel* reported that removing diseased trees as well as street trees that were planted too
densely could provide valuable firewood. This was the case even if, as Tiegs cautioned,
the wood of linden, horse chestnuts, and poplars was unsuitable for heating purposes.
On the other hand, *Der Tagesspiegel* also cautioned, based on Tiegs's expertise, that
it would be dangerous to tolerate large-scale deforestation on city streets and in the
Grunewald as urban trees were vital for the urban microclimate. Without trees the
aquifer would sink and Berlin would risk desertification and soil desiccation. For Tiegs
it was clear that Berlin's Tiergarten needed to be reforested, a desire harbored by many
Berliners and begun during the blockade in spring 1949.[24]

Indeed, it was the Tiergarten where the blockade's battle of trees culminated on
17 March 1949, on the occasion of the tree-planting ceremony near Großer Stern
along Hofjägerallee. Mayor Ernst Reuter, chairperson of Berlin's Social Democratic

Party Franz Neumann, and city planning director Paul Bonatz planted a linden tree to inaugurate the Tiergarten's resurrection as a park after wartime destruction and post-war tree cutting. Although elaborate reconstruction plans for the former royal hunting ground were still pending, the timing of this political act was well chosen. After a first winter of isolation with much disputed tree felling, the planting signaled to Berliners in both East and West a new beginning. It showed West Berliners that care was taken for their city's reforestation and that their perseverance and staunch resistance against Soviet propaganda would be rewarded. It also demonstrated West Berlin's resilience to the world, in particular to East Berlin and the Soviet military government. To this effect, Mayor Reuter proclaimed in his inaugural remarks: "We Berliners will not be defeated." Indeed, as the *Telegraf* trumpeted on the occasion of the tree planting, "Berlin [was] ris[ing] again." The first 150,000 to 200,000 trees for Tiergarten's refor-estation were flown in during the Berlin Blockade from nurseries in the northern state of Schleswig-Holstein. By May 1951, thirty-five cities and the Bundestag had donated money and altogether 346,882 trees. One of the largest deliveries came from Lower Saxony, which donated 77,000 trees, mostly two- to three-year-old beeches and oaks.[25]

Newspapers in both parts of the city did not fail to mention the planting of the first linden tree of Tiergarten along Hofjägerallee, even if they did so in very different ways. While *Der Tagesspiegel* reported the facts of the ensuing replanting and design competition for the Tiergarten, the *Berliner Zeitung* merely commented by publish-ing a satirical cartoon. Entitled "Reuter plants firewood," the cartoon ridiculed the event (figure 5.11).[26]

As the replanting of Tiergarten and the first trees along Hofjägerallee showed very clearly, Berliners and their city government valued their street trees. While they were deforesting their city, they were also beginning to reforest it.

Aware of the harm done to the urban living environment, the British military government had included in its order of October 1948 that plans be drawn up for the reforestation of cleared areas. Not only was this considered a necessary part of the city's reconstruction as an aesthetically pleasing, green city, it was also thought to be particularly pertinent to the health of the city's inhabitants, given the amount of rub-ble and barren areas that in their bare state were unable to bind dust, especially during the dry, warm summer months. Making a virtue out of necessity and with the help of emergency workers, street trees were replanted, and many ruined small lots between buildings were turned into park and play spaces.

## Trees against Dust and Desert

One of the major concerns brought forward by promoters of tree conservation and planting on both sides of the city was trees' climatological effect and their role in slow-ing down winds and in binding dust—a big problem in the city given the amount of rubble and ruins. There was no question that urban trees, whether along streets, in forests, in groves, or in plantations on rubble areas, were important in the creation of a comfortable and healthy urban climate. As early as 1946, a directive of the Greater

**Reuter pflanzt Brennholz**

„Oberbürgermeister" Reuter hat gestern die „Aufforstung" des abgeholzten Tiergartens damit begonnen, daß er die erste Linde pflanzte.

„Hör mal, Ernst, Kollege Füllsack meint, wir sollten die Linde mit Kartoffelkloßmehl düngen, damit wir im Winter wieder was zum Heizen haben."      Zeichnung: Schmitt

FIGURE 5.11.
"Listen, Ernst, colleague Füllsack [Senate director P. Füllsack] is suggesting we should fertilize the linden tree with potato flour so that we will have something to keep us warm next winter." ("Reuter pflanzt Brennholz," *Berliner Zeitung*, 18 March 1949, 2)

Berlin Magistrate's Department for Building and Housing (Abteilung für Bau- und Wohnungswesen) ordered the city's reforestation to improve, among other things, its microclimate. However, reforestation, as Berlin's director of open-space planning Reinhold Lingner argued, was not only a matter of making Berlin beautiful and healthy; it would also reestablish a "balanced landscape" on a regional scale, saving it from desertification and forming steppes. In 1946 the Soviet military administration had begun a reforestation program for its occupation zone, a program that Lingner wanted Berlin to participate in.[27]

The program was modeled upon earlier wide-reaching plans that had been drawn up for the Soviet Union and that ultimately culminated in the 1948 Great Stalin Plan for the Transformation of Nature. The Great Stalin Plan was an attempt to reorganize Soviet forests into protected zones and exploited zones and to afforest the steppes. Eight enormous shelterbelts were planned to combat drought and increase agricultural production. Similar to these wide-reaching plans in the Soviet Union, the plans for the Soviet occupation zone (SBZ) were to include the creation of large forested areas, and Berlin had its part to play in them.[28]

Lingner, who had Communist convictions and a long-standing interest in the Soviet Union, was aware of the developments leading up to the 1948 Great Stalin Plan, which he referenced in several lectures throughout the late 1940s and early 1950s as well as in his 1952 book *Landschaftsgestaltung* (Landscape Design). A welcome means of socialist propaganda in the GDR, the plan's afforestation initiatives were not only well known to landscape architects and other specialists, they were also widely publicized, for example, through articles in East German newspapers and in the weekly news summaries produced by DEFA (Deutsche Film-Aktiengesellschaft), the first German postwar film-production company founded in the SBZ. Envisioning a reunited Communist Germany, Lingner used the devastating consequences of deforestation and dust storms in the capitalist oppressor country the United States as examples of the dire results of a lack of attention paid to the environment. In the burgeoning Cold War, the United States was also causing deforestation in West Germany, where, according to Lingner, U.S. authorities were unabashedly felling trees to construct military bases and training sites. In contrast, the socialist economy and society of Communist Germany provided the necessary conditions to "cure" the landscape, prevent any further damage, and render it most productive. In the postwar years both Lingner and his colleague Georg B. Pniower, who had assumed the professorship of the Institute for Garden Art and Landscape Design (Institut für Gartenkunst und Landschaftsgestaltung) at Humboldt University, therefore became preoccupied with plantations both in and outside cities.[29]

The wide-ranging postwar lumber scarcity affected all military occupation zones. In the rural northwestern regions, which lacked large forest cover, multifunctional arboricultural practices were revived to create a sustainable, productive cultural landscape. As in East Germany, trees were planted along roads, railroads, riverbanks, and ponds, as well as in former open-pit mines and industrial areas, to increase lumber production, protect crops against dry wind, and enhance the landscape's aesthetic appearance. For these purposes, in February 1947 the regional government of Schleswig-Holstein supported the foundation of Lignikultur, an association for lumber production outside of forests. At the same time in the Franken region of southern West Germany, former transportation engineer Rudolf Eichhorn proposed a different type of production. He argued that the reconstruction of country roads and the revival of fruit tree planting along them would not only create the desperately needed jobs but also feed the people.[30]

Despite Lingner's interests concerning the design of the larger landscape and although he believed that only comprehensive landscape design addressing "the entire area between Elbe and Oder" in collaboration with similar planning activities "in adjacent climate zones" could tackle the problems of drought and flooding, in his capacity as Berlin's director of open-space planning he promoted afforestation measures in the urban realm. Adopting the idea of shelterbelts for the increase of agricultural production in the city of Berlin, Lingner argued that hedges should be planted on barren land and around allotment gardens, and that escarpments and slopes should be afforested to prevent topsoil, sand, and rubble from being blown through streets and being lost in rivers and lakes. "Only if protected by woody plants will our fields and gardens grow fruit," he explained. "Only trees and shrubs will protect against the dust storms that are blowing through our walls and ruins. Where trees and shrubs are missing, birds are missing and pests multiply." For Lingner, the city was part of a larger organic system, a system that had been severely damaged and required treatment on the basis of scientific analysis. Using Lingner's ideas with regard to urban development and the city's reforestation, in 1948 botanist and Humboldt University professor Kurt Hueck drew up a phytosociological map of Berlin accompanied by lists of native plant species for use in the different urban areas (plate 9). These documents were intended to provide the district Departments of Gardens with a foundation and guide for their work.[31]

In the occupied and later divided city, around 80 million cubic meters of rubble had to be removed and the development of dust clouds had to be prevented. Although the dust caused by wartime destruction was a new problem that Berlin and other war-ravaged cities had to tackle, other types of air pollution—caused by smoke from coal fires and industry—and their effects on people and trees had been a topic widely discussed already in early twentieth-century Germany.

By that time German scientists had begun to research the role of trees in the urban climate more thoroughly, thereby joining their colleagues abroad—in particular in Britain, where as early as the seventeenth century polymath John Evelyn had famously pointed to the role of vegetation in mitigating urban air pollution. In early twentieth-century Berlin, Otto Behre had highlighted the role that trees played in maintaining urban citizens' health through climate mitigation. In 1908 Behre therefore argued for land-use planning that would secure certain urban areas as parks planted with deciduous trees. The positive effects that Behre and others attributed to trees were first and foremost related to their gas exchange and to their role in mitigating climate extremes through shade and evaporation. Only later studies paid attention to the role that trees could play for binding dust and other aerosols. Alfred Löbner, of the Institute of Water, Soil, and Air Hygiene in Dahlem, mapped the dust data he had collected in Berlin between 1935 and 1936, and he discovered higher amounts of dust in the air near industry, railway stations, and commercial centers as well as in densely built-up neighborhoods. In contrast, in Tiergarten and other parks and their surrounding areas, dust concentrations were relatively low (figure 5.12).[32]

A few years later, engineer Heinrich Reisner from Essen found that it was trees' porosity that turned them into effective dust binders. Because tree branches and trees standing in rows slowed down air currents rather than blocking them like a wall, dust could quickly settle in trees and their vicinity. Thus, scientists and engineers had begun to prove the age-old assumption that trees cleaned the air, providing the basis for several new investigations in the immediate postwar years.[33]

After World War II, scientists considered rubble dust a health hazard for two reasons. First, the particular quality of the dust particles with their sharp angles and edges could easily cause conjunctivitis and other harm to eyes as well as various respiratory diseases. In addition, rubble dust was considered a vector of infectious diseases as it was often mixed with organic material that could include bacteria. Besides these potential more or less immediate effects on human health, the other negative effect

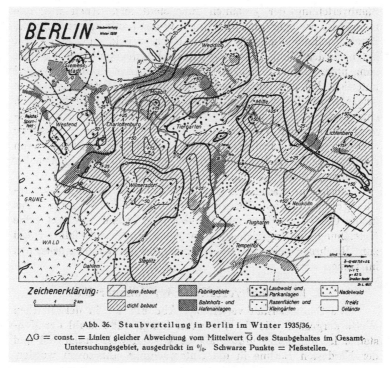

Abb. 36. Staubverteilung in Berlin im Winter 1935/36.

△G = const. = Linien gleicher Abweichung vom Mittelwert $\overline{G}$ des Staubgehaltes im Gesamt-Untersuchungsgebiet, ausgedrückt in %. Schwarze Punkte = Meßstellen.

Abb. 37. Staubverteilung in Berlin im Sommer 1936.

△G = const. = Linien gleicher Abweichung vom Mittelwert $\overline{G}$ des Staubgehaltes im Gesamt-Untersuchungsgebiet, ausgedrückt in %. Schwarze Punkte = Meßstellen.

FIGURE 5.12.
Alfred Löbner's maps showing the distribution of dust in Berlin, winter 1935–36 and summer 1936.
(A. Löbner, "Vergleichende Untersuchungen über den Staubgehalt der Großstadtluft im Winter
und Sommer," *Kleine Mitteilungen für die Mitglieder des Vereins für Wasser-, Boden-,
und Lufthygiene e.V.* 13, nos. 6–8 [1937]: 181–200)

was more indirect and of a psychological nature. Dust particles could change the urban microclimate through facilitating the obfuscation of the atmosphere and the development of fog and mist. Scientists, planners, and designers believed that these climatic conditions negatively affected the human psyche, a concern that seemed especially relevant during the hardships of the immediate postwar years. Managing Berlin's dust from rubble areas and ruined grounds by planting trees that could hold and bind dust was one of the declared objectives of the Greater Berlin Magistrate. Although Berlin's industry had caused much dust before World War II, Alfred Löbner measured a 50 to 70 percent dust increase in Berlin's inner city after the war.[34]

The problem of tackling rubble dust was closely related to questions that arose around rubble disposal and recycling in general. As architect and urban designer Fritz Schumacher explained in 1949, rubble was the new building ground for urban reconstruction. Its treatment was part of the necessary earthwork, a sophisticated area of construction that would contribute to the city's future form. Only part of Berlin's large amount of rubble could be recycled by using it in the construction of roads and sports grounds like the Poststadion in Moabit and the new sport stadium at Lochowdamm in Wilmersdorf. Rubble was also used to fill fire-protection ponds and air-raid protection trenches that had been built during the war. But these uses did not suffice to remove all the rubble that was covering streets and obstructing movement through them. One of the first postwar directives sent to the directors of the district Departments of Gardens, therefore, was to identify those spaces that could accommodate ground modulation. Not only did the debris have to be sorted to separate recyclable materials, its unusable remains had to be dumped in locations where they would least hinder reconstruction. Rubble hills developed, and consequently Berlin's topography changed. The hills either were amassed in existing parks and forests, giving them new contours— as in Humboldthain, Friedrichshain, Hasenheide, and Grunewald (Teufelsberg)—or they formed new park spaces, as in the cases of Schöneberg's so-called Insulaner (plate 10), Prenzlauer Berg's Oderbruchkippe, Biesdorf mountain, and the park at Columbiabad in Neukölln. Creating a new topography and planting it functioned as another sign of Berlin's resurrection.[35]

## Rubble Greening

While piling up rubble hills was a logistical task, planting them with trees was a scientific challenge that both parts of the city had to contend with. Yet planting trees to turn barren rubble hills into permanent green luscious parks and to bind dust on cleared prospective building sites appeared as the most effective solution to the rubble question. While it cleaned the city, it also provided new public open space and nature. As Berlin gardener Karl Schmid commented in 1952, "Tree and shrub will at some point have covered up the vestiges of the harrowing event out of which these mountains have grown." For better or for worse, trees would with time quite literally grow over the rubble and let the dust settle, exemplifying what cultural geographer Stephen Daniels has called the "duplicity of landscape." The unprecedented *Trümmerbegrünung* (rubble greening), a neologism that entered the German language in the postwar years, and the newfound interest in dust and in the climatological aspects of city building led to the implementation of various test sites and research projects. The planting experience most similar to rubble greening before World War II was the greening of open-pit mines undertaken in East Germany's Lower Lausatia in the first two decades of the twentieth century and in West Germany's Ruhr region in the 1920s and 1930s. But these projects had occurred outside of cities and on even vaster scales. The experience with inner-city rubble greening and the new research into dust and urban climate therefore attracted attention in the specialist press in the East and West alike.[36]

In Berlin's West sectors in 1949, researchers at the Robert Koch Institute began representing the data they collected with the help of seventeen dust-collection stations on Berlin's rooftops in dust maps, with particular attention paid to plazas, intersections, and parks. On the basis of collected dust samples they determined the amount of fly ash from furnaces, dust from streets, dust from rubble, and dust caused by traffic. In the same year, the Department of Open-Space Planning established a test site for rubble greening in the rear of the Charlottenburg palace gardens, which had been heavily damaged by bombs and grenades. The planting experiments on the site, which also functioned as a spatial barrier and noise buffer between the palace gardens and the Spandau light rail line, made clear that the best tree species for the greening of Berlin's rubble sites were those that thrived on limestone and produced much foliage so that they quickly created a humus layer. Suitable species included ash, mountain ash, black locust, poplar, Norway maple, sea buckthorn, black cherries, and varieties of hawthorn.[37]

One of the challenges that cities in both East and West Germany had to tackle in terms of rubble greening and planting in general was the lack of topsoil. In Berlin fallen foliage and pruned limbs from street trees therefore became a highly sought-after commodity for composting. As early as February 1946, the city had established a composting committee. Soon the first composting facilities were constructed in the exterior districts, for example, on 30,000 square meters along Argentinische Allee in Zehlendorf. The Departments of Waste Management and Open-Space Planning began to compost waste to create humus. Although the Allies did not agree to pass the ordinance for the protection of humus (Verordnung zum Schutz der Muttererde) drafted by the Magistrate's Department of Building and Housing, a directive with a similar content became binding for all building projects in Berlin. The existing humus layer of any type of construction site had to be saved and heaped into a pile up to 3 meters wide and 1,20 meters high so that the fertile soil could be reused on the site or elsewhere.[38]

To further alleviate the situation and provide the necessary humus for rubble greening, garden architect Alwin Seifert suggested that the city should pump sludge from the rivers Spree and Havel onto rubble areas. According to Seifert, this humus layer would enable cultivation in the subsequent spring and could be protected against the dry summer winds by vertically erected reed mats. Indeed, in 1947 excavated sludge from Lake Rummelsburg was to be used together with trash and topsoil to cover the rubble hill in Friedrichshain the following spring. When this plan ran the risk of failing because the light railway for the transport of rubble that could have moved the sludge to Friedrichshain was used for other purposes, the Department of Open-Space Planning made a new plan to employ the sludge on a sports ground in Lichtenberg. Later experience in West Berlin showed that saving and then mixing humus with finely ground rubble provided a good substrate for the successful planting of rubble hills. Furthermore, the planting of trees along the contours of the new rubble hills and shelter plantings was to prevent erosion (figure 5.13).[39]

Opinions about the necessity of topsoil in rubble greening diverged, not least because the rubble's quality differed even within the same city, requiring site-specific treatment. Nevertheless, in both parts of Germany, the city of Kiel soon served as a model for rubble greening (figures 5.14 and 5.15). There, trees had been planted successfully directly into the rubble, thus bolstering the argument of the forester of the

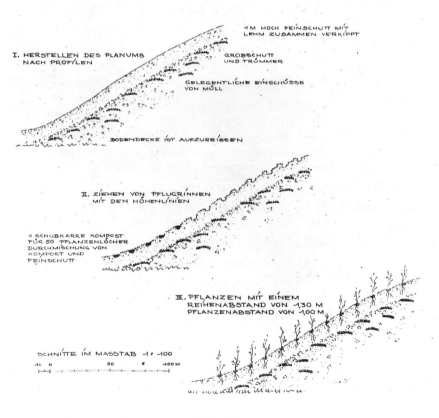

FIGURE 5.13.
Section drawings used to give instructions for planting rubble hills.
(IRS Erkner, Nachlass Johann Greiner, C10, K54, 001)

FIGURE 5.14.
Rubble along Gerhardstraße looking toward Beseler Allee, Kiel, August 1947.
(Stadtarchiv Kiel, 2.36 Trümmerräumung in Kiel)

FIGURE 5.15.
The same area after rubble clearing and greening with young saplings, September 1949.
(Stadtarchiv Kiel, 2.36 Trümmerräumung in Kiel)

FIGURE 5.16.
Schoolchildren of Hebbel Schule
planting saplings—including
alders, elms, black locusts, and
willows—on rubble areas in Kiel,
October 1948. (Bundesarchiv
Koblenz, Bild 183-2005-0707-508)

Lausatian coal mines Rudolf Heuson, who had shown that trees could thrive on crude rubble and were more stable than trees that were planted on sites with humus. In his studies of the greening of dumps, dunes, and wastelands, which he had begun in the early twentieth century, Heuson had dealt with soil conditions that came close to those that landscape architects and urban planners had to confront in the postwar years. Consequently, his 1947 book *Die Kultivierung rauher Mineralböden* (The Cultivation of Rough Mineral Soils) became a widely read resource among practitioners of rubble greening.[40]

Kiel's mayor, Andreas Gayk, had asked citizens to replant before rebuilding. Following his directive, schoolchildren were instructed to plant trees on some of the city center's rubble areas for four hours each day, which quickly turned the sites into informal tree nurseries (figure 5.16). By 1949, trees from the inner-city forest were being transplanted and used for replanting the city's parks and gardens.[41]

Over and above the science and techniques of rubble greening was its aesthetic, although there were some who argued that questions of design were secondary. In 1954 Berlin gardener Eberhard Fink published a prototypical planting design for a grove of black locust and silver birch on rubble hills (figure 5.17). He had observed

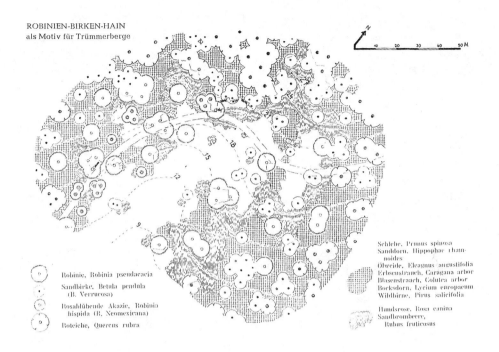

ROBINIEN-BIRKEN-HAIN
als Motiv für Trümmerberge

Robinie, Robinia pseudacacia

Sandbirke, Betula pendula
(B. Verrucosa)

Rosablühende Akazie, Robinia
hispida (R. Neomexicana)

Roteiche, Quercus rubra

Schlehe, Prunus spinosa
Sanddorn, Hippophae rhamnoides
Ölweide, Elaeagnus angustifolia
Erbsenstrauch, Caragana arbor
Blasenstrauch, Colutea arbor
Bocksdorn, Lycium europaeum
Wildbirne, Pirus salicifolia

Handsrose, Rosa canina
Sandbrombeere,
Rubus fruticosus

FIGURE 5.17.
Eberhard Fink's prototypical planting design for a grove of black locust and silver birch
on rubble hills, Berlin, 1954. (Eberhard Fink, "Die Gemeinschaft der Robinie als
Motiv der Trümmerschuttbepflanzungen," *Das Gartenamt* 3, no. 1 [1954]: 3)

how the appearance and practical value of black locust had remained underappreciated, although the tree was one of the most suitable and successful pioneer species on rubble thanks to the nitrogen-fixing bacteria on its root system, which facilitated its fast growth in poor and dry soil. Other benefits were easily summed up: its bark, flowers, and overall habitus were ornamental; its loose branching allowed for undergrowth; it resisted air pollution and cold; it was a honey plant for bees; and its wood was durable and valuable. Studies even reported that black locust fence posts lasted three times as long as those made from oak. Fink therefore concluded that the tree should play a leading role in planting plans for rubble areas. To capitalize on the species' capacity to prepare the ground for future cultivation, he proposed that three times more black locust trees than ultimately needed for the design should be planted. They should be mixed with silver birches, which were ideally suited to mediate between black locust and the third species, pine trees. The naturalistic planting design for the dry southern slope of Fink's imaginary rubble hill was to be complemented with an undercover of blackthorn, Russian olive, Siberian pea-tree, bladder senna, European nightshade, willowleaf pear, sea buckthorn, and bramble-berry. These would protect against soil erosion and underscore the dry atmosphere through their physiognomy. On the open areas, Fink suggested planting creeping baby's breath and rock soapwort as grass substitutes, or a colorful mix of wild perennials, including toadflax, catmint, and bigleaf lupine.[42]

In East Berlin and the GDR as well, rubble greening—also called "instant greening" (*Direktbegrünung*)—became a necessary practice and a research interest in addition to the topics of urban climate and dust pollution. As Potsdam horticulturalist Günter Bickerich observed in the early 1950s, in the summer particularly Berlin's inner-city areas that had been cleared of coarse rubble created dust, requiring the wearing of protective glasses. Garden architect and dendrologist Hans F. Kammeyer at the Department of Garden and Landscape Design of the Research Center for Horticulture in Dresden-Pillnitz (Abteilung Garten- und Landschaftsgestaltung an der Versuchs- und Forschungsanstalt für Gartenbau Dresden-Pillnitz) differentiated between three different types of rubble greening: it could be used to complement spontaneous vegetation growth, to provide a temporary plant cover as "interim" or "primitive" greening on prospective building lots, or to establish more permanent parks and gardens.[43]

East Berlin's later version of the second, temporary type of rubble greening was the so-called Wild Grass Program (Wildgrasprogramm/Wildrasenprogramm) that was begun in the winter of 1961 to improve the visual appearance of the inner city, bind dust, and prevent rubble areas from becoming sources of contagion (plate 11). Rubble areas were cleared, covered with up to ten centimeters of topsoil, treated with herbicide, and then seeded with a drought-resistant grass mixture that had been developed especially for the purpose. Subsequently, many lots were planted with trees, especially with fast-growing poplars. Most wild grass areas were located in the Mitte district, which had been hit very badly by the wartime bombing raids, but other more peripheral districts, like Pankow, also expended a significant amount of money on the program throughout the 1960s. Given the special urgency to reconstruct the city center so that it would adequately represent the East German capital, the wild grass campaign played a particularly important role in this part of the city. Yet it also required the education of the citizens—it appeared curious to many East Berliners that the seemingly wild growth of "weeds" should be the outcome of a highly curated program that employed considerable effort and money.[44]

In the early 1950s, Kammeyer's department undertook a survey of the rubble greening efforts in twenty-eight East German cities with more than fifty thousand inhabitants. The cities considered the preparatory labor for rubble greening to be the most challenging. This included the removal of building debris, the filling of basements to prevent hollows, and the destruction of ground floors to reconnect the plant layer with the water table. Many cities found that topsoil needed to be mixed into the available substrate. Although the suitability of species varied depending on the material quality of the rubble, in particular birches, black alder, black locust, tree of heaven, elderberry, dog rose, and blackberry were thought to be among the most successful plants for rubble greening.[45]

Besides these empirical observations of rubble greening, a dissertation project undertaken by Dieter Hennebo at the Institute for Garden and Land Culture (Institut für Garten- und Landeskultur) of Humboldt University assessed the role of tree plantings, parks, and gardens in binding dust from different types of sources, including rubble. The dust-binding capacity of street tree rows in urban environments was not measured, but studies of different types of hedgerows and dense plantings along country roads revealed various degrees of effectiveness. Hennebo, who published his study in 1955, concluded that measurements in parks and gardens and their surroundings revealed reduced dust levels; that urban vegetation acted as an air filter for dust rather than a source of it; and that the microclimate created by parks and gardens could direct air currents to flow above and past them, a reason for their relatively low dust levels.[46]

Hennebo's study was one of three dissertation projects advised by Georg Pniower in the 1950s that focused on urban trees. Pniower, who had recently moved from a professorship in West Berlin to Humboldt University, was establishing his new chair of garden design, and the studies of his disciples were intended to serve East Germany's reconstruction based on the newly adopted Sixteen Principles of Urban Planning (16 Grundsätze des Städtebaus). The Sixteen Principles, drafted by East German architects after a 1950 trip to Moscow, followed Soviet planning guidelines. In an effort to distance the GDR's plans from the "dispersed city" paradigm and related garden city ideas used in West German reconstruction planning, the Sixteen Principles declared it impossible to turn the city into a garden. While it was clear that cities needed to be provided with greenery, the principle that life in cities was more urban and life in the country was more rural was not to be subverted. Hennebo's 1955 study on the dust-binding capacity of trees and parks was part of this effort to create a new healthy urbanism. As he explained, urbanity needed to be sustained and all negative aspects of the urban living environment, such as dust, removed.[47]

Rejecting zoning, the separation of functions, and any type of urban fragmentation, which were considered character traits of capitalist cities, the Sixteen Principles also stressed the importance of a central hierarchical organization that was given expression in a city center with administrative and cultural institutions, representative architecture, and plazas based upon the city's history. Although the Sixteen Principles did not explicitly include climatological concerns and mentioned matters of landscape and "greening" only in passing, the urban climate became an important consideration in East German reconstruction efforts. To complement the Sixteen Principles, therefore, the Research Institute for Urban Design (Forschungsinstitut für Städtebau) commissioned meteorologist Wolfgang Böer of the GDR's meteorological and hydrological service in Potsdam to establish guidelines for the consideration of climate in urban planning and design. After all, the leading voices in East German planning at the time argued that urbanization did not necessarily have to cause climate deterioration. Instead, the climate could be designed and managed through intelligent urban design. For example, in his attempt to interpret the Sixteen Principles for landscape architecture, Lingner listed the improvement of the urban microclimate as the first of three main points, the others being the provision of open space for sports and recreation and its artistic design.[48]

Of course, climatological concerns preoccupied planners, architects, and landscape architects in East and West Germany alike. Already before the war in 1937, German meteorologist Albert Kratzer had argued that urban planning could mitigate urban climate. "More than generally assumed," he posited, "cities are co-designers of their own climate." After the war, publications abounded. Discussing urban heat island effects, Erich Kühn, professor of urban design and landscape planning at the Technical University Aachen, argued in 1957, "Every dispersal of the city also decentralizes the source of warmth." But East German planners wanted to illustrate that "socialist city building" did not need the dispersal of buildings to respond adequately to climatological concerns. Hennebo even claimed that it had been impossible for cities characterized by capitalist land ownership to consider climatological concerns. In contrast, plans for new cities in the Soviet Union—Stalingrad and Magnitogorsk among them—had taken climate into consideration. In his early postwar arguments for a socialist land reform, Hennebo's advisor Georg Pniower had included the description of a prototypical diagram of an ideal city based upon climatological concerns. The city would be positioned perpendicular to the main wind directions, it would be surrounded by forests, and its area would be interspersed with afforested wind corridors.[49]

## Trees along East Germany's First "Socialist Street"

The first big opportunity in East Berlin to integrate climatological concerns into a representational reconstruction project was the laying out of Stalinallee, East Germany's first "socialist street." Construction along this new East Berlin showcase boulevard began in 1952, but the planning and design process had been long and complicated. The preparation of its planting had begun in 1950 when a directive from the director of open-space planning Lingner asked for increased attention to the cultivation and care of trees for Stalinallee. At the time, adequate street trees were hard to come by as most tree nurseries had been used for the cultivation of vegetables during and after the war. Stalinallee's final urban design was assembled on the basis of five prize-winning competition entries and subsequent consultation with Soviet architects. They had advised that the buildings on both sides of the boulevard should form an ensemble that included commercial space on the ground floors facing the street. As the most important propaganda project in East Germany at the time, Stalinallee needed to express the joy of life. The Soviet architects had advised that the boulevard should have an uplifting effect on East Berliners. The landscape design of the public open space played an important role in the creation of this purportedly happy urban ensemble even though, as Lingner later criticized, landscape architects were employed for the open-space design only once building masses and urban spaces had already been determined.[50]

Most building blocks had neoclassicist-inspired facades with varying offsets in depth so that they "moved" back and forth along the street. But more important for the design of the public open space was the fact that the building blocks lining the street on the north were set back further from the street than the buildings on the south. The urban design and the subsequent tree planting and landscape design responded to climatological considerations regarding the east-west orientation of Stalinallee (fig-

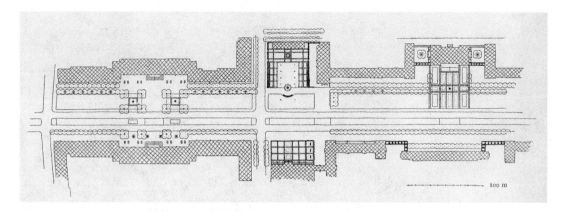

FIGURE 5.18.
Landscape design for part of Stalinallee by the Kollektiv Reinhold Lingner, 1953.
(Frank Erich Carl, *Kleinarchitekturen in der deutschen Gartenkunst*
[Berlin: Henschel, 1956])

ure 5.18). Basing his work on the first ideas developed by Lingner in 1951, landscape architect Helmut Kruse designed a linear ornamental park space for promenading along the northern side of the boulevard, which would get the most light and sun. Altogether it was around forty to fifty meters wide. In early 1953, a double row of silver linden trees was planted at a distance of eight meters from the buildings (figure 5.19). This ten-meter-wide tree promenade stood far enough away from the building facades so that the trees would not prevent light from entering through windows. At the same time, the trees provided a shady walkway in the summer that contrasted with the open twenty-meter-wide lawns with low ornamental plantings that reached toward the boulevard's roadway and accompanying bike path. Along the south side of Stalinallee, the design was limited to a generous sidewalk and a row of trees along the street. Although the tree rows were a response to climatological concerns, they also provoked criticism. It was clear that in due time the tree canopies would conflict with the tall monumental streetlights designed for the new boulevard. Lingner himself, who had provided the initial ideas for Stalinallee's landscape, ultimately questioned the length and schematic rigidity of the tree rows, which countered the architects' idea of the more flexible building line, which jumped back and forth. According to him, it was not sufficient for tree rows to be interrupted at various points to highlight important buildings. Instead, trees should also have been used to create spaces in the broad areas leading toward the thoroughfare.[51]

In all, 450 silver linden trees were planted along Stalinallee's first construction section. They mitigated climate and their rows unified the right-of-way while providing color and seasonal change. The trees structured the wide open space, in particular along the northern side, and they added a human scale to the monumental boulevard. The tree planting was also ideologically motivated, in this sense serving as urban cosmetics. On the one hand, designers used the trees in ornamental ensembles flanking

FIGURE 5.19.
Chargesheimer's view of Stalinallee from the terrace of Café Warschau, ca. 1953.
(Repro: Rheinisches Bildarchiv Köln, Chargesheimer, rba_L010538_32)

and highlighting important buildings like the German Sports Hall. On the other hand,
they realized the trees' potential to hide buildings that by the early 1950s no longer
corresponded to the new urban design paradigm preferred by the Socialist Unity Party.
In the immediate postwar years between 1949 and 1951, the first new modernist apartment buildings along Stalinallee had been constructed in the spirit of the dispersed
city paradigm that shaped Hans Scharoun's reconstruction planning along what was
then the easternmost section of Stalinallee (then Frankfurter Straße/Frankfurter Allee;
today Karl-Marx-Allee between subway stations Frankfurter Tor and Weberwiese).
Given the 1950s turn toward the Sixteen Principles of Urban Planning, the architecture of this first phase of reconstruction along Stalinallee no longer seemed appropriate. It appeared too close to reconstruction initiatives in the West. Following the suggestion of consulting chief architect of Moscow Alexander W. Wlassow, therefore,
trees were soon planted in front of what the first secretary of the Socialist Unity Party,
Walter Ulbricht, had decried as "American egg boxes" to cover them from view.[52]

Despite the ideological claims to climate control through tree planting, the improvement of the urban climate preoccupied planners, architects, and landscape architects in the East and West alike. During the war, Berlin's garden director Josef Pertl
had already requested that in the case of newly planned streets running east-west, side-

walks on the north should accommodate street trees and landscaping and be wider than on the south side. In east-west streets of average width, only the northern sides were to be planted with street trees. Viennese architect Camillo Sitte had made a similar request in the early twentieth century, noting the happy coincidence that street trees along northern street sides received the necessary light and gave pedestrians the desired shade. Pertl's later directive reverberated not only in the 1944 guidelines for the construction of urban streets but also in various postwar instructions and deliberations about street tree planting.[53]

In 1953 Berlin gardener Eberhard Fink noted in the West German journal *Das Gartenamt* that in narrow east-west streets only the northern side should be planted with street trees. Most notable for East Berlin was Alfred Hoffmann and Karl Kirschner's study of urban street tree planting that formed the basis of Hoffmann's 1954 doctoral dissertation at Humboldt University. Advised by Pniower, the study was part of a larger research project conducted in 1953 on the fundamentals of green planning under the auspices of Lingner at East Berlin's Research Institute for Urban Design. Although Hoffmann's work synthesized existing knowledge and expertise and could hardly claim innovative content, it was part of an attempt to create a new socialist foundation for landscape design. In the work, which listed suitable street trees, studied street trees' various hygienic functions, and examined best management practices and street tree care, Hoffmann also elaborated on tree planting along east-west streets. Besides the bioclimatic functions, he drew attention to the needs of the trees themselves. Tree growth and health depended, among other things, on access to light and therefore on the street side and its orientation (figure 5.20). Hoffmann's work was also published in West German landscape journals.[54]

The time was ripe for further scientific study of how trees could mitigate climate and airflows in built-up areas. Research on this subject in cities was still lagging in Germany at the time.

In the United States in the 1950s, American landscape architect and researcher Robert F. White studied the role of trees in the climate control of inhabited areas at the Texas Engineering Experiment Station (figure 5.21). His results, which informed Victor Olgyay's seminal 1963 book *Design with Climate,* were relayed in West German journals in the late 1960s. Undertaking airflow studies around real trees outside and model trees in the laboratory, White had shown how tree and hedge planting near buildings could be used as air-conditioning. He had expanded upon earlier shelterbelt studies carried out to enhance agricultural production in rural regions to serve human habitation in the urban realm. As West German urbanist and professor Erich Kühn pointed out in 1957, landscape architects could use trees and urban green space to regulate airflow and the intensity and direction of winds in the city.[55]

The effects of war in German cities had made it clearer than ever before that trees and parks were needed to mitigate the urban climate. Street trees had been a point of contention during the Berlin Blockade, furthering conflicts of the Cold War. They

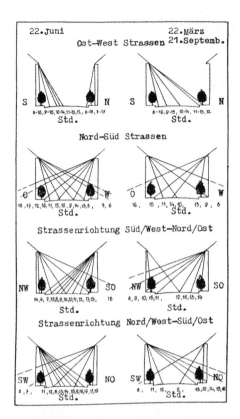

FIGURE 5.20.
Light conditions for trees in east-west streets, drawn by Alfred Hoffmann after W. Grocholskaja. (Alfred Hoffmann, "Eine ökologische Betrachtung zum Problemkreis des Straßenbaumes," *Das Gartenamt* 4, no. 10 [1955]: 182–86)

FIGURE 5.21.
Roland Chatham's photograph of a tree model made of moss shows the breeze pattern when positioned in the wind channel at the Texas Engineering Experiment Station. (Robert F. White, "Effects of Landscape Development on the Natural Ventilation of Buildings and Their Adjacent Areas," Research Report 45 [College Station: Texas Engineering Experiment Station, 1954])

were needed to make the dust settle. But they also continued to be a means of competition between East and West Berlin in the decades that followed. Once the blockade was over, the street tree battle between East and West Berlin carried on, albeit in a different form: now it was about afforestation and the replacement of trees that had been destroyed during the war and the blockade, rather than about questions of deforestation. The competition was on: would East or West Berlin earn the title of the greenest city?

# 6
# Greening Trees
## *Replanting East and West Berlin*

The Berlin Blockade had only aggravated the plight and poverty of the people in the bombed city. Many lost the jobs they had only just found. Street tree work helped to temper unemployment and revive both parts of the city. Largely sponsored in West Berlin by the European Recovery Program and in East Berlin by the National Construction Program (Nationales Aufbauprogramm Berlin, and then Nationales Aufbauwerk), the district departments' emergency programs for unemployment relief included the planting and maintenance of new street trees and the extremely laborious removal of former tree stumps. For example, in West Berlin, the Tempelhof district Department of Gardens, which had just planted two thousand new street trees, anticipated that it could employ workers to weed the tree pits, plant them with grass, water them three times a day, and even them out. In East Berlin's Weissensee district, twenty workers were to be employed in the removal of tree stumps in parks and along streets, and ten to augment the regular tree gang. In Pankow, twenty-five hundred new street trees were to be planted along the main streets and thoroughfares with the help of relief workers.[1]

In West Berlin in 1949 a fluctuating number of between two thousand and four thousand emergency workers were used by the Departments of Open-Space Planning to rebuild and replant Berlin's urban green; in the period 1950–52, the numbers fluctuated between five thousand and fifteen thousand for most of the time but reached a high with up to nineteen thousand emergency workers in the summer of 1950. Women made up 40 to 60 percent of this workforce. As the new garden director of West Berlin, Fritz Witte, reported in 1952, his department, in contrast to the engineering and construction departments, was fortunate in that little special equipment was needed for the labor. The work mostly entailed rubble clearance, the construction of playgrounds and small parks, and street tree planting and maintenance. The activities confirmed that the city could indeed receive "help through green" (*Hilfe durch Grün*), as the motto of the 1952 GruGa, the Big Rhineland Horticultural Show (Große Ruhrländische Gartenbauausstellung), and of following West German horticultural shows aptly suggested.[2]

The reconstruction years were a period when interest in the various functions of street trees peaked in both East and West Berlin. This was only logical given the trees' importance for climate mitigation, their role in spatial structuring and aesthetic improvement, and the new requirements mandated by increased automobile traffic. While their ornamental and spatial role continued to be of major importance, most attention was given to the functions that benefited public health. In East Berlin, Dieter Hennebo sought to better understand the dust-binding capacities of trees. In West Berlin, researchers studied trees' capacity to buffer sound. But although street trees were attributed with health and aesthetic functions in East and West Berlin and although research projects were undertaken in both cities to scientifically prove previous pseudoscientific assumptions, the emphasis on these functions and the methods to achieve them varied. East Berlin and East German landscape architects emphasized street trees' materialist functions and their efficient planting and maintenance, leading them to develop tree-planting concepts. In West Berlin, as in other parts of West Germany, the opportunities offered by reconstruction inspired landscape architects to discuss the street tree aesthetic and consider alternate street tree–planting paradigms. In East and West Berlin as in East and West Germany in general, however, the postwar years marked the beginning of a new phase of scientific street tree management. East and West Berlin landscape designers and tree experts continued developments that had begun in the late nineteenth century when the first street tree standards had been implemented, and began to conduct research and apply best practices. Finding ways to measure the various street tree functions was the only way to further improve their effectiveness and legitimize their planting in an environment that was increasingly dominated by the private automobile and concerns for its space and safety requirements.

## Socialist Street Tree Planting in East Berlin

The declared goal of East German officials in the postwar years was to turn Berlin into a green socialist capital city serving Germany in its entirety. "Socialist city greening" was a new term used in the 1950s and 1960s for this purpose. In true socialist spirit, East Berlin landscape architects and city officials argued for the equality of architecture and landscape. In a state in which history and the accomplishments of particular periods and people were co-opted for political and social goals, the landscape creations of nineteenth-century Prussia were considered useful in legitimizing landscape architecture's status. For East Berlin landscape architects, the collaboration between architect Karl Friedrich Schinkel and landscape gardener Peter Joseph Lenné under the Prussian reign was a local precedent that could engender national pride and emulation. Following this nineteenth-century legacy and its heritage—while overlooking the aristocratic character—architecture and landscape architecture were to complement one another to form one urban ensemble.[3]

Although the reality, particularly during the early years of the GDR, was quite different, landscape architects argued vehemently that architecture, parks, and gardens had to be considered of equal importance in the "socialist city." Green open space was no luxury in socialist city building—it was a necessity. Consequently, an even distribution of trees in all streets was seen as a means to turn Berlin into a "socialist metropolis." To make sure Berlin would no longer be the "stony desert" it had typically been referred to since the late nineteenth century, trees were also to be planted along streets that had not been planted before the war. As several landscape architects, tree experts, and scientists pointed out, tree planting was a fundamental contribution to the creation of "an environment worthy of socialist society." East Berlin's director of gardens Helmut Lichey argued that street trees could fulfill their hygienic and aesthetic functions in the loose new urban designs of socialist city building better than ever before.[4]

Although contradicting East Berlin's comprehensive street tree–planting objectives, which projected trees along all city streets, landscape architect Harri Günther, like many of his Western colleagues, wanted to avoid the monotony of regular lines of street trees. In socialist East Germany, he easily found an ideological argument to dismiss regular street tree rows if they did not share the urbanistic importance of boulevards like Unter den Linden. After all, regular street tree rows were a conceptual "relic of the feudal epoch that had actually been overcome." On the emotional level, "the beauty of plants" was thought to help citizens overcome a city's "capitalist asphalt atmosphere." Informed by the ideologies of the Cold War, the notion was that Communists would use street trees, parks, and gardens to create happy cities symbolic of life, whereas capitalists would drown their cities in asphalt. Street trees contributed to the city's image and East Berlin's government considered a well-maintained streetscape a mirror of the Soviet sector's economic progress.[5]

The East Berlin press, quick to understand the value of street trees in the ideological battles of the burgeoning Cold War during the Berlin Blockade, continued to pay attention to street trees in the years that followed. Now it reported on East Berlin's street tree progress. Between 1949 and 1951, 10,000 street trees were planted in East Berlin, and by April 1952, East Berlin boasted 90,000 street trees overall, suburban Köpenick appearing as the greenest district with 18,009 trees. Most of the newly planted trees were linden, black locust, mountain ash, maple, and plane trees.[6]

Street tree planting and maintenance in East Berlin soon began to be informed by interpretations of Marxist theory and by the new socialist efficiency principles the state was seeking to introduce. Paradoxically, after the East Berlin press had ridiculed and condemned street tree use as firewood in the Western sectors during the Berlin Blockade, in 1953 the GDR government considered street trees and other roadside trees merchantable wood that could yield considerable profit. In a directive of 19 March, the government announced that the wood from street trees was a valuable resource that had not yet been recognized as such. Plans for the maintenance and use of street and roadside trees were to be prepared each year and confirmed by the Ministry of Forestry and Agriculture. Contracts for the respective sales quantities from street and roadside trees were to be arranged with the local state forestry operations.

So scarce had wood become in Berlin that as early as June 1951 the Magistrate had ordered district Departments of Gardens to report available lumber quantities; in May 1952 it ordered that all scrap wood from workshops, woodshops, and building sites, including from the departments' tree-pruning and maintenance activities along streets and in parks, was to be centrally collected so that it could be sold and reused. While workshops were allowed to reuse the wood for their own purposes, under no circumstances were workers allowed to take it.[7]

This new materialist ethos and the state's new economic goals had also contributed to the dissolution of East Berlin's Department of Gardens (Hauptamt für Grünanlagen) in 1950. Like all other tasks regarding urban green, street tree care and production were transferred to the Departments of Municipal Economy (Abteilung Kommunale Wirtschaft) and Agriculture (Abteilung Landwirtschaft). Despite their lack of expertise, these departments would remain in charge of street tree care, planting, and production until in 1960 all work regarding parks and gardens, including street trees, was united again in a central East Berlin Department of Gardens (Stadtgartenamt) and its district offices (Bezirksgartenämter).[8]

The 1953 directive addressing the country's lumber shortage and the transitional organizational restructuring was in line with the formation of the GDR's centrally planned economy and with Marxist theories regarding the relationship between humans and nonhuman nature. With their nineteenth-century belief in the infinite reproduction and resources of nature, Karl Marx and Friedrich Engels had considered this relationship as one of human power over nonhuman nature. For the two philosophers it was a matter of course that humans, on the basis of their knowledge and technological progress, would modify and make use of nonhuman nature according to their needs. Although Marx and Engels lacked a theory of nature per se, after the GDR's foundation, landscape professionals began to mine their writings for use in journal editorials and other official statements, seeking to politically legitimize their work and couch it in socialist rhetoric. But it was not only the transformation and use of nature that could be framed as appearing in line with Marxist thought. The opposition against a reckless exploitation of nature could similarly be bolstered by Marxist arguments. As garden architect Walter Funcke explained in 1950, if the socialist system was to eliminate the exploitation of humans, it was only logical that the elimination of any exploitation of (nonhuman) nature should follow.[9]

Given that the new materialist approach to trees and to nature in general could hardly be accommodated in the 1935 Reich Nature Protection Law (Reichsnaturschutzgesetz), the GDR replaced it in 1954 with the Law for the Conservation and Care of Native Nature (Gesetz zur Erhaltung und Pflege der heimatlichen Natur). Besides protection, the new legislation also emphasized research in the natural sciences and accommodated the use of natural resources, thus tolerating the human design and alteration of nature. The preamble emphasized that nature was protected out of care for the well-being of the people and for the sensible use of its resources. Nature was to be protected so that the economic goals during the formative years of socialism could be reached.

Some decades later, in 1981, the GDR's new tree ordinance created a mandate for cities and citizens to protect their trees while at the same time assuring the state's maximum profit if trees did have to be felled, for example, on the occasion of new construction projects. Cities were required to offer any lumber they accrued to the state forestry units for purchase. Based on this 1981 state ordinance, East Berlin's tree ordinance of the same year allowed for the clearing of trees if they obstructed the use of properties, the use of land for allotment gardening, the renewal of tree stands, and the planned economic use of the trees. Furthermore, permission to cut a tree was not required if it endangered human life and health, socialist property, or people's personal property, or if its removal was required as a result of the tree's health. As determined in the GDR's ordinance, lumber that Berlin's district offices accrued through tree removal had to be offered first to state forestry units for purchase.[10]

Yet to achieve these declared socialist goals, tree planting and the design and construction of expansive parks and gardens had to be made economically feasible and rendered efficient. Material, money, and labor had to be employed as sparingly as possible and to maximum effect. In addition, close ties between science, industrial development, and technological progress as well as between research and practice were to be forged, following the Soviet model. The Working Group for City Greening (Arbeitsausschuß "Stadtbegrünung") of the Berlin Association of Building Trades (Fachverband Bauwesen) in the Chamber of Technology (Kammer der Technik) therefore proposed to form committees to strengthen the ties between academic research and professional practice and to promote the mechanization of all work and maintenance processes through the consideration of technological progress, new design methods, and administrative organization. The Working Group argued that research conducted at the Institute for Garden and Land Culture (Institut für Garten- und Landeskultur) and the Department of Green Planning (Abteilung Grünplanung) at the Institute of Urban Design (Forschingsinstitut Städtebau) of Humboldt University should be made readily accessible to all professionals dealing with the city's streets, parks, and gardens. Obviously, forging close ties between research and practice made sense only if adequate research was being produced, something that appeared questionable at the time. University assistant Dieter Hennebo and horticultural scientist Robert Zander therefore sought to tackle the dearth of serious research in 1956 with a small manual, *Anleitungen zur Grundlagenforschung in Grünplanung und Gartenkunst* (Instructions for Basic Research in Green Planning and Garden Art), perhaps the first book ever written in any language on research methods in landscape architecture. Landscape architects and horticulturalists at the German Building Academy (Deutsche Bauakademie) and the universities reacted. They sought to produce a theoretical and scientific grounding for landscape design within the framework of the socialist state.[11]

## Industrializing Trees

It became clear that as in architecture, industrialization, mechanization, and typification could be used to render the practice of landscape architecture more efficient and effective. As landscape architect Johann Greiner stressed in 1966, garden architecture needed to follow the typification and industrialization in housing construction and develop efficient processes of design, implementation, and maintenance. Some of the work that fed into ideas for landscape architecture's standardization was produced at the universities and at the German Building Academy.[12]

In the realm of street tree planting, industrialization included the industrial production of street trees, an effort undertaken in the GDR's tree nurseries, which were turned into state-owned People's Enterprises (Volkeigene Betriebe [VEB]) and People's Estates (Volkseigene Güter [VEG]). For example, the VEG Saatzucht Baumschulen in Berlin was preoccupied with cultivating large numbers of fast-growing poplars that were needed to turn the new residential districts green as fast as possible. Since the immediate postwar years, the growth of poplars in and outside forests had received particular attention in forestry science, arboriculture, and the nursery industry as trees and wood were needed for multiple reconstruction purposes. Cities and forests that had suffered during the war needed to be reforested as fast as possible, prompting the establishment of a poplar cultivation program.[13]

As demand for larger and stronger trees with a caliber of more than eighteen centimeters increased, a system of tree-classification numbers was introduced to determine the future need for trees of different sizes and facilitate their timely cultivation in the nurseries. Different tree-classification numbers were attributed to different dimensions and types of open space in new urban developments so that a total anticipated number of trees could be calculated for each projected development and cultivated in the nurseries in advance. A rule of thumb was that any new urban development would be planted with 20 percent young poplars and 8 percent "big trees" (Großbäume), defined as having a caliber of over thirty centimeters. Furthermore, 50 percent of the trees were to be of a "leading" species and 22 percent of an "accompanying" species. Among the leading and accompanying species, the ratios between young and more mature trees with calibers of eighteen to twenty and twenty to twenty-five centimeters were determined as well, thus making it possible to provide nurseries with detailed purchase lists of trees years before the actual design and its implementation.[14]

Nevertheless, nurseries struggled to produce the necessary trees, and plantings in new urban housing developments and replacement plantings in old urban districts lagged behind. In order to facilitate the creation of instant landscapes and in response to a request from traffic planners for tall and full-canopy street trees, Dresden landscape architecture professor Harald Linke in 1966 began studies to economize and maximize the vegetative propagation of large-caliber trees in so-called tube containers. These trees could be transplanted year-round without limiting what would today be called their ecosystem services. The city of Dresden sponsored the continuation of the project so that by 1975, Linke and his collaborator Hans Prugger were able to file a patent for a new method of vegetative propagation of large trees. According to Linke and Prugger, their method enabled the production of sturdy three-meter-tall trees without multiple transplantations, which typically damaged root systems. The trees could also be produced in only six months instead of the usual twelve years. By the time Linke and Prugger had filed their patent, the method was also thought to benefit the GDR's 1973 public housing program. Linke's advisee Wolfgang Fischer continued the studies of the fast production of large-caliber trees through his dissertation research in the 1980s. Fischer explained in socialist jargon how this research could serve the "intensification of society's reproductive processes" by facilitating a qualitative increase of tree production. Another focus was added to tree production once it had become clear that what was needed was pollution- and pest-resistant tree species with fastigiated shapes or small crowns that developed only above a certain height.[15]

Industrialization went hand in hand with mechanization, which involved not only standardized planting and maintenance procedures but also the use and development of new technology and machinery for tree planting and maintenance. Following Georg Pniower's call for the standardization of the transplanting of large trees, in 1955 Alfred Hoffmann and Karl Kirschner published a first analysis of the respective work process and thoughts on how transplanting in tree nurseries could be rendered more efficient. Dividing the process into three steps—digging out, transportation, and plantation—would allow for the independent preparation of trees to be transplanted, for the transportation of numerous trees at the same time, and for continual planting at the new location. Hoffmann and Kirschner argued that the goal should be to develop technologies that would enable such a work process throughout the entire year, regardless of the season.[16]

In the 1960s, landscape architect Hans Georg Büchner, inspired by architects Hans Schmidt and Konrad Wachsmann, explored the application of industrial building methods and efficiency principles in landscape design. Because it was impossible to fully standardize living plants, Büchner argued that in landscape design attempts at standardization could be undertaken only with regard to tools and methods. Of particular importance would be new machines and their employment according to the Mitrofanow method, which implied the increase of efficiency through the temporal and spatial consolidation of work processes. For example, if tree groves and playgrounds with sandpits were positioned next to each other, both tree pits and sandpits could be dug out with the same machine during one operating sequence. Efficiency could similarly be increased if trees in nurseries were planted at a distance that facilitated subsequent care with hack machines, or if a new hack machine was produced that was adapted to the distance needed for the optimal development of trees.[17]

Once the 1961 state plan New Technology (Neue Technik), which aspired to increase labor productivity through the promotion of scientific and technological progress, had been published, the East Berlin Department of Gardens sought to further increase the mechanization and standardization of its maintenance procedures. In a 1964 directive for the department's five-year plan for 1965–70, director Lichey mandated that the care and maintenance of street trees was to be rationalized by the formation of special brigades that were equipped with the latest technological tools. In 1972 his department sought to rationalize the care of street trees and of all other plants and elements of public parks and gardens further by implementing standards

(*Mustertechnologie*) for all care operations carried out by the district offices. Young trees, like perennials, lawns, park trees, and other elements the offices cared for, were divided into care categories. Depending on their respective care category, young trees would, for example, be watered thirty, twenty, ten, or only five times per year.[18]

Besides the implementation of these operational standards, efficiency could also be augmented by the invention of new machines. A case in point was the development of a machine for the removal of tree stumps, one of the biggest challenges of all tree work. The Institute for Technology and Mechanization (Institut für Technologie und Mechanisierung) at the GDR Building Academy finally invented such a machine in the early 1980s. What in the past had required hard menial labor or the use of explosives by 1982 necessitated only the special attachment of "LBA 7/1" to any hydraulic shovel. LBA 7/1 was the appropriation of a machine for the production of posts made out of site concrete. Mounted onto a hydraulic shovel, it could be used to dig out stumps without damaging surrounding earthworks and pavements.[19]

Despite these efficiency principles, Büchner and others also argued for an aesthetic standard that was not to be compromised by the use of new technologies and machines, nor by designs and planting procedures that curtailed tree health and growth. Thus, garden director Lichey had rejected a 1971 proposal to reduce the size of all tree pits regardless of their location. The proposal had been submitted by a group of workers as part of the state-driven *Neuerer* system, which encouraged innovative ideas to elevate productivity. Instead, the general objective for Lichey, as for many of his colleagues, was to achieve attractive designs economically and efficiently. In this vein, Klaus-Dietrich Gandert, horticultural researcher and later professor at Humboldt University, argued that modern technologies and economic demands could even elevate aesthetics. With relatively little effort, tree planting was to achieve the maximum effect, including the emotional benefit attained through seasonal rhythm and colors.[20]

But as much as technology was to be adapted to plant morphology and growth, the plant selection and the planting designs were supposed to facilitate the plants' care in the first place. The trees to be planted were to be site-specific, able to withstand harsh treatment with machinery, and resilient against the chemicals used to keep insects at bay. "Typification" was another means to achieve this goal. Regarding street tree planting, typification meant that planting typologies should be developed for different types of streets, neighborhoods, and cities. The most important effort to this end was the development of comprehensive tree-planting concepts for reconstructed cities or districts and for new housing developments.[21]

## Comprehensive Tree-Planting Concepts

The interest in standardizing street tree planting within a rationalist urban planning and construction process and the desire to treat architecture and landscape as equally important components in urban design led East German landscape architects to develop comprehensive tree-planting concepts. In 1966 Berlin professor of horticulture

Klaus-Dietrich Gandert and his student Klaus Ostwald contended that trees had to be considered integral components of the city plan, not ornamental afterthoughts. Trees, they argued, were as important as buildings. Like their colleague Johann Greiner, the researchers believed that cities should be divided into recognizable and characteristic planting units.[22]

Similarly, Reinhold Lingner and his university assistant Hans Georg Büchner suggested in 1968 that trees could provide cities and their neighborhoods with characteristic identities. At Humboldt University's Institute for Garden Design (Institut für Gartengestaltung) they had embarked on a project to develop a method for tree-planting concepts using Halle-Neustadt as a case study. Basing their study on previous elaborations by Greiner and on the first tree-planting concepts developed in the postwar years by various colleagues, Lingner and Büchner explained the objectives: tree-planting concepts should be drawn up to guarantee a unified look that stressed the respective site's landscape qualities, provide for the trees' adaptability and health, and ensure their timely provision. The concepts were thought to optimize aesthetic and what Gandert and Ostwald called "biological-technical" street tree functions. Established on the basis of an analysis of the respective site's ecology and use, tree-planting concepts should become integral to both land-use plans and more detailed area development plans. On the scale of land-use plans, they determined the characteristic plant palette for the various areas of a city or district on the basis of aesthetic, functional, and economic criteria. Borrowing a representational technique from Walter Funcke and Heinz Karn's 1959–60 tree-planting concept for Potsdam-Waldstadt (figure 6.1), Lingner and Büchner produced an abstract schematic diagram (figure 6.2) in which different horizontal hachures indicated areas with various ecological conditions. Different vertical hachures stood for areas with diverse functions, like residential areas and city centers. Denser hachures indicated that more intense tree care would be required. A number was used to describe the combination of tree species, including the ratio of the various species to be used for each functional unit. In addition to this information, more detailed tree-planting concepts drawn up to accompany area development plans should determine actual tree locations and the trees' aesthetic quality (figure 6.3). The rational, minimalist graphic representation of the tree-planting concepts supported the planar treatment of urban areas and the understanding of them as large vegetated areas or forests. More or less unwittingly, therefore, Lingner and Büchner—as well as their colleagues who had provided the early examples—anticipated the conception of urban forestry that would in the following years become a declared objective of city governments in the United States.[23]

The first comprehensive tree-planting concept was drawn up for the new industrial city of Stalinstadt. Renamed Eisenhüttenstadt in 1961, Stalinstadt was planned in the early 1950s as a model socialist town for thirty thousand inhabitants. While garden architects considered the tree-planting concept a necessary aesthetic component of urban design, it was also meant to contribute to the creation of an amenable bioclimate.

Garden architect Walter Meißner's 1952 tree-planting concept was based on a

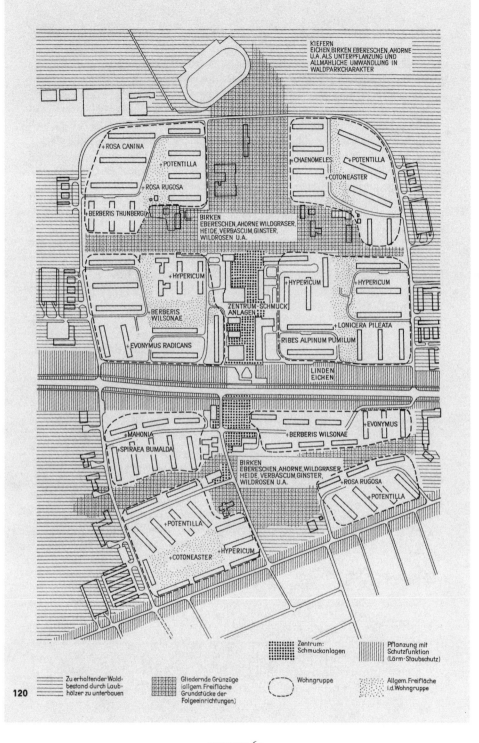

KIEFERN
EICHEN,BIRKEN EBERESCHEN, AHORNE
U.A. ALS UNTERPFLANZUNG UND
ALLMÄHLICHE UMWANDLUNG IN
WALDPARKCHARAKTER

+ROSA CANINA

+POTENTILLA

+ROSA RUGOSA

+BERBERIS THUNBERGII

CHAENOMELES       +POTENTILLA

+COTONEASTER

BIRKEN
EBERESCHEN, AHORNE WILDGRÄSER,
HEIDE, VERBASCUM, GINSTER,
WILDROSEN U.A.

+HYPERICUM

+HYPERICUM      +HYPERICUM

+BERBERIS
WILSONAE

ZENTRUM-SCHMUCK
ANLAGEN

+LONICERA PILEATA

+EVONYMUS RADICANS

RIBES ALPINUM PUMILUM

LINDEN
EICHEN

+EVONYMUS

+MAHONIA

+SPIRAEA BUMALDA

+BERBERIS WILSONAE

BIRKEN
EBERESCHEN, AHORNE, WILDGRÄSER,
HEIDE, VERBASCUM,GINSTER,
WILDROSEN U.A.

+ROSA RUGOSA

POTENTILLA

+POTENTILLA

+COTONEASTER      +HYPERICUM

Zentrum:
Schmuckanlagen

Pflanzung mit
Schutzfunktion
(Lärm-Staubschutz)

Zu erhaltender Wald-
bestand durch Laub-
hölzer zu unterbauen

Gliedernde Grünzüge
(allgem. Freifläche
Grundstücke der
Folgeeinrichtungen)

Wohngruppe

Allgem. Freifläche
i.d.Wohngruppe

120

FIGURE 6.1.
Walter Funcke and Heinz Karn's tree-planting concept for Potsdam-Waldstadt,
1959–60. (Johann Greiner, *Grünanlagen für mehrgeschossige Wohnbauten*
[Berlin: VEB Verlag für Bauwesen, 1966])

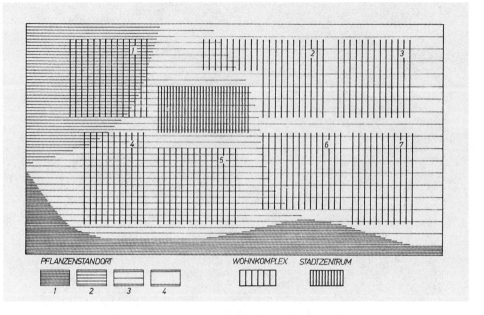

FIGURE 6.2.
Reinhold Lingner and Hans Georg Büchner's diagram for use in the representation of tree-planting concepts. (Reinhold Lingner and Hans Georg Büchner, "Bepflanzungskonzeptionen für Städte," *Wissenschaftliche Zeitschrift der Humboldt-Universität zu Berlin, Math.-Nat. R. 17, no. 2 [1968]: 215–30)*

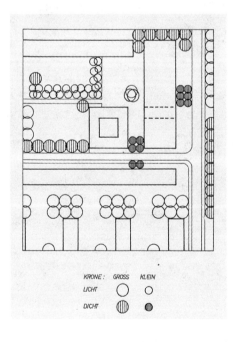

FIGURE 6.3.
Reinhold Lingner and Hans Georg Büchner's example of a detailed tree-planting concept to accompany area development plans. (Reinhold Lingner and Hans Georg Büchner, "Bepflan-zungskonzeptionen für Städte," *Wissenschaftliche Zeitschrift der Humboldt-Universität zu Berlin, Math.-Nat. R. 17, no. 2 [1968]: 215–30)*

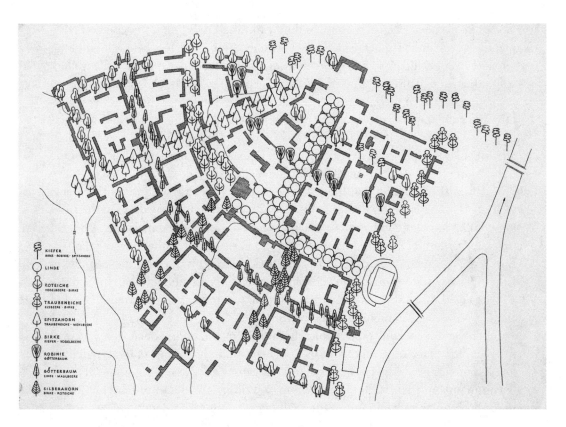

FIGURE 6.4.
Walter Meißner's tree-planting concept for Stalinstadt, 1952.
(Staatsbibliothek Berlin, Nachlass Walter Funcke, plan 75)

hierarchy of streets and the tree species growing in the region. It determined which streets were to be planted with rows of trees and which with irregular planting patterns (figure 6.4), thus providing a tool to emphasize main street axes and important buildings along a street.

For example, Meißner planned regular rows of lindens along the central north-south and east-west axes. Other areas of the city were to be planted in more irregular patterns and with a specified "leading" species, accompanied by one or two other species of trees. Meißner developed ten different tree combinations for this purpose. For example, streets along the northern and southwestern edges of the city were to be planted with birches leading. Pine trees and rowan berries could also be mixed into these plantings. A street running from the north to the southwest was to be lined with Norway maple, which could be accompanied by durmast oak and whitebeam. In other areas, silver maples were to dominate, accompanied by birches and northern red oak.[24]

Although garden architect Walter Funcke had succeeded in making the case for landscape architecture in Stalinstadt in the first place and had obtained its landscape design commissions, and although his colleague Reinhold Lingner had hailed the landscape plans in socialist rhetoric as the first in the GDR that were based on comprehensive scientific studies, Funcke was demoted in 1954. Spaces between buildings remained largely unkempt and the realization of the street tree concept and any landscape designs languished. One of the few exceptions was a tree-planting campaign involving citizens in the fall of 1955. "1,000 Trees for Stalinstadt" brought the city its first, albeit incomplete, green veneer. One of the challenges the landscape architects had faced in the new city had been the lack of available trees. Over the years, random and irregular tree plantings occurred, neither following the species suggestions of the initial plan nor creating any meaningful spatial effects.[25]

It was this situation that the graduate student Klaus Ostwald sought to tackle with plans for Eisenhüttenstadt that he drew up as part of his 1965 thesis at the Institute for Garden Design of Humboldt University. Following the initial plans, Ostwald suggested determining a "leading species" for each street and introducing regular tree lines along the main streets. With his advisor Gandert, Ostwald used the example of Eisenhüttenstadt to argue for the equal importance of trees and buildings in land use and area development plans. Trees, they argued, had to be included as fundamental and binding design elements in these plans.[26]

The attention landscape architects paid to tree-planting concepts resulted in the inclusion of a requirement for these concepts in the 1966 draft of the Guidelines for Comprehensive Master Plans for Cities (Richtlinie zur Ausarbeitung von Generalbebauungsplänen der Städte) by the German Building Academy. Although it would never become an official decree, the draft became a guiding document. As part of the new GDR Urban Design Regulation (Städtebauordnung), considered a revision and refinement of the Sixteen Principles of Urban Planning, the guidelines on tree-planting concepts aimed to guarantee that existing trees were integrated into development plans as far as possible; that trees were used for the spatial structuring and aesthetic enhancement of new mass-housing projects; that their planting would occur well ahead of time—if appropriate, before the construction of buildings; and that planting designs would facilitate efficient tree planting and maintenance. The requirement also became a component of the draft for sheet 10 of TGL 113-0369, the standard specifications for the content of maps and comprehensive master plans, but with regard to street tree planting, the urban master plans were finally only to include indications of relevant existing and planned street tree lines. But even in the last years before German reunification, tree-planting concepts and any cartography that could inform them, like tree surveys and phytosociological maps, were still seldom available to the urban master planners.[27]

By the late 1960s, relevant comprehensive tree-planting concepts had already been drawn up for several new cities and urban developments, including Potsdam-Waldstadt, Glauchau, parts of Schwedt, and parts of East Berlin. Although they often turned out to be hard to realize given the bottlenecks of tree supply, which they were intended to alleviate, tree-planting concepts remained an important planning method until reunification. They were considered part of "socialist city greening." Their comprehensiveness was what made them inherently socialist, Lingner and Büchner argued in a rhetorical effort to distinguish socialist from capitalist design practice and further their own cause as garden architects in the GDR. They claimed that given the diverging interests of individual property owners, comprehensive tree-planting concepts were impossible in capitalist cities. This rationale nonchalantly disregarded not only the late nineteenth and early twentieth-century history of comprehensive park system planning in the United States but also Peter Joseph Lenné's 1840 plan for ornamental greenways in Berlin, much closer to home.

In East Berlin, at the conference "The Street Tree in the Metropolis," organized by the Department of Gardens in January 1966, garden architect H. Wilcke presented a draft tree-planting concept (*Entwurf für einen Leitplan für Straßenbaumbepflanzungen*) for the center of the East German capital. The concept showed that the number of street trees in the district of Mitte, where there were then around fifteen hundred street trees, was to be increased by ten thousand. The planting plans ultimately included four rows of maple trees along Alexanderstraße, linden trees along Mollstraße and the extension of Karl-Marx-Allee, and plane trees along Hans-Beimler-Straße.[28]

Other resolutions determined at the conference were to draw up a map of existing trees by the end of the following year; to formulate street tree–planting guidelines; to carry out an annual tree survey conducted by the district Departments of Gardens to document all street trees and assess their condition; to test the feasibility of a tree ordinance; and to collect and document best management practices in Germany and abroad. East Berlin's director of gardens Helmut Lichey, who had particular interest in street trees, wanted to take action because he considered the planting instructions in the general Guideline for City Streets (Richtlinie für Stadtstraßen), which both East and West Germany had inherited from the war period, insufficient. Although the guideline determined that a garden architect or landscape designer was to be in charge of all plantings along streets, it was first and foremost drawn up to guarantee a secure traffic flow rather than tree-lined streets and adequate living conditions for plants.[29]

But despite the important role attributed to street trees in "socialist city greening" and the Magistrate's objective to plant all streets with trees, the adequate planting and care of street trees in East Berlin turned out to be challenging. Nurseries could not provide the number and quality of trees needed. Street tree surveys did not exist, and tree planting was often undertaken by the citizens themselves, who were asked to volunteer on Saturdays but lacked the necessary expertise.[30]

In 1974 Wolfgang Reckling, advised by Klaus-Dietrich Gandert, concluded the first comprehensive street tree survey of East Berlin. In this work for his thesis in horticulture at Humboldt University, Reckling found that East Berlin was losing eighteen hundred trees annually to disease, insect damage, pollution, and natural death. A gradation of tree age could be observed throughout the city, with a higher number of younger trees in the central districts and more older trees in Berlin's peripheral districts. The treatment of street trees and the importance attributed to them varied considerably across the eight district Departments of Gardens. Whereas Mitte, Friedrichshain, and Lichtenberg had achieved an increase of street trees between 1971 and 1973, Prenzlauer Berg, Treptow, Köpenick, and Weissensee had only felled trees and were lacking any type of future planning documents. According to Reckling, the bad condition of street trees was the result of a lack of prospective planning, management, and care, including the neglect of watering, fertilizing, and the prophylactic use of insecticides. Reckling's specific recommendations included the establishment of greater unity among the street tree plantings of Friedrichshain and Treptow, where many streets were lined with various species of different age groups. His general recommendations included the use of a larger variety of species and of more mature trees, those with a caliber of eighteen to thirty centimeters rather than nursery saplings with sixteen-centimeter calibers; the identification of tree damage with the help of infrared aerial photography; the continuation of a computerized tree survey; and the formation of special tree brigades, each in charge of some five thousand trees and each of which would maintain a current tree survey.[31]

The idea of tree-planting concepts was promoted repeatedly throughout the decades. With the goal of protecting existing trees and tree groups, determining the main and accompanying tree species and the number of "big trees" to be planted, tree-planting concepts were especially drawn up for new housing developments on East Berlin's periphery. By the latter half of the 1970s, Berlin's Department of Gardens was working with tree-classification numbers that determined that 55 percent of newly planted trees would have a caliber below sixteen centimeters, 33 percent would have a caliber of eighteen to twenty centimeters, 7 percent twenty to twenty-five centimeters, and 5 percent would be big trees. These percentages were further differentiated between leading and accompanying species, and 20 percent of the saplings were to be poplars. The tree species planted with calibers of eighteen to twenty-five centimeters included various maples and lindens, ailanthus, black locus, tree hazel, red oak, and London plane.[32]

In 1972 Walter Meißner and K. Rubel developed a tree-planting concept for East Berlin's new housing development Fennpfuhl in the Lichtenberg district in which the choice of species emphasized the site's landscape features (figure 6.5). Fennpfuhl had been slated as an area for development in the immediate postwar years, but its future had remained uncertain for two decades. Its core area had been the focus of the only all-German postwar urban design competition, held in 1956–57. But the winning design, by Ernst May's team from Hamburg, was never realized as this would have appeared too big a concession to West Berlin and the Federal Republic. Instead, in the early 1960s, the area was used as a test site for the erection of the first prototypical building of the GDR Building Academy's standardized P 2 housing series. Fennpfuhl's urban design for mass housing was finally developed and realized from the 1970s until 1986. Detailed tree-planting concepts based on Meißner and Rubel's comprehensive 1972 plan were included in various area development plans (figure 6.6).[33]

In the early 1980s Hubert Matthes, a long-serving Berlin government official who had served as the director of open-space design in the Magistrate's Office of Urban Design (Büro für Städtebau) before he was appointed professor at the Weimar School for Architecture, Engineering, and Construction (Hochschule für Architektur und Bauwesen Weimar), returned to Berlin's phytosociological map as the basis for draw-

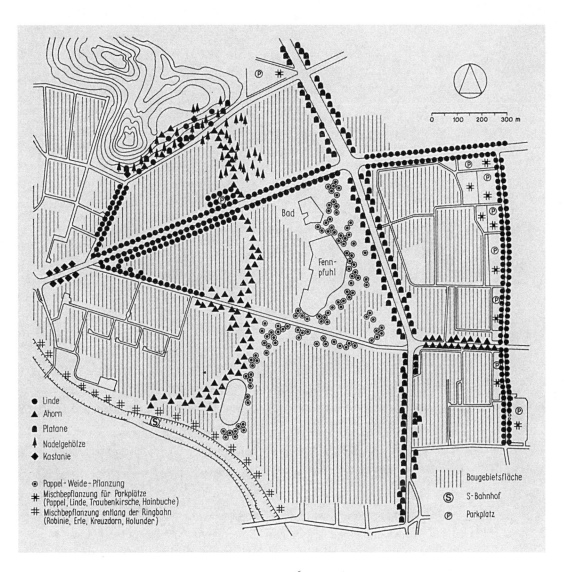

FIGURE 6.5.
Walter Meißner and K. Rubel's tree-planting concept for the new housing develop-
ment Fennpfuhl in Lichtenberg, 1972. (Johann Greiner and Helmut Gelbrich,
*Grünflächen der Stadt* [Berlin: VEB Verlag für Bauwesen, 1972])

ing up planting concepts for the city's various neighborhoods. Berlin's 1979 master
plan (*Generalbebauungsplan*) had demanded the planting of allées and promenades
based on ecological conditions. Urban ecology was a surging new research area at the
time. Berlin's street tree care was inadequate, but decisions had been made to increase
the number of trees, not least in expectation of the tenth Socialist Unity Party convention to be held in the capital in April 1981. The mandate was to plant eleven thousand
new trees in Berlin's new urban neighborhoods. Following what by this time were well-
known guidelines for tree-planting concepts, Matthes argued that every neighborhood

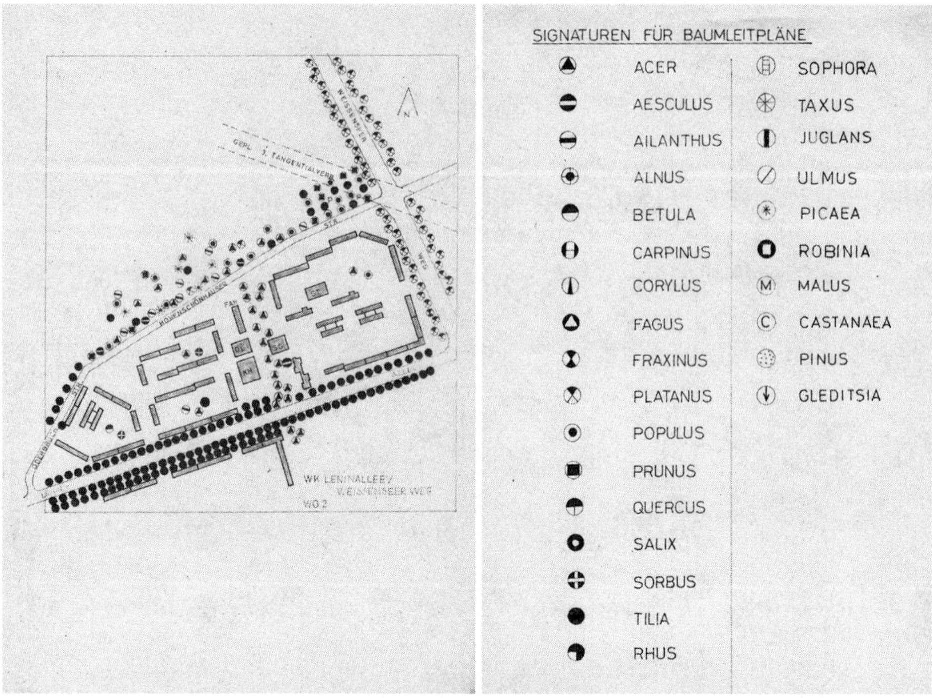

SIGNATUREN FÜR BAUMLEITPLÄNE

| | | | |
|---|---|---|---|
| ⬤ | ACER | ⊞ | SOPHORA |
| ⬤ | AESCULUS | ✳ | TAXUS |
| ⬤ | AILANTHUS | ◉ | JUGLANS |
| ⊕ | ALNUS | ⊘ | ULMUS |
| ⬤ | BETULA | ✳ | PICAEA |
| ⊖ | CARPINUS | ⬤ | ROBINIA |
| ⬤ | CORYLUS | Ⓜ | MALUS |
| ⬤ | FAGUS | Ⓒ | CASTANAEA |
| ⊗ | FRAXINUS | ⬤ | PINUS |
| ⊗ | PLATANUS | ⬇ | GLEDITSIA |
| ◉ | POPULUS | | |
| ⬛ | PRUNUS | | |
| ⬤ | QUERCUS | | |
| ⬤ | SALIX | | |
| ⊕ | SORBUS | | |
| ⬤ | TILIA | | |
| ⬤ | RHUS | | |

FIGURE 6.6.
Detailed tree-planting concept for the housing complex between Leninallee and
Weissenseer Weg in the new housing development Fennpfuhl in Lichtenberg, n.d.
(Studienarchiv Umweltgeschichte, STUG 002-48, "Materialsammlung Baumschutz 1,"
K.-D. Gandert, "Über die Verwendung von Bäumen in Berliner Grünanlagen,"
sheet nos. 53–54)

should be planted with a characteristic plant palette. He added a topographic layer to Kurt Hueck's 1948 phytosociological map and echoed garden architect Willy Lange's early twentieth-century nature garden idea on a citywide scale when he suggested that the plant selection for each neighborhood should be based on its respective "autochthonous species," complemented by non-native species that had a similar physiognomy. As one of the advantages of this approach he noted the reduction of necessary plant care.[34]

Despite these efforts, one of the biggest problems in the landscape design of Berlin's new housing developments was the lack of street trees. Extensive conduits below sidewalks prohibited their growth. Landscape architect Johann Greiner, already recognizing the problem in the 1960s, proposed planting trees along streets' northern side and running conduits along their southern side north of the buildings. In east-west running streets, this street side would not be lined with trees to allow adequate sunlight to reach apartments. Yet it took the Magistrate officials until the early 1980s to sign a white paper with which they sought to tackle the lack of street trees in new urban neighborhoods and take another step toward the realization of the equality of architecture and landscape. The 1984 Basic Regulation for the Design of Local Streets in Residential Neighborhoods (Grundsatzregelung für die Gestaltung der Anlieger-straßen im komplexen Wohnungsbau) determined that in the streets of residential neighborhoods, space needed to be set aside for both conduits *and* street trees. The regulation, which complemented others governing the distance between trees and various construction elements more generally, determined the necessary space between trees and underground conduits. The 1984 regulation also served the district departments that were in charge of street trees, parks, and gardens in the old neighborhoods where, by the 1980s, many trees that had been planted at the end of the previous century were nearing the end of their life.[35]

## A New Street Tree–Planting Paradigm for West Berlin?

In the Western sectors, tree-planting activities also picked up quickly after the war. This area of the city had counted 268,323 street trees in 1939, but had lost around 100,000 to 110,000 trees during the war. In the postwar years, around 11,000 additional trees were lost due to road widening and subway construction. By 1952, West Berlin could boast 189,250 trees, a number that further increased to 200,150 by 1959 but stagnated around that level until the mid-1970s. An initiative to fill "tree gaps" along streets where possible and the provision of 9 million marks for this street tree program led to a further increase; 234,104 street trees were reported in 1985.[36]

Shortly after new young linden trees had been planted along Unter den Linden in East Berlin, efforts were undertaken in the Western district Departments of Gardens in Wilmersdorf and Charlottenburg to replant the showcase street Kurfürstendamm. However, at that time the street's reconstruction plans had not been finalized yet, and in early 1951, further discussion among officials in the district departments and the Department of Gardens (Hauptamt für Grünflächen und Gartenbau) questioned the choice of species. Should Kurfürstendamm be lined with the eight hundred London plane trees that had been shipped to Berlin from West Germany for the purpose, or should it rather be lined with Berlin's famous linden trees, thus creating a West Berlin Unter den Linden? Although the native linden species lost its foliage unfavorably early in the autumn, some argued that its more compact growth was better adapted to Kurfürstendamm's heavy traffic. Regardless of this, in April 1951, the street's few remaining old and crippled trees were finally uprooted to make room for the initially intended London plane trees.[37]

The question regarding the suitability of species for Kurfürstendamm highlights a relevant point in the general postwar discussion about street tree planting: How "natural" could and should street trees be? Although the big size and majestic character of mature London plane trees were well suited to turn Kurfürstendamm into West Berlin's central showcase boulevard, could their natural growth be accommodated in this urban condition?

Postwar reconstruction offered an opportunity to improve and reconceive street tree planting in big cities. In Berlin's Western sectors, the Magistrate determined in

1949 that street trees were not to be planted in rows if sidewalks were less than five meters wide; they were to be planted at minimum intervals of sixteen meters, at least eight meters from street lights and eighty to one hundred centimeters from the curb. In addition, standardized tree grates were to be used in paved sidewalks. The West Berlin Neukölln district Department of Gardens had reacted to the loss of street trees during the Berlin Blockade by selecting new locations for them so that they would have the necessary space to develop and thrive without blocking windows in narrow streets and conflicting with urban life in general. Although the street tree ordinance prohibited street tree planting on sidewalks less than five meters wide—a requirement that would have left Neukölln treeless—the district department decided to plant street trees. However, this was to be undertaken in an irregular pattern in front of window-less gables and industrial buildings, and in the spaces between buildings.[38]

Neukölln's street tree planting was not dissimilar from what was happening in other German cities at that time, nor was it a new idea. In 1949 Düsseldorf's garden inspector Helmut Schildt proposed a street tree–planting paradigm similar to the one that had been put forward for Neukölln in Berlin. Schildt argued that the reduction of the number of street trees in a row and site-specific replacement plantings not only improved the cityscape but also saved labor and costs. According to him, it was not necessary to replace every street tree that had been destroyed during the war and the ensuing deforestation for firewood. Instead, fewer trees given more space could con-tribute more successfully to the creation of an aesthetically appealing street. In addi-tion, he promoted cooperation between city governments and property owners so that trees could be planted in private front yards if sidewalks were too narrow.[39]

Although presented in the guise of German reconstruction, like the street tree planting followed in Neukölln, Schildt's planting paradigm, knowingly or not, fol-lowed similar ideas regarding street trees and front yard treatment put forward in the first three decades of the twentieth century by garden architects and urbanists like Camillo Sitte, Harry Maasz, Joseph Pertl, and Michael Mappes. Unease with ordered street tree rows had begun to develop even earlier, in the last decades of the nine-teenth century, affecting ardent street tree promoters and modern city builders like Joseph Stübben and Carl Hampel. They had queried if trees should indeed be planted in monotonous rows. While it was hard to argue that a systematic serial street tree planting did not have benefits—it could unify disharmonious facades along a street and provide an even line of shade in the summer, and it was relatively easy to manage—unease with the systematic and rather unartful planting of street tree rows had been mounting among experts. To prevent monotony and a potential "corridor effect," Stübben and Hampel had suggested that trees could be planted asymmetrically along a street, and that different species could be used in the various street sections. For ex-ample, Stübben reported on the successful breaking up of long rows of elms by horse chestnuts at street crossings. However, in the design of street tree planting, Stübben advised his readers, the variety of the trees' life cycles was as important to keep in mind as their various appearances. Along Hamburg streets, oak trees and mountain ash had successfully been planted in an alternating pattern. But as soon as the oaks had developed a wide canopy, it was important to fell the faster-growing mountain ash as the aesthetic effect would otherwise have become too unwieldy. Berlin garden architect Carl Hampel added the effect of leaf forms and canopy colors to the discussion, arguing that although variety was desired, tree differences should be subtle.[40]

In 1900 Austrian architect Camillo Sitte, an ardent promoter of picturesque city building, went a step further. He wanted to see the loosening of the rigid, monotonous staccato of closely planted rows of trees along streets. Tree planting should create shade where necessary and frame buildings or cover them, responding to specific circumstances and need. Instead of the standardized street tree treatment that ruled contemporary modern town planning and disregarded the streets' orientation and character, a site-specific treatment was required. Even more important, Sitte drew attention to the aesthetic role and beauty of individual trees in the urban ensemble. He lamented, "The *motif of the single tree*, which is so rewarding at such a slight cost," had been "almost totally ignored in modern town planning." In his estimation it was still worthy of attention as it could stimulate the imagination and had a beneficial psychological effect besides functioning as a signifier of national and local folklore and history.[41]

Supporters of the German *Heimatschutz* movement—a movement to protect the homeland—including garden architect Harry Maasz, in whose office Schildt had trained in the early 1930s, promoted similar ideas. In his 1927 book *Das Grün in Stadt und Land* (Green in City and Country) Maasz highlighted the aesthetic role of single trees and of street tree rows in the urban ensemble. In an effort to avoid monotonous tree lines, Berlin's garden director Erwin Barth and Düsseldorf's garden director Walter von Engelhardt also promoted a site-specific treatment and variety in street tree planting. To achieve a more varied streetscape, in the late 1920s, Berlin advised its district Departments of Gardens to cease replacing street trees in old streets with front yards and to refrain from planting trees in newly laid-out streets with front yards. This way, the city argued, plant growth in the front yards in general and single trees in particular would be able to unfold their beauty more fully. The vegetation of private front yards was "the best and cheapest ornament" in these types of streets.[42]

The same ideas were applied during the Nazi regime in the 1930s when Berlin's district mayors were advised to instruct their Departments of Gardens to promote the use of single trees in the urban streetscape. In a 1941 directive to all district mayors, Berlin's director of gardens Josef Pertl promoted the replacement of "strict and straight tree rows" in streets with front yards with loose and irregular tree groups. If tree growth was restricted to front yards, sidewalks could be broadened, their paving saved from damage from tree roots, and labor for the department reduced. According to Pertl, the replacement of trees was also inconspicuous, and therefore easier, if they were planted in loose tree groups. Like many of his colleagues, he stressed the importance of single trees in the streetscape. Only a few years before, Pertl and garden architect Michael Mappes had begun a campaign to visually merge the space of private front yards and public streets through their naturalistic planting and the elimination of fences and walls. This treatment was to give a spatial and aesthetic expression to the Nazi ideal of a community's collective spirit as well as to its blood and soil ideology. At the same time, streets laid out in a more naturalistic way were thought to contribute to air-raid protection measures.[43]

During the following war years, discussions about a suitable planting paradigm for street trees continued. The single-tree motif not only bolstered Nazi ideology, which used old oak and linden trees as signifiers of mythical nature worship and of the people's ties to the land, it was also practical because lumber was desperately needed. The cutting of old street trees did more than provide wood; it also offered an opportunity to dispense with their replanting or to replace them in a different, more fitting location, or with a more appropriate species. As Detmold's director of gardens Th. Falck noted, single trees were considered most successful both in enlivening a quiet streetscape and in calming a disturbed one.[44]

In 1949 the suggestions by Berlin's Neukölln district Department of Gardens and Düsseldorf's garden inspector Schildt continued along these lines. They also stood in the context of Bernhard Reichow's ideas about the "organic art of city building" (*Organische Stadtbaukunst*), published in 1948. Reichow suggested that the city be designed as a living, flexible organism—a city landscape that was understood as a loose urban fabric embedded in green open space. This paradigm shift in urban design built on a combination of anti-urban thought, political ideologies and policies, and the necessities of protecting against air raids that had developed before and during the war years. Although focused on comprehensive urban design schemes that translated the organism metaphor quite literally into urban form, Reichow did not fail to address trees as an integral design element of his city landscape. In the same way in which the organic, healthy, and defensive city of the future was to grow naturally, following natural patterns, trees as integral components of this city landscape were to be given the opportunity to follow their natural growth patterns. A diagram published in *Organische Stadtbaukunst* illustrated the growth phases of four native trees. Reichow argued that organic city building took into account the trees' life cycles, not only their mature state, as the "static art of city building" had done (figure 6.7). While Reichow addressed urban trees in general, Schildt applied similar thoughts to street trees, proposing that they should be given the opportunity to unfold their whole beauty and become design elements in the city. Some years later, in 1957, West Berlin tree expert Rudolf Kühn similarly suggested that the new urban design paradigm, with its loose order of housing slabs and individual high-rise buildings, required "strong tree characters" and irregularly shaped groves and groups of trees. "The melody" of the times also had to be found in a new street tree aesthetic, Kühn argued. While provocatively declaring allées as elements of the past, he did concede that new residential streets without traffic would provide opportunities to reestablish "real allées and promenades."[45]

## Intangible Trees

Despite differences between the ideologies and the various aesthetic and spatial concerns in the street tree discourse among professional designers in East and West Germany at the time, the trees' role in improving the urban climate was undisputed. Trees

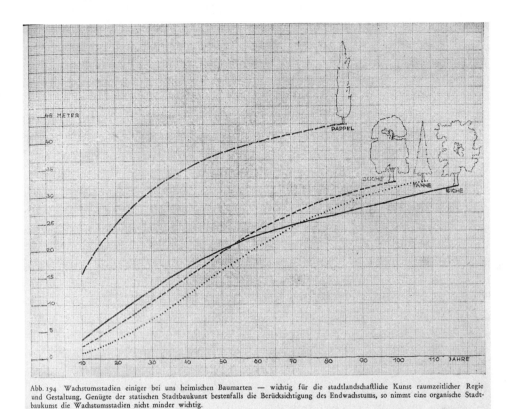

Abb. 194  Wachstumsstadien einiger bei uns heimischen Baumarten — wichtig für die stadtlandschaftliche Kunst raumzeitlicher Regie und Gestaltung. Genügte der statischen Stadtbaukunst bestenfalls die Berücksichtigung des Endwachstums, so nimmt eine organische Stadtbaukunst die Wachstumsstadien nicht minder wichtig.

FIGURE 6.7.
Diagram illustrating the growth phases of four native trees. (Bernhard Reichow,
*Organische Stadtbaukunst* [Braunschweig: G. Westermann, 1948])

had an impact on the temperature, humidity, and flow of urban air, and they could bind dust. But they also had psychological functions that were in part closely related to these microclimatic benefits. Trees could change what it felt like to be and live in the city. Urban designers and landscape architects especially pointed to these other qualities of trees. When in 1931 engineers J. Goldmerstein and Karl Stodieck at the Technical University Berlin showed in a widely cited study that Berlin's carbon dioxide was not assimilated by the trees in its parks, as had so often been assumed, but instead was merely transported higher into the atmosphere above the vegetated areas, urban designer Herbert Göner was irritated. He feared that the use of the Berlin engineers' positivist mathematical study as a basis for urban design decisions would lead to a disregard of the many other functions of urban trees. Their scent and their aesthetic value, especially on glaring summer days, when their green color and shade would relieve the strain on the eyes, were cases in point. In addition, he claimed that mathematical calculations could not capture "the psychological effect in the form of joy about beautiful tree shapes, about their growth and becoming, and the distraction from everyday life through a feeling of closeness to nature."[46]

It was certainly true that trees brought color and seasonal change into an otherwise relatively inert physical environment. As studies in the 1950s began to show, they
also had the capacity to buffer sound. Although these character traits were in part
ephemeral and difficult to measure and package into the data of hard science, they
appeared essential to human well-being.

In the first two decades after the war, East and West Berlin's urban officials,
planners, designers, and tree experts began to construe the microclimate trees could
produce as a bioclimate, that is, a climate understood in relation to life in general and
human life in particular. The bioclimatic and related psychological effects of street
trees were increasingly recognized and discussed. The science of bioclimatology, which
gained ground in this time period, bridged the disciplines of medicine and meteorol-
ogy. While it had always been clear that trees had an impact on human well-being,
systematic and scientific studies to explain this phenomenon had begun only in the
nineteenth century. In Germany, studies on urban climate and psychology as related
to the urban environment multiplied in the first decades of the twentieth century.
They were part of a more general surge in urban studies in which scientists sought to
address the challenges and dangers of urban life, often in an attempt to overcome the
strong anti-urban bias shared by many intellectuals of the time.[47]

Psychologist Willy Hellpach, often referred to today as the founder of environmental psychology, was one of the early twentieth-century scientists who argued for
the production of scientific knowledge to improve urban life. Committed to strict
scholarship and objectivity, his writings remained an important reference point for
planners and designers in the postwar years despite—or perhaps, rather, because of—
the environmentally determinist bent of his psychological theories. In 1911 Hellpach
had published *Geopsyche*, which would appear in six editions by 1950. In *Geopsyche*
he sought to shed light on the psychological effects of weather, climate, soil, and
landscape. In 1939, at the beginning of World War II, his book *Mensch und Volk der
Großstadt* (Man and People of the Metropolis) applied some of these ideas to the
urban environment. The book's second edition in 1952 returned the author and his
work to the attention of urbanists and landscape architects in East and especially West
Germany.[48]

Hellpach finally provided them with a seemingly legitimate psychological rationale for urban green. Of particular importance to street tree planting and the design
of urban parks and gardens were Hellpach's observations regarding the psychological
effects of landscape in general: its colors, light, smells, sounds, movement, atmosphere,
and "physicalness," by which he meant the perception of its wind, heat, and cold. As
Dietrich Hieronymus noted in *Das Gartenamt* and his East German colleague Alfred
Hoffmann referred to in his investigations on street trees in the early 1950s, Hellpach
claimed that the shorter bandwidths of light between green and violet had a calming
effect on the senses. Cities often lacked the colors green and blue, which had a particularly soothing effect on nerves and mind; they lacked large parks and open blue skies.
Instead, the urban environment was dominated by exciting red and yellow artificial

lights and by the gray tones of buildings, which had a rather depressing and paralyzing effect. In contrast, moving leaves in the wind, the seasonal color changes of foliage, the smell of flowering linden trees, and different tree shapes and outlines were character traits of trees that, as Hellpach confirmed, could elicit joy. Street trees, parks, and gardens could therefore turn the city into a pleasant, healthy, and happy living environment.[49]

The role of trees in buffering noise, in particular traffic noise, also began to assume growing importance in the reconstruction years. Traffic was on the increase and it was clear that its air and noise pollution could harm human physical and psychological well-being. And yet West Germany favored the "dispersed city" paradigm that built on private motor transportation, and the GDR was building large, majestic boulevards — the so-called Magistralen — as central ordering devices in new cities and those under reconstruction. Berlin's building inspector Georg Pinkenburg had analyzed the city's street noise already at the turn of the century, drawing attention to its potentially harmful effects on the nervous system and identifying means of prevention and protection. Nevertheless, in the 1950s noise was quickly becoming "a new area of knowledge" and "a public problem with technical, medical, and juridical tasks," even warranting a new journal entitled *Lärmbekämpfung* (Noise Abatement).[50]

In the early 1950s, studies by the West German engineers Franz Josef Meister and Walter Ruhrberg confirmed the assumption that plants could absorb, reflect, and divert sound, depending on their branch and leaf structure. West Berlin's Zehlendorf district Department of Gardens was among the first public offices to undertake sound studies along some of its streets and public parks, seeking to understand better how particular plants and planting designs could buffer traffic noise. In 1958 district gardener R. Dittmann reported that densely growing shrubs that had reached a height of over one and a half meters along Potsdamer Chaussee provided good noise protection. In contrast, rows of street trees whose crowns often only began at that height hardly buffered traffic noise (figure 6.8).[51]

But as one of the earliest studies on the soundproofing of buildings in France by engineer and sociologist Abraham Moles and a 1960s study in Berlin showed, street trees could indeed buffer noise, and quite effectively. Their capacity to absorb and deflect sound depended upon their leaf size, orientation, and density as well as on the height and distance of the sound source. Particularly effective were the large-leafed species, species with leaves oriented perpendicular to the sound waves, and species with many dry interior leaves and branches. Sycamores and little-leaf linden, for example, turned out to have an especially big buffer effect during the summer months (figure 6.9). To come to this and other conclusions, between 1962 and 1965, Gerhard Beck of the Institute of Landscape Technology (Institut für Technik im Landschaftsbau) at the Technical University Berlin had undertaken various measurements of sound waves traveling through trees and shrubs in different seasons and weather conditions. Beck's dissertation project, advised by garden architect and professor Hermann Mattern, was sponsored by Berlin's Senate Department for Economics (Senator für

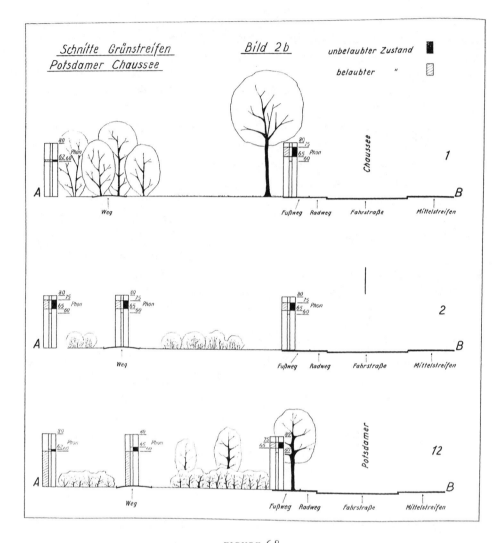

FIGURE 6.8.
Sketches illustrating the sound protection provided by different types of plantings along Potsdamer Chaussee. (R. Dittmann, "Öffentliche Grünanlagen, ein Faktor zur Lärmbekämpfung," *Das Gartenamt* 7, no. 8 [1958]: 175–80)

Wirtschaft). Building on the studies by Moles, Meister, and Ruhrberg, Beck aimed to find out which plant species and which types of planting designs were most effective in buffering sound and which could save the city money.[52]

Due to street trees' location along traffic arteries, many tree experts considered them more effective than park trees when it came to abating noise, binding dust, and ameliorating climate. To increase street tree density, by the 1970s, one hundred trees per street kilometer was cited in West German specialist literature as an ideal value, a goal aspired to by West Berlin's Senate.[53]

To counter occasional arguments that trees did not belong in cities, where they

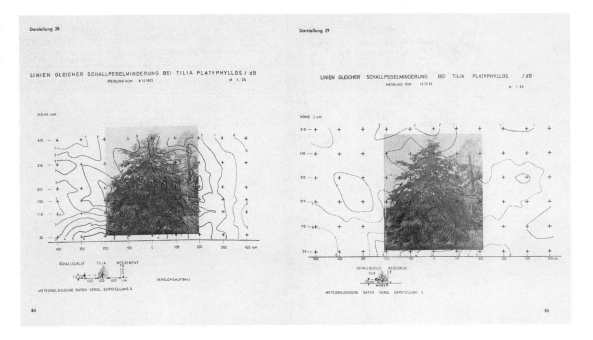

FIGURE 6.9.
Sectional sound contours representing the data gathered at two sound measurements
behind a largeleaf linden in October 1963. (Gerhard Beck, *Pflanzen als Mittel zur
Lärmbekämpfung* [Hannover, Berlin: Patzer, 1965])

disturbed and endangered traffic, researchers had already in the postwar years begun
to elaborate on street trees' roles in traffic guidance. In East Berlin, Alfred Hoffmann
had explored trees as elements that structured transportation space and could provide
visual guidance to automobile drivers and protection to pedestrians. In cities built
according to the GDR's Sixteen Principles of Urban Planning, transportation space
was to accommodate all types of movement, and trees provided a valuable structuring
element that could increase security as long as they were not planted too close to each
other, obstructing sightlines and views, and as long as they were maintained properly.
As Hoffmann showed, the distance between trunks should depend upon the expected
traffic speed (figure 6.10). Along thoroughfares where higher speeds were anticipated,
a fifteen-meter distance between trunks was considered ideal for drivers to have sufficient range of view over the traffic space ahead of them. Hoffmann's East German
elaborations, which built on data provided by Soviet colleagues, in turn informed the
recommendations published in the 1970s by landscape architect Aloys Bernatzky for
a professional West German audience. According to Bernatzky, the minimum space
between trees was to be ten meters, a value that complied with the minimum density
believed necessary for the trees' benefits for public health.[54]

When in 1937 nationalist landscape architect Max Bromme had described the urban
site conditions as an urban steppe (*Großstadtsteppe*) that required the least demanding

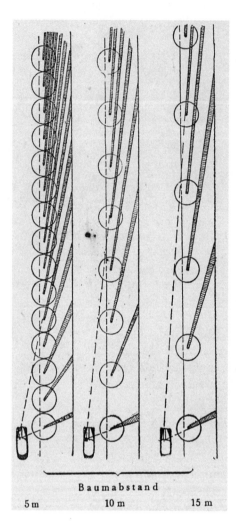

FIGURE 6.10.
According to Alfred Hoffmann, the distance between tree trunks along roads had to depend on the expected speed of traffic. (A. Hoffmann with K. Kirschner, "Die Straßenbepflanzung in Städten in ihren Beziehungen zur Lufthygiene sowie zu den verkehrs- und stadttechnischen Verhältnissen," in *Fragen der Grünplanung im Städtebau*, ed. Johann Greiner and Alfred Hoffmann [Berlin: Henschel, 1955], 62–91)

Baumabstand

5 m        10 m        15 m

of tree species, like black locust and tree of heaven, he would have been unable to imagine the devastated German cities only a decade later, after the war. Trees were used to help restore this new barren landscape and its rubble hills. They were planted to bind dust, buffer sound, provide color and contrast, and enliven the streets of the city. The reconstruction period, with its unforeseen opportunities, inspired new research into the manifold roles and functions of trees in the city. Despite the everlasting importance of trees' ephemeral and seasonal characteristics, the postwar years marked a renewed scientific interest in their traits and potential. Besides this scientific research, phenomenological approaches to street trees also continued. In the early 1960s, the West Berlin landscape architect Rudolf Kühn categorized urban streets according to the sense perception of their street tree plantings. Kühn differentiated between "the green street" (*die grüne Strasse*), "the glowing street" (*die leuchtende Strasse*), "the fragrant street" (*die duftende Strasse*), "the singing street" (*die singende Strasse*), "the grand street" (*die Prachtstrasse*), and "the graceful street" (*die zierliche Strasse*).[55]

These were the characteristics that the public, consciously or subconsciously, perceived as being threatened by increasing pollution, traffic, and urban development in the 1970s. Citizens and tree experts alike, therefore, rallied to protect Berlin's street trees from destruction.

# 7
# Shades of Red

*Art, Action, and Aerial Photography for a Green Berlin*

I n 1985 West Berlin hosted the Federal Horticultural Show (Bundesgartenschau [BUGA]). Held in a different West German city every four years, the show provided urban governments with a tool for urban reconstruction and renewal. In West Berlin it led to the construction of a new park in the city's southeast district of Neukölln, an area still largely deprived of park space at the time. The show also promised an opportunity for significant local and national media coverage, a situation that Berlin's tree experts and tree lovers welcomed as it was an opportunity to draw attention to the dire state of the urban environment. Every second street tree in West Berlin at the time was reported to be suffering traffic-related air and soil pollution that had resulted from the extensive and unrestrained use of road salt in the 1960s and 1970s.

Thousands of visitors to the show were greeted with colorful planting patterns of perennials, water fountains, and luscious large-scale and more intimate garden spaces. They were also confronted with a row of trees that in the early 1970s had been used in a study of different tree species' resistance to de-icing salt. Furthermore, visitors encountered a critical installation that involved live and dead street trees. Street trees from Neukölln and other parts of the city had been transplanted to the new park site as early as 1976, when many trees along Germaniastraße had to be moved for the construction of motorway A 100. In 1985 some of these trees which, due to poor transplanting practices and compacted soil, had shown only minimal growth, became the subject of an installation called *Street Tree* by landscape architects Falk Trillitzsch and Edelgard Jost. Their installation was both a shrine and a monument to Berlin's dead street trees. It was also a moralizing landscape that sought to attract people's attention to the dire situation of Berlin's street trees in the interest of a greener environmental politics. The designers built a street near the trees to simulate their original street-side location (figure 7.1). In addition, a zigzagging pathway circumnavigated each tree, which was signposted with its biography. The installation's centerpiece was a tree sur-

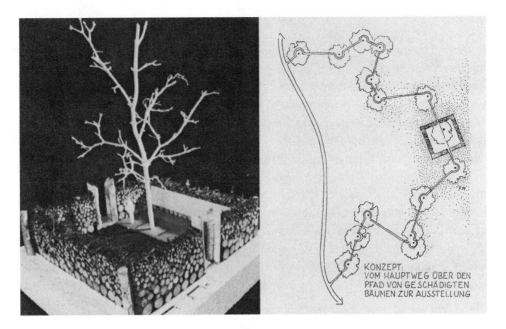

KONZEPT:
VOM HAUPTWEG ÜBER DEN
PFAD VON GESCHÄDIGTEN
BÄUMEN ZUR AUSSTELLUNG

FIGURE 7.1.
Falk Trillitzsch and Edelgard Jost's model and plan of the *Street Tree* installation for
the 1985 Federal Horticultural Show in Berlin. (Falk Trillitzsch and Edelgard Jost,
"Dying Trees as Exhibition Pieces," *Garten und Landschaft* 95, no. 4 [1985]: 60)

rounded by a quadrangular enclosure built out of the lumber of dead and diseased
street trees that had been felled in the preceding winter. Many of the rootstocks,
trunks, and branches used were labeled with the street trees' original locations in West
Berlin.[1]

The 1985 street tree installation at the Federal Horticultural Show in the Britz
locality of Neukölln was neither the first nor the last of its kind. Fueled by local observations of the decay of street trees and by the increasingly frequent nationwide discussions and alarm about forest death, in the 1970s, several artists in West and East Berlin
had begun to focus their attention on street trees. A new genre of street tree art developed that, together with citizen activism, induced the cities' urban governments to
increase their care of street trees and the urban environment in general. Street tree
plantings were a way for citizens in both East and West Berlin to oppose the state and
its environmental politics. In East Berlin, oppositional street tree activism provided
cover to evade state control and surveillance and created space for other adversarial
activities. Given their relative "neutrality," street tree–planting campaigns were tolerated more easily by the state, although it followed these campaigns closely. In West
Berlin, oppositional street tree activism inspired experts to employ tools of surveillance like aerial infrared photography in the development of a more comprehensive
and accountable street tree management in the first place.

# Street Tree Art

At the eighth spring convention of Berlin art galleries at the Academy of Arts (Akademie der Künste) in 1976, West Berlin gallerist, sculptor, and environmental activist Ben Wargin exhibited twenty trees. Collaborating with the district Departments of Gardens, he had chosen species able to survive urban conditions and sought to find citizen caretakers for each tree, which would then be planted in the respective citizen's neighborhood. The installation at the Academy of Arts was part of a larger project entitled *Baumpate / Grün ist Leben* (Tree Guardian / Green Is Life), which involved a series of events that Wargin organized together with a group of West German artist friends. To promote their tree-care concept and draw citizens' attention to environmental issues more generally, the artists toured West Berlin and other West German cities in old converted BVG buses, performing a series of tree-related happenings along the way (figure 7.2). Their thirty happenings in Berlin included decapitating a tree with a guillotine on one occasion, and on another simultaneously setting a tree and Cologne artist Dieter Reick on fire (the latter in a protective asbestos suit) in front of

FIGURE 7.2.
Ben Wargin's *Baumpate / Grün ist Leben* (Tree Guardian / Green Is Life): Wargin's sculpture advertising his activist tree planting and one of the BVG buses he used for his activism parked in front of Schöneberg's town hall, 8 June 1976. (Photos by Karl-Heinz Schubert. Landesarchiv Berlin, F Rep. 290 [09] Nrn. 0191182, 0191184)

the Reichstag. At Gedächtniskirche, the installation *Welthölzerbrand* (Global Wood Fire) was ignited, and on Hohenzollerndamm a tree was set in concrete. Artist Oskar Blasé's *Autogärten* (Auto Gardens) featured green installations on car roofs, symbolically replacing the vegetation that had been killed for road construction, and his Düsseldorf colleague Bernd Flemming eradicated all green open space on city maps.[2]

Wargin's tree work neither began nor ended with the eighth spring convention of Berlin art galleries in 1976. He had already planted his first tree in front of Gedächtniskirche in 1967 and had called attention to the perils of urban trees in his famed 1975 mural *Weltbaum* (World Tree), which he painted on a facade near S-Bahn station Tiergarten together with artists Peter Janssen, Fritz Köthe, Narenda Kumar Jain, and Siegfried Rischar. The remains of the mural can still be seen and have been celebrated as Berlin's first street art. Wargin and artist friends also planted the parking space in front of the S-Bahn station with trees sponsored by the European Parliament, and in 1978 he raised money to plant twenty ginkgo trees along Kurfürstendamm. The activist planted more trees on West Berlin's biggest rubble hill, Teufelsberg, and in the 1980s he began planting trees along the Western side of the Berlin wall. Since then, although the locations and contexts have changed, Wargin's tree plantings have been ongoing.[3]

In the 1970s, the activist-artist also mounted a series of exhibitions shaped by his environmental ethics in the Berlin Pavilion in Tiergarten. From *Der Baum bist Du/sind wir* (The Tree Is You/Us) to a 1979 show that warned against the negative effects of automobiles and technological development on the environment more generally, all exhibitions had an environmentalist agenda. The latter show was accompanied by an outdoor installation that included wild grasses and four ginkgo trees sponsored by the Council of Europe.[4]

Ginkgos, especially the male trees and the more recently produced seedless cultivars, were and still are considered suitable street trees. Wargin had first encountered a ginkgo tree on a visit to East Berlin in the 1950s. There, in front of Humboldt University, he came upon a specimen that had been planted in 1851. He chose the species for his planting campaigns because of its unique beauty and symbolism. As the oldest known living plant species, which has populated the earth for the past 250 million years, the ginkgo symbolizes survival, resilience, longevity, peace, and human oneness with nature. Like the ginkgo trees that had survived the atomic bomb at Hiroshima, a couple of Berlin ginkgo trees had survived the ravages of World War II. The ginkgo's qualities inspired Wargin's environmental and peace activism, which was informed by both deep ecology and the Romantic humanism given expression by Johann Wolfgang von Goethe's often-cited 1815 ginkgo poem. For Wargin, whose work in general was informed by his preoccupation with an environmental Armageddon caused by the unforeseen results of progress in technology and science, the ginkgo therefore appeared as a suitable choice.[5]

In East Berlin in the same period, graphic artist Manfred Butzmann's attention was similarly drawn to the fate of street trees in his city. In 1978 he designed a poster entitled *Ein Platz für Bäume* (A Place for Trees) that won him the 1979 best poster award assigned by the GDR Association of Visual Artists (Verband Bildender Künstler der DDR) (figure 7.3). A space between words in the title and a series of photographs of unoccupied curbside tree pits of different shapes and sizes and in different conditions called attention to the voids left behind by vandalized, damaged, and felled trees—and by trees that had never been planted. Following up on this work seven years later, in 1985 Butzmann produced a second poster entitled *Kein Platz für Bäume* (No Place for Trees) (figure 7.4).

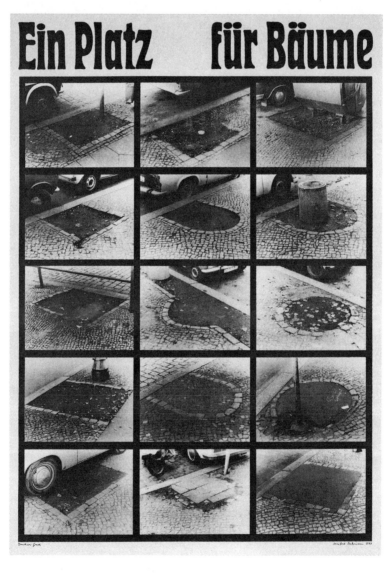

FIGURE 7.3.
Manfred Butzmann, *Ein Platz für Bäume* (A Place for Trees), 1978.
(© 2016 Artists Rights Society [ARS], New York/VG Bild-Kunst, Bonn)

FIGURE 7.4.
Manfred Butzmann, *Kein Platz für Bäume* (No Place for Trees), 1985.
(© 2016 Artists Rights Society [ARS], New York/VG Bild-Kunst, Bonn)

This poster featured a series of photographs taken in August 1985 of defoliated, poorly growing, and unhealthy street trees along Schönhauser Allee. Taken together, the posters illustrated the paradoxical condition of East Berlin's street trees: on the one hand there was space, but on the other hand there was none. By highlighting both the lack and the decay of East Berlin street trees, Butzmann also drew attention to the degradation of public space and the environment more generally.

Like much of his work, Butzmann's street tree posters indicted the state's neglect of environmental concerns and public open space, the provision of which was part of its self-proclaimed larger socialist values. In an ironic twist, therefore, his street tree work subversively sought to encourage citizens to take ownership of the neglected socialist public space by planting trees in it. Butzmann's preoccupation with street trees' absence and decay in East Berlin's inner-city districts was part of a new environmental activism that began to surge in East Germany in the 1970s. The poster medium allowed him to work with photography, not considered an artistic medium in East Germany at the time. Due to the documentary character of photography, the state saw the medium as running counter to the objectives of socialist realist art, which should merge fact and fiction, the present and an imagined future.[6]

But Butzmann's work included more than graphic art. Although on a smaller scale than Wargin, Butzmann also became a "street gardener." In the early 1980s, he began to turn the new concrete lamp posts that the city had used to replace the old ornamental cast iron posts into new types of trees, planting ivy and wild vine at their bases. With these small interventions Butzmann sought to draw his fellow citizens' attention to the loss of history and the slow and creeping degradation of the public realm.[7]

Given the widespread concerns about the state of the environment at the time, Trillitzsch, Jost, Wargin, and Butzmann were not the only designers and artists engaging street trees or individual trees in their work. Street tree art and activism could be found in many cities. In 1969 Robert Smithson had exhibited *Dead Tree* at a gallery show in Düsseldorf. Showing an uprooted tree lying horizontally in a white exhibition space like a corpse, *Dead Tree* was the culmination of a series, *Up-Side-Down Trees*. Smithson had "planted" trees upside down, their roots taking the place of the canopy in New York State, Florida, and Yucatán. Highlighting the mirrored morphology of trees' rootstocks and canopies, Smithson was drawing attention to matters of displacement, dislocation, and death. While *Dead Tree* stood for the death of nature, it also criticized art's prevailing preoccupation with artifacts.

Dissolving the boundary between nature and art was also among the concerns of the artist couple Christo and Jeanne Claude. A few years before Smithson, they had staged single trees lying horizontally in museums and galleries, turning them into anthropomorphic sculptures. With their root balls wrapped in burlap and crowns covered in polyethylene, the trees resembled both packed and preserved nursery stock and mummified corpses. They appeared to be precious objects, but also looked harrowed, weak, suffering, and old. In 1966 and 1969 Christo and Jeanne Claude sought to draw

FIGURE 7.5.
Wrapped trees along Donauwörther Straße, Augsburg, September 1985.
(Photo by Daniel Biskup)

attention to the state of living trees by wrapping trees near the City Art Museum in Forest Park, St. Louis, and the street trees along the Parisian showcase boulevard Champs-Élysées. The projects were meant to turn the trees into living sculptures. *Wrapped Trees* intended to draw attention to street trees' value, their architectural and spatial qualities, and their vulnerability. Adding a protective layer of artificial gauze over the trees also highlighted their artificial environment and the labor needed to maintain them in cities. Although it took until 1997 for a version of *Wrapped Trees* to be realized in Berower Park in Riehen near Basel, Switzerland, a similar project had already been carried out in Augsburg, Germany, in September 1985.

A group of young urban horticulturalists inadvertently used Christo and Jeanne Claude's principle of "revelation through concealment" when they wrapped a series of linden trees along Donauwörther Straße to draw attention to street tree death and disappearance (figure 7.5). Banners between the ghostly trees satirically asked passing motorists and pedestrians whether a gray future was to be a vision or reality. The young people distributed flyers that provocatively informed about street trees' negative influences on public urban life: street trees obscured light, cost money, occupied space that could otherwise be used as parking lots, caused traffic deaths, and attracted insect pests, and their leaves in the fall increased the danger of skidding.[8]

It was precisely these arguments that some Kassel citizens had used to criticize a street tree art project in their city. Performance artist Joseph Beuys's 7000 *Eichen—Stadtverwaldung statt Stadtverwaltung* (7,000 Oaks—City Forestation Instead of City Administration) was the street tree project that achieved the biggest international attention in the decades characterized by the environmental movement. Perhaps inspired by Wargin's Berlin-based tree-planting activism, which had also led to an offshoot in Kassel, the project was realized on the occasion of the 1982 international art show Documenta 7. During the show, Beuys began to plant seven thousand oak trees along Kassel's streets, in parks and playgrounds, and near public buildings. Citizens, institutions like the Dia Art Foundation, international artist friends, and Beuys himself sponsored the trees.

The planting project continued over a period of several years, leading to spin-offs and exhibitions in other German cities; it was also presented in 1987 at the Tree Show at the Massachusetts College of Art in Boston. Through East Berlin performance artist Erhard Monden, who included a working paper on 7,000 Oaks by Beuys in his own performance Sender-Empfänger (Sender-Receiver), 7,000 Oaks also reached the other part of Germany.[9]

7,000 Oaks was administered by a group of people brought together by Beuys in a nonprofit organization that included two landscape architects and numerous volunteer workers. They raised the funds and directed the planting work in collaboration with the city's Department of Gardens and in coordination with the design and planning departments. Seven thousand solid granite stones from a nearby quarry, which were stored in the form of a temporary sculpture on Friedrichsplatz, marked every planted tree as a permanent point of reference. The granite stone marking a tree's location would also act as a scale for measuring the tree's growth and maturation, and it would remain after its death. The shifting of proportions that became visible in the development of the trees, in the movement of the granite stones, and in the change of permeable versus impermeable ground cover in the city as more and more asphalt surfaces were replaced with gravel to accommodate the oaks and other tree species was a central concern for Beuys. With 7,000 Oaks he challenged authority and criticized the nontransparent, unwieldy, and detached bureaucratic procedures of urban administrations. He wanted to empower citizens to choose their own tree locations throughout the city. Beuys saw in the slow-growing oak trees a potential power to transform society and its environment. He considered 7,000 Oaks the beginning of a new cultural layer enveloping the earth and its biosphere. For him, 7,000 Oaks was a means to criticize capitalist development and reconceive capital as the inner values of humankind.[10]

Like the projects by his colleagues Wargin, Smithson, and Christo and Jeanne Claude, Beuys's street tree activism sought to extend the definition of art. He created "social" and "time sculptures," artworks that were realized interactively with citizen participation over periods of time. Yet many Kassel residents did not accept the art status of the project. Given that the project hardly differed from the city's regular street tree planting, Beuys and his collaborators also had to tackle all challenges commonly associated with regular street tree planting, including tree vandalism. Many Kassel residents questioned the nature of the artwork. The biggest matter of contention was the large triangular sculpture of initially seven thousand loose granite stones on Kassel's

Friedrichsplatz, which would gradually become undone as more and more trees were planted. Many citizens viewed the temporary sculpture as an eyesore and dump that disfigured the plaza and its open space.[11]

## Autumn in the Summer

As in Kassel and other cities, the East and West Berlin street tree art of the 1970s and 1980s was inspired by strong concerns about the environment in general; about forest death, which caused anxiety and reached a particular notoriety through the media in the 1980s; and about the state of Berlin's street trees in particular. In 1970s West Berlin, street trees produced an autumn scenery already in the summer months. As early as July and August, they carried colored leaves and began to lose them due to the winter road salt application. In 1971 the number of West Berlin street trees damaged by road salt was estimated at between ten thousand and twelve thousand. By the end of the decade, as a result of the particularly harsh winter of 1978–79, West Berlin's tree experts reckoned that forty thousand to fifty thousand trees had been damaged by road salt, a damage that was translated into a monetary value of 750 million marks. Road salt had begun to be used in the winter of 1962–63, but its effects were often protracted as soils stored it and trees accumulated the chlorides in their branches and leaves before reacting through coloring, defoliation, and finally death.[12]

But street trees' ecological functions were valued highly, and in 1971 the Berlin Senate's Department of Economics commissioned scientist Hans-Otfried Leh at the Federal Biological Institute for Agriculture and Forestry (Biologische Bundesanstalt für Land- und Forstwirtschaft) in Dahlem to study the effects of road salt on West Berlin street trees and to find ways to conserve and cure affected trees. Leh's objectives were to make recommendations for future species selection and for the improvement of street trees' site conditions. To cure and prevent the premature death of the many existing linden, maple, horse chestnut, and London plane trees that were very susceptible to the chemicals, Leh recommended severe pruning and the application of humus-rich fertilizers. Among the species he suggested for future planting were oak, red oak, black locust, and the Japanese pagoda tree, species that would ultimately also be added to the official list of recommended street trees for West German city governments. By 1980, this list would be expanded to include more foreign species, especially from the Netherlands and the United States, for example, *Sophora japonica* 'Regent,' the thornless and upright *Gleditsia triacanthos* 'Shademaster,' and the pyramidal *Gleditsia triacanthos* 'Skyline.' While these varieties of the Japanese pagoda tree and the honey locust were considered to be relatively resistant to environmental pollutants and insects, they also fit into increasingly crowded spaces.[13]

Leh's species recommendations for Berlin were first applied on Kantstraße, where the Charlottenburg district Department of Gardens planted black locusts in 1975. Other measures that began to be employed were the use of planting containers for small trees in narrow streets and the construction of elevated planting beds. A concerted effort was also undertaken to remove snow mountains from tree pits before the thawing season. The West German working group of the Municipal Garden Directors' Conference (Gartenamtsleiterkonferenz [GALK]) discussed the challenges of road salt application at its twenty-second workshop in 1979. It passed a resolution demanding the limitation of road salt application in general and its prohibition on sidewalks.[14]

West Berlin had prohibited road salt application on sidewalks already the year before and extended the prohibition the following year, although the city still allowed exceptions for heavily trafficked and important thoroughfares. At the same time, seventy-five kilometers of West Berlin's streets were being used for a research project undertaken in collaboration with the Federal Environment Agency (Umweltbundesamt) in Berlin that experimented with an alternative and more effective de-icing substance consisting of 80 percent common salt and 20 percent calcium chloride. Although only slow street tree recovery was visible in the first half of the 1980s, by the end of the decade, the data regarding the effects of road salt on street trees reported by the German Environment Agency showed a distinct improvement. Excluding the data for the trees along the two major thoroughfares Avus and Heerstraße, the natrium and chloride content in West Berlin's street tree leaves was reported to be reduced by an average of 75 percent.[15]

Although trees were suffering in East Berlin as well, the situation was different. By the 1960s, large parts of the inner city had not yet been reconstructed, and there was less traffic and fewer trees. Whereas West Berlin in 1973 had around 200,000 street trees on an area of about 488 square kilometers, Wolfgang Reckling's 1973 survey of East Berlin street trees counted 78,944 in a city area of some 403 square kilometers. Until the mid-1960s, most streets in East Berlin were treated only with sand and grit. Once increased traffic did require chemical solutions, road surfaces were treated with liquid magnesium chloride, a waste product in potash mining that the GDR considered less damaging to roads, plantings, and automobiles than the chloride salts used in cities of capitalist opponents West Germany, Austria, Switzerland, Scandinavia, and the United States.[16]

However, the outcome soon appeared to be very similar. When Reckling assessed the street trees of Unter den Linden in the late summer of 1973, he observed a situation not dissimilar from that of West Berlin. All trees, except for the young replanted ones, showed significant road salt damage. In the same year, the Institute of Forest Science (Institut für Forstwissenschaft) had undertaken measurements of chloride in the dry leaf matter of Unter den Linden's famous street trees. The results revealed a chloride content of 3.05 to 3.45 percent, which was up to ten times higher than the 0.35 percent that was measured in the dry leaf matter of the lindens standing in lawns near the Brandenburg Gate.[17]

As in West Berlin, East Berlin's district departments' first response to the soil pollution was the use of planting containers that heavily compromised the many climatic and aesthetic functions that trees growing in the ground could provide. This response was also used in the 1980s, when trees began to suffer from soil pollution caused by gas leakages occurring as a result of the switch from city gas to the dryer natural gas deliv-

ered from the Soviet Union. A second response was the plan to plant forty-five thousand new trees along streets and in parks in the years 1976 to 1979.[18]

## Street Tree Activism in West Berlin

As many tree experts noted in the early 1980s, the public at large was quickly learning
about trees and other environmental matters. Already in the early 1970s, the dire state
of Berlin's street trees could hardly go unnoticed by the public. Given the slow reaction of the urban government, West Berliners decided to take matters into their own
hands. What followed were protests by informed citizens who stood up for trees against
the economic interests of what they perceived to be an egoistic, profit-reaping urban
government.[19]

In the early 1970s, a group of concerned citizens founded the nonprofit Association for the Protection of Berlin Trees (Gemeinschaft für den Schutz des Berliner
Baumbestandes e.V.) and quickly began to issue what became an annual newsletter
entitled *Der Baumschutz* (Tree Protection). Endorsed by tree experts such as landscape architect Aloys Bernatzky, the association declared its mission to be to monitor
urban tree care and maintenance activities by the district Departments of Gardens;
to draw attention to endangered trees; and to distribute the latest scientific findings related to tree health and care. From its beginning, the Association for the Protection of
Berlin Trees protested the destruction of trees as a result of road salt and air pollution
exacerbated further by industry and motor traffic in East Berlin; street tree felling that
resulted from road-building activities; the destruction of permeable mosaic sidewalk
paving and its replacement or covering with impermeable asphalt; the insufficient
expertise of the city's climbers and pruners, who "butchered" trees; and the lack of
comprehensive tree surveys.[20]

The association supported other citizen initiatives that pursued similar objectives. After the particularly long and snowy winter of 1978–79, it teamed up with the
Working Group Green Tempelhof (AG Grünes Tempelhof), the Initiative for the Salvation of Berlin Street Trees (Initiative zur Rettung Berliner Strassenbäume), and a
number of other environmental nonprofit organizations and citizen initiatives to hold
a first "Tree Hearing" on the condition of street trees and initiate a campaign to stop
the use of de-icing salt. While this campaign, called "Aktion Tausalzstopp," further
pressured the West Berlin city government to rethink its winter snow policies and ultimately contributed to the prohibition of de-icing salt on most roadways and all sidewalks, other protests were less successful.[21]

In 1977 the Association for the Protection of Berlin Trees reported street trees'
dry and curled-up leaves and premature leaf loss in both East and West Berlin to the
district Departments of Gardens and the director of the West Berlin Office of Plant
Protection (Pflanzenschutzamt). Although the symptoms reappeared in the following
years and tree expert Aloys Bernatzky had voiced his belief that the trees were reacting
to air pollution, the association found governmental agencies and city offices evasive
and unresponsive. Neither Berlin botanists and plant pathologists nor politicians and

the media appeared interested in seriously addressing the issue. The association accused the city of neglecting to commission air-quality studies. The wall could not prevent the air from passing between East and West Berlin, so why could the two sides not talk about what had turned Berlin into one of the three worst air-polluted cities in the world besides Chicago and Osaka? Berlin was, after all, an "affected area" as defined by section 44 of the 1974 Federal Law for the Protection against Emissions (Bundesimmissionsschutzgesetz). To the Association for the Protection of Berlin Trees the West Berlin Senate appeared fearful of the results of an air study that would likely have contradicted what was promoted as a "green city" with "air like champagne." Extensive lobbying by the association and other environmental groups finally led the Berlin Senate in 1978 to implement a smog regulation that would force industry and citizens to adjust their behavior with regard to the use of private automobiles and the burning of fossil fuels that emitted sulfur dioxide.[22]

But underground the air was not so good, either. Gas leakages were damaging street trees throughout the city. Although the Berlin gasworks (GASAG) had known that the switch from city to natural gas would lead to leaking gas pipes due to the different chemical compositions of the gas, the GASAG had not taken precautions and had neglected the renewal of the piping system. While the association's critique of the city and its economic interests never relented, the politicians certainly knew how to fire back. In the early 1970s, the senator for building and housing Klaus Riebschläger derisively called the association a "Back-to-the-Woods-Movement" (Zurück-auf-die-Bäume-Bewegung).[23]

While the Association for the Protection of Berlin Trees, other nonprofits, and citizen initiatives considered it their mission to draw attention to grievances affecting urban trees, they did not seek to interfere in the city's actual planting and maintenance work until the hot summer of 1989. Due to work overload, it was impossible for many district departments to keep up with watering street trees, so the Foundation for Nature Protection (Stiftung Naturschutz) proposed that citizens could be assigned trees to take care of. The idea was greeted with skepticism by some city officials, although citizens in Marienfelde und Lichtenrade were already practicing it. Some years later, after the fall of the wall, several newsletters of various environmental opposition groups like Baumschutz and Berliner Luftzeitung encouraged citizens to do their part by reporting any damage and maltreatment of trees on a form to be sent to the respective district Department of Gardens (figure 7.6).[24]

## Street Tree Activism and Citizen Opposition in East Berlin

When Manfred Butzmann's poster No Place for Trees featured the decaying trees along Prenzlauer Berg's Schönhauser Allee in 1985, public interest in conservation issues in the East Berlin district was increasing. This included the conservation of trees. For almost two decades street trees in the district had been struggling. In the early 1970s, 5 percent of all existing street trees were lost, among them some old majestic linden, London plane, and horse chestnut trees that had contributed to the distinctive

**An das Bezirksamt**_____
**Naturschutz- und Grünflächenamt**

Betr.: Anzeige wegen Baumschäden bzw. -beeinträchtigungen durch Baumaßnahmen

Sehr geehrte Damen und Herren,

hiermit zeige ich Ihnen den untenstehenden Sachverhalt an. Ich bitte Sie, mich vom Eingang meines Schreibens sowie von den durch Sie ergriffenen Maßnahmen und ggf. dem Ausgang des Verfahrens zu informieren.

Mit freundlichen Grüßen

Unterschrift, Datum

STANDORT DES BAUMES
Straße (vor Hausnr.):
Baum-Nr. (falls vorhanden):
Ortsteil:

SCHADSTELLE
(bitte in der Skizze markieren!)
O Astwerk
O Stamm
O Wurzeln

BEEINTRÄCHTIGUNG
O Lagerung von Baumaterial
O Beparken
O ...

evtl. SCHADENSEINTRITT
Datum

evtl. VERURSACHER
Firma
Pol. Kz. des verursachenden Fahrzeugs

Weitere/r Zeuge/Zeugin

Diese Seite kann und soll kopiert werden - oder als Infoblatt bei der Baumschutzgemeinschaft anfordern!

**8**  *Baumschutz-Sondernummer 19/95*

FIGURE 7.6.
Report sheet for street tree damage, 1995. (*Baumschutz-Sonderheft*, no. 19 [1995]: 8)

identity of Hufelandstraße, Kastanienallee, Schönhauser Allee, and Dimitroffstraße (today Danziger Straße). The street tree situation in the early 1980s mirrored the more general state of affairs in this old and most densely populated district of East Berlin. Its housing stock was crumbling. When in 1983 architects Herbert Pohl and Wolf Dietrich Werner published their design ideas for the district's urban renewal, including extensive street tree plantings, in the industrial design journal *Form und Zweck*, they lost their jobs. Their analysis of the existing conditions of the district's urban structure and its architectural and natural heritage had been too critical.[25]

For a long time the urban government had neglected the conservation, let alone the renovation, of the district's natural and architectural heritage, which, to socialist leaders, epitomized capitalist urban development. Instead, the city had invested much of its energy and effort in the construction of prefabricated public housing units in new residential districts on greenfield sites along East Berlin's periphery. They were part of the public housing program adopted in 1973 under the new party leader Erich Honecker to solve the GDR's housing shortage by 1990. But as political opposition and the criticism of neglect mounted in the 1980s, Prenzlauer Berg finally became the location of both grassroots citizen initiatives that led to the greening of derelict land and courtyards, and government action that struggled to keep these citizen actions under control while itself undertaking some first steps toward restoration and urban renewal.[26]

In 1982, in an attempt to increase the district's number of trees, calm and control Prenzlauer Berg's inhabitants in the process, and prepare for the city's 750-year anniversary celebrations in 1987, the urban government launched a tree-planting campaign that sought to involve citizens in the planting of nineteen thousand to twenty thousand trees along district streets and in parks. In the following years, around five thousand trees were planted in as many streets as possible; these were pollution-resistant species with small crowns like bird cherry 'Schloss Tiefurt,' Turkish hazel, fastigiated hornbeam, and various maple species adapted to growth in narrow streets. In the mid-1980s, the journal *Architektur der DDR* reported that existing mature trees in single or double rows lined altogether thirty-four kilometers of district streets and that new trees had been planted along forty-two street kilometers. Further plans were laid out for future tree planting in all electoral wards throughout the district, and the planting campaign was complemented with a first tree survey of all mature trees undertaken by engaged citizens, many of them members of church, school, and student groups as well as the GDR's Cultural League (Kulturbund der DDR). Although the Cultural League was a mass organization that served the Socialist Unity Party (SED) in furthering a socialist society, it also provided those members interested in local history, conservation, nature, and the environment a space that was less closely tied to party politics than that of other organizations.[27]

Prenzlauer Berg's planting campaign was reminiscent of one of the first activist street tree–planting campaigns, undertaken in 1979 by a church youth group in Schwerin in collaboration with the local people's enterprise VEB Grünanlagen, which had provided trees and support. A successful collaboration between an environmentalist church group and a governmental agency, Schwerin's activist street tree–planting campaign had soon become the beginning of a countrywide tree-planting and environmental movement. It became the precedent for similar activities in cities across the GDR, including in Berlin's Prenzlauer Berg. By the late 1980s, Prenzlauer

Berg's citizens and schoolchildren were to plant around three thousand new trees per year on the occasion of the National Economic Mass Initiative (Volkwirtschaftliche Masseninitiative) in an activity called "Beautify Our Cities and Municipalities— Participate!" (Schöner unsere Städte und Gemeinden—Mach mit!), and then care for the trees in the years to come. Street tree care in the first three years after the trees' planting comprised their regular watering and the loosening of the soil every two to three weeks (plate 12).[28]

In contrast to citizen initiatives in West Germany, those in East Germany, like Beautify Our Cities and Municipalities—Participate!, were neither grassroots activities nor political protests. Rather than being bottom-up, they were top-down and often coercive. As historian Mary Fulbrook has shown, East German citizen initiatives were part of a "participatory dictatorship" in which citizens took on functions in party and state organs for a variety of reasons that did not always include political conviction. The state saw in the initiatives a means to incite in citizens a socialist patriotism. Also described as events in a "socialist people's movement," such activities were organized by the GDR's National Front, an association of all GDR political parties and mass organizations intended to bring citizens into party lines, promote cohesion between party and citizen interests, and use citizen labor for those public works and services that the state was unable to supply and that were not part of the planned economy. The most common Participate! efforts involved the renovation of old housing stock for seniors and young couples, the collection of recyclable materials, and city and village beautification measures. The latter included the construction, care, and maintenance of public parks, front yards, and courtyards, and street tree planting and care. For the purposes of Participate! citizens were organized into neighborhood associations (*Hausgemeinschaften*), street associations (*Straßengemeinschaften*), which represented all neighborhood associations along one street, and district associations (*Wohnbezirksausschüsse*). For their work, groups of citizens and individuals—whose participation in the purportedly voluntary initiatives was carefully registered and observed—could earn awards. But for many citizens it may have been more important that their participation could earn them certain favors such as a promotion, the ability to travel abroad, or a bigger apartment.[29]

Organized "voluntary" citizen labor had already been requested on the occasion of one of the earliest street tree–planting campaigns in Prenzlauer Berg, in the fall of 1952. Some fifteen hundred trees had been newly planted along the streets of the district, which had lost 40 percent of its street trees during the war. In December, the director of the Garden Unit (Referat Grünflächen) in the district Department of Reconstruction (Abteilung Aufbau) suggested that citizens should adopt and care for the street trees in front of their houses by watering them and loosening the soil. Unit staff alone would be unable to care for the trees. Prenzlauer Berg's Garden Unit was responding to the East Berlin Magistrate's effort to improve the city's image. Earlier that year, the Magistrate had ordered the planting of altogether forty-seven hundred new street trees throughout East Berlin. In a related prescription, it had demanded that all citizens be encouraged to participate in this and other urban greening and tidying-up measures. Building caretakers (*Hauswarte*) as well as house and street leaders (*Haus- und Straßenvertrauensleute*) were to facilitate and control their realization.[30]

However, it took until 1959 for East Berlin to begin placing official care contracts with neighborhood and street associations for the purpose of urban embellishment and improvement. Sixty percent of the citizens' labor was recompensed, whereas 40 percent was used by the city to provide the necessary material, equipment, and tools. The associations' earnings were paid into a communal account and could be used to renovate community facilities or pay for social gatherings.[31]

In the mid-1960s, East Berlin director of gardens Lichey demanded that by 1970 at least 50 percent of all residential green open space should be maintained by the people. He proclaimed that maintenance contracts with citizens were "a big step toward socialist thought, a big contribution to the function of the new city organism." For Lichey and others, the care contracts and Participate! initiatives in landscape construction and horticulture were further legitimized by the Russian model of volunteer landscape work. In Moscow and other cities, such as Stavropol, Tambov, Poltava, Kharkov, Baku, and Tallinn, residents who had organized in Associations for the Promotion of Greening and Nature Conservation not only paid the association membership dues but also volunteered their time and labor on weekends and national holidays to build and care for public urban parks and gardens. East Germans soon began to call their Saturday "voluntary" service along streets and in front yards, parks, and gardens *subbotnik*, a derivation of the Russian word for Saturday—*subbota*. The term had been used in the Soviet Union since 1919 to describe the unpaid organized Saturday employment of labor. Subbotniks were hailed as the recipe for "socialist construction" and the basis of large-scale "socialist competition" that, according to Stalin, would transform labor from a burden into "a matter of honour, matter of glory, matter of valour and heroism."[32]

Beginning in November 1969, East Berlin's new city regulation (*Stadtordnung*), adopted to ensure the capital's "order, salubrity, and hygiene," provided an explicit framework and basis for Participate! initiatives and care contracts, also regulating the competitions held on their occasion. The Central Committee for Dendrology and Garden Architecture (Zentraler Fachausschuss Dendrologie und Gartenarchitektur) of the GDR's Cultural League bolstered the Participate! initiatives, too, not least because they encouraged a sense of collective ownership among citizens that could enhance the care of the environment.[33]

Both the 1981 GDR tree ordinance, which was signed into law in January 1982, and in its extension East Berlin's own tree ordinance finally codified street tree care as citizen labor. The state tree ordinance required city, district, and municipal councils to ensure that street tree care and maintenance became integral components of Participate!. In Berlin, this responsibility fell to the district governments. To mobilize citizens, they used a series of posters that graphic designer Hajo Schüler had originally created for the city of Erfurt's Department of Gardens (plate 13).[34]

East Berlin's first director of gardens, Reinhold Lingner, had backed this type of citizen participation, but in 1953 he had also cautioned that rallying this kind of support would require time. Indeed, the placement of care contracts in 1959 and in the 1960s advanced slowly. As landscape architect E. Jaenisch would explain some years later, an awareness had to be fostered among citizens that landscape work in public parks and along streets provided the same health benefits as leisure activity in allotment gardens or private backyards. Jaenisch considered it important that the organization and control of concerted mass initiatives by the local state organs, the committees of the National Front, and the respective city officials not compromise the impression of the activities' voluntary status. But the contrived and forced nature as well as the real intention of Participate! initiatives could hardly, of course, escape most citizens.[35]

The official Participate! success stories often differed from the reality on the ground. The official report of Prenzlauer Berg's 1974 spring Participate! planting efforts, for example, highlighted the district inhabitants' "vital interest and active initiative" in environmental beautification. Prenzlauer Berg's district director of gardens Reinhild Zagrodnik explained that "despite their future responsibilities for the trees' or shrubs' care, citizens enjoyed planting because they were contributing in a concrete way to the increase of urban green." In her official report Zagrodnik also pointed out that plant material and machines had been supplied on time, and that the organization of expert supervision of the planting work had been exemplary. More than a decade later, in 1985, the report of the equivalent spring initiative highlighted a successful 85 percent growth of newly planted trees along streets.[36]

But in 1989 Holger Brandt, in the third issue of *Arche Nova*, one of the first publications of the GDR's oppositional environmental movement, noted that although tree-care contracts could help the trees in their early growth phase, their effects were limited in scope. In 1990 during the GDR's dissolution, acting director of gardens in East Berlin Hans Georg Büchner questioned the value of Participate! and care contracts. He criticized the low growth rate of newly planted trees and maintained that this was due to the frequent lack of expertise and motivation among the "volunteers" of the mass initiatives, their deficient supervision by experts, and the delivery of poor nursery stock. Another challenge citizens often had to face was the delayed delivery of planting material by the nurseries. In the early 1980s, for example, Pankow's district Department of Gardens repeatedly reported on citizens' inability to plant the stipulated number of trees due to delivery shortages. On the other hand, one of Büchner's colleagues, the long-serving director of the Prenzlauer Berg district Department of Gardens before and after the fall of the wall, Wolfgang Krause, considered care contracts with citizens a benefit of East Berlin's administration that would have been worthwhile maintaining after German reunification.[37]

By the 1980s, there was perhaps no area and part of the Participate! initiatives that was easier for oppositional environmental activists to co-opt or build upon than street tree management and care. Conversely, activist street tree–planting campaigns were also relatively easy for urban governments to control or take over, and they were therefore tolerated as long as they were carried out by affiliates of church groups. Trees quite literally provided the space that the government was willing to cede to environmental activists. Even though activist street tree–planting campaigns were re-actions to the environmental degradation that the government had allowed, they first appeared neutral enough to give the activists an outlet without seriously compromis-ing the state's power and its laissez-faire environmental politics. The state terminated cooperation and began to criminalize tree-planting activists only once it realized that the nascent environmentalism could not be tamed and that related political interests could not be neutralized.

## The Value of Trees

In the 1970s and 1980s for East and West Berliners alike, street tree planting, care, and protection was a means to oppose the state and its environmental politics. While mo-tivations and forms could differ, street tree activism developed out of the same deep and affective appreciation that many people had not only for the material and eco-logical values of trees but also for their aesthetic, ephemeral, variable, and intangible qualities—qualities that were difficult to assess. While the political economy of the materialist values attributed to trees differed in East and West Berlin, their intrinsic, immaterial values were the same for Berliners on both sides of the iron curtain. In East and West Berlin, it became clear that to protect trees and urban nature as a whole, the value of trees had to be monetized, or at least quantified. Attempts were made in both parts of the country to devise a calculation method that would satisfac-torily render street trees' immaterial benefits quantifiable, a feat that soon appeared rather impossible.

Werner Nohl, professor of landscape architecture and urbanism in Hannover and then at the Technical University Munich, undertook a first attempt to quantify the psychological benefits and the aesthetic qualities of diverse urban trees in the early 1970s. At a time when environmental psychology increasingly attracted researchers' attention, street trees were measurable elements in the urban environment, standing for both a real material and a culturally and socially produced metaphorical nature. Nohl explored the experiential value of trees in a study that he published in 1974. He quantified the psychological effects felt—and the value judgments made—by trained designers and an untrained public when asked to assess two different trees. Nohl's premise was that the reactions and the reception of the public needed to be studied and understood so that designers could base their designs on empirical data rather than exclusively on their own subjective impressions and intuitive understanding. Through his study, he began to describe and quantify an urban tree aesthetics that he found to depend very much on the respective socialization and education of his subjects.[38]

Nohl was tackling an area that had remained elusive in all discussions about tree value so far and that complemented a study carried out in the same years by the young researcher Dieter Boeminghaus at the RWTH Aachen University. In contrast to Nohl's concerns, Boeminghaus's interest was in trees along country roads. In light of increasing traffic deaths at the time and with the intent of providing design guidelines that could contribute to their reduction, Boeminghaus had set out to explore whether and how roadside trees influenced motorists' psychology of perception. He found that trees contributed significantly to the identification of roadways and helped drivers assess the speed and distance of vehicles, and that they could therefore indeed reduce accidents.[39]

Discussions about the value of trees had begun in Germany in the early 1950s when Nuremberg tree expert Michael Maurer had returned from war captivity in Siberia. Maurer had trained as a tree surgeon in the United States. When he returned in 1935, he brought back with him the idea of the value assessment of trees as well as new tree surgery and transplanting methods, which he introduced at Nuremberg's Nazi Party rally grounds in 1938. In the 1950s he began to apply his knowledge of tree care, maintenance, and evaluation in West German cities more widely. In Germany at the time, the monetary value assigned to trees was based solely on the value of their lumber. Their shade and aesthetic functions as well as their role in forging a collective memory and identity and their "time value," as tree expert Werner Hoffmann described the factor of age, were not accounted for.[40]

The situation in the United States was different. There, it had already been an accepted fact in the early twentieth century that trees had value besides the worth of their lumber. Trees could raise property prices. In the late 1920s, recently retired New York state entomologist Ephraim Porter Felt, who directed the Bartlett Tree Research Laboratories, had developed a method and table for the calculation of tree values. The assessment method had been distributed widely through a talk on the New York radio station WEAF and through publications in forestry and horticultural journals. Trees were attributed a base value for their trunk diameter at breast height. This base value could be reduced by various factors, depending on the species, its location, and the tree's physical condition. For example, American elms were rated higher than Norway maples, which in turn were rated higher than American lindens. The tree's value could be further reduced or increased depending on the worth of the residential land it was standing on. Land that was worth between $500 and $2,000 per acre or lot reduced the tree value, whereas prices between $2,000 and $28,000 could increase the tree value by up to 300 percent, turning street trees into precious objects. Once the United States had entered World War II, Felt argued that the war conditions augmented trees' values even further. Given their use in protective concealment against air raids, their value was particularly high if they concealed an important military installation. But Felt, like many of his colleagues, considered all trees, regardless of whether they were growing along streets, in parks, or on private property, valuable shelter trees against both air raids and the sun, and thus they contributed to national welfare and security. He admitted, however, that it was doubtful whether a satisfactory formula could be developed that accounted for their air-raid protection value. Still, he made a strong argument for street tree planting overall. Looking "down upon the treeless streets of a large city like New York" sufficed "to note the harsh outlines of the closely placed buildings and the accompanying dark shadows" that could be broken up by street trees, in particular if they were planted in irregular groups. Like his colleagues in Germany, Felt promoted the planting of "irregular groups of trees with foliage presenting a dissimilar texture or with variable branching habits" for the purposes of air-raid protection.[41]

The American expertise in monetizing trees that developed before and during World War II informed German discussions about trees' value in the postwar years. Ultimately, tree experts considered the attribution of monetary value to street trees the only way to protect them against their destruction by street enlargements, mechanical injuries by cars, and environmental pollution. Assigning street trees a monetary value was the only way to guarantee their replacement in cities that had to deal with an increasing amount of traffic and a decreasing amount of open space. In West Germany throughout the 1950s and 1960s, many cities followed Maurer's lead and began to adopt the American calculation method, which soon became known as the Maurer-Hoffmann method. But calculation methods continued to vary in different cities, and the Maurer-Hoffmann method did not win the upper hand, possibly due to legal concerns and the relatively high restitution costs entailed in its application. For example, a 1968 survey showed that while West Berlin and Frankfurt applied the Maurer-Hoffmann method in the case of a tree's total destruction, Düsseldorf and Hannover based their calculation on the replacement costs for a tree of the same species and size, or for the largest available tree of the species.[42]

To set a unified standard for calculating tree value, in 1969 a working group was founded at the Municipal Garden Directors' Conference of Deutscher Städtetag. Spearheaded by Mannheim garden director Heinrich Wawrik, the group included tree surgeon Maurer and agricultural and horticultural assessor Werner Koch from Stuttgart. In less than two years, the committee drafted a new calculation method to be adopted throughout West German cities and municipalities. Aware that it was difficult to monetize trees' intrinsic values, the tree and real estate experts adapted a method used for the calculation of real estate and property values to street trees. The new tree-value guidelines (*Baumwertrichtlinien*), which would become known as the Koch method, were decided at the 1971 Garden Directors' Conference in Cologne and published the following year. They were based on the replacement and maintenance costs of the damaged or destroyed street tree and further bolstered by what would become known as the horse chestnut tree verdict (*Kastanienbaumurteil*). This 1975 ruling by the federal court, prompted by an automobile accident that destroyed a forty-year-old horse chestnut tree standing on the median of West Berlin's Meraner Straße, determined that the accused had to pay for the initial planting and care of a five-year-old substitute tree in addition to thirty-five years of annual care and maintenance costs and 5 percent annual interest on the basis of these costs.[43]

But given the Koch method's disregard for trees' immaterial hygienic and aesthetic qualities, attempts to assess trees' true value remained insufficient. Only shortly after the guidelines had been published, the Berlin Senate Department for Building and Housing (Senator für Bau- und Wohnungswesen) commissioned a group of young engineers at Berlin Technical University's Department of Landscape Economy to develop a method that could be easily applied, delivered higher quotes, and took into consideration trees' qualitative values. Referencing some of their East German colleagues at the Technical University Dresden, the authors of the study argued that tree canopy cover and its seasonal changes had to be factored into the calculation as it was the tree's canopy that provided its various hygienic, aesthetic, and spatial functions. Besides the trees' physical condition, the authors wanted to see the trees' uniqueness and historic value taken into account. Furthermore, trees in urban districts with few street trees and parks should be assigned a higher value, they argued. The researchers surmised that in West Berlin, street trees were particularly valuable in the districts of Neukölln, Kreuzberg, Schöneberg, and Spandau. Despite this and various other attempts at improving the Koch method, however, it has remained the most widely used system for the monetization of trees in Germany.[44]

Similarly, in East Germany a group of tree experts charged with drawing up the GDR's first tree ordinance was preoccupied with the question of monetizing trees. Existing large trees were indiscriminately felled, particularly in areas slated for new mass housing. The only way to save them appeared to be their monetization. Although the Maurer-Hoffmann method had also been applied in East Germany, and the state's capital based its 1974 tree-evaluation draft on this method, East German tree experts soon declared it too subjective and capitalist. In their eyes the Maurer-Hoffmann method furthered financial speculation. Unquestionably, therefore, the socialist economy needed a different method. After all, Marxist ideology determined that street tree values were purely materialist and therefore subject to the Marxist theory of labor. Thus, their calculation had to be based on the effort employed in the trees' production and on their societal use value, which meant that the trees were valuable only insofar as they fulfilled required human needs. While early attempts to calculate tree value dismissed trees' aesthetic, hygienic, and bioclimatic functions as too subjective, later methods did seek to accommodate "facultative functions" such as trees' capacity to bind dust, cast shadow, shield against wind, buffer sound, foster identity, contribute to the definition of space, and guide traffic.[45]

In 1966 W. Zipperling of the Prenzlauer Berg district Department of Gardens had suggested that a tree's value was the sum of expenses for its removal, its replacement, and the maintenance efforts required during its life span, from which the tree's worth as lumber was to be deducted. Some years later, in 1973, professor Harald Linke and his student Friedbert Klein at the Technical University Dresden added trees' purportedly subjective "social-hygienic and cultural values" to the equation. In fact, these values lay at the basis of their method, which used the trees' canopy as reference parameter rather than the diameter of their trunk, as the Maurer-Hoffmann method and many other methods had done. As Linke and Klein pointed out, it was the trees' canopies, not their trunks, that protected against wind, noise, and dust, and provided many of the other generally accepted tree benefits. Given the researchers' observation that trunk diameters could vary between trees with similar crown sizes but that trees with the same trunk diameter varied considerably in canopy size, the canopy appeared to Linke and Klein to be the better reference point. Klein's 1974 dissertation, which presented a method for the economic evaluation of trees, provided the guidelines used in the early drafts of what would become the GDR's 1981 tree ordinance. Beginning in November 1973, Klein, a landscape architect at the VEB Garden and Landscape Design Leipzig, had led the working group that drafted the first versions of the tree ordinance to complement the 1970 GDR's environmental law (Landeskulturgesetz), which determined how the state's natural resources were to be protected, improved, and effectively used. These early drafts included an economic tree value assessment based on Klein's dissertation work.[46]

Klaus-Dietrich Gandert, a member of the working group and president of the Central Expert Committee on Dendrology and Garden Architecture (Zentraler Fachausschuss Dendrologie und Gartenarchitektur) of the Cultural League who actively supported the effort to draw up a state tree ordinance, also supported the canopy idea. He agreed with his colleagues that a tree's monetary value—which was also its societal use value—should be calculated on the basis of its canopy volume, its functions, and its condition. Nevertheless, the greater practicality of using trunk diameters as a reference parameter meant that the calculation method of tree value appended to East Berlin's 1976 tree ordinance was based on trunk diameter, species, locality, and condition. While this method did not account for the land value, the remaining factors were the same as in the initial American value calculation method. The GDR's first tree ordinance, signed into law in May 1981, ultimately omitted any kind of tree-evaluation method on the assumption that the civil code and indemnification law were sufficient to rule in cases of tree damage and destruction. It disregarded the multiyear efforts undertaken by tree experts and conservationists to devise a new tree-evaluation method that sought to quantify trees' immaterial values and protect them more successfully.[47]

## The Risk of Trees

Discussions about the value of trees invariably highlighted the many risks and uncertainties connected to the growth and development of trees in an urban environment. People and material property like cars could easily be harmed by falling trees and branches, and German law required cities and municipalities to regularly assess street trees' "traffic security" (Verkehissicherheit). In the 1980s, therefore, research projects were undertaken and new instruments were developed to assess tree health and static and with this the risks of trees harming people and property. The arborscope, an endoscope for trees, was invented to look inside the hollows of tree trunks and assess their level of decay. Less invasive was the use of computer tomography, which was tested throughout the 1980s in a research project sponsored by the Volkswagenwerk Foundation and undertaken by professor of physics Adolf Habermehl and his research group for medical physics at the Radiology Center of Philipps-University Marburg. The measurement of tree static became another full-fledged area of research and a new business opportunity developed by garden architect Günter Sinn. Soon, however, his method, which required boring holes into the trunk, stood in competition with Lothar Wessolly's "elasto-method," hailed as a noninvasive way to calculate a tree's tensile resistance.[48]

In the academy, tree static became an area of research at the universities of

Stuttgart and Tübingen under the leadership of architect Frei Otto, well known for his web- and tent-like roof structures: for example, the Munich Olympic stadium and the roof of the German Pavilion at the 1967 Montreal Expo. Ever interested in the morphology and anatomy of plants and animals, Otto had assembled humanities scholars, architects, engineers, and biologists in a special research cluster sponsored by the German Research Foundation and called Natural Constructions: Lightweight Construction in Architecture and Nature (Sonderforschungsbereich 230—Natürliche Konstruktionen: Leichtbau in Architektur und Natur). Two projects specifically dealt with the static of plants informing Sinn's and others' arboricultural applications. One, led by paleontologist Adolf Seilacher, addressed the morphology of plants' structural systems. The other project, by engineers Gallus Rehm and Rainer Blum, measured the various densities, bending tensile, and compressive strengths of wood from different tree species, and studied how trees and their branches break.[49]

Whereas for Otto and his collaborators in architecture and engineering, the objective was to learn from trees and nature in general in order to naturalize architecture and create heretofore impossible lightweight structures, for Sinn and others the goal was to render trees more secure and controllable, turning them into what Heinrich Wawrik from Mannheim's Department of Gardens called a *Kunstbaum*, an artificial tree.[50]

German arboricultural research was further inspired by the idiosyncratic work of the charismatic American biologist and tree pathologist Alex L. Shigo, who for many years worked for the U.S. Forest Service. Shigo's visit to the State Horticultural Training and Experiment Station in Heidelberg in 1984 led to inspired discussions among German tree experts. Continuing education tree seminars for landscape workers, gardeners, and landscape designers had been held in Heidelberg since 1966, but only since 1977 had there been an annual three-part seminar on tree surgery, planting, care, and the survey of trees that concluded with an exam to become a certified tree carer (*Baumpfleger*). After Michael Maurer, whose teachings had been followed in Heidelberg and by most West German firms offering arboricultural services, Shigo was the second American-trained tree expert to introduce new, or amended, aboricultural practices. German tree experts decided to follow some but not all his recommendations. For example, whereas Shigo advised the filling of cavities for cosmetic reasons, German tree experts did not want to see cavities filled as this would prevent easy control of further interior decay. Shigo considered wound-closure agents superfluous or even harmful. In contrast, German tree experts were in favor of wound-closure agents because they had not been proven to be ineffective or harmful. While new studies and technologies changed the immediate arboricultural practices when it came to the care of single street trees, they also affected the management of the entire urban forest.[51]

## Shades of Red

After the death of Berlin's street trees had led to the first upsurge in citizen concern, complaints, and environmental activism in the late 1970s, West Berlin's Senate Department for Building and Housing initiated the first infrared aerial survey to assess the health and condition of West Berlin's trees on a citywide scale. Among the many things that the Association for the Protection of Berlin Trees had criticized in its first issue of *Der Baumschutz* in 1976 had been the lack of aerial infrared surveys. It took the city a year to secure the necessary permissions from the Allied occupying forces, but finally in late August and early September 1979 the first aerial infrared survey was flown by a pilot provided by the Western Allies. It would take almost another year for the film material to be cleared by the Allies so that it could be put to use.[52]

Despite arguably having the biggest tree canopy among German cities, West Berlin lagged behind other German metropolises, which had begun infrared surveys for the inventory and analysis of green space and street trees in the early 1970s, after Amsterdam had set an early example in 1969. It took East Berlin even longer to initiate a comparable survey, although East German landscape architects had been intently observing the developments in West Germany and other countries. Parts of Fried-richshain were finally surveyed in 1986 with infrared Spektrozonalfilm produced in the Soviet Union as an equivalent to Kodak's Aerochrome Infrared film, which was in wide use on the Western side of the iron curtain.[53]

The Eastman Kodak Company had developed infrared film during World War II to allow American reconnaissance missions to identify camouflage schemes. Photo interpreters were able to detect false trees and false vegetation more easily on infrared than on panchromatic film because leaf cells reflect infrared light waves differently than painted artificial materials. In the 1970s, infrared photography became a sought-after tool in urban street tree management because it could be used to detect trees' varying health conditions caused by physiological and morphological changes in their leaf cells. It could render visible to the human eye what regular vision and black-and-white aerial photography could not. For example, regular terrestrial field analysis could often detect the effects of de-icing salt only years later when foliage developed necrosis. In contrast, infrared photographs could reveal changes in the leaves' cellular structure much earlier because these morphological changes also caused changes in the leaves' reflective properties. In aerial infrared photographs, healthy trees appeared in a dark red tone, whereas reduced vitality in trees was indicated by lighter red and gray foliage.

By the time a second aerial survey was conducted in West Berlin in the late summer of 1985, it was clear not only that between 20 and 30 percent of West Berlin's street trees were damaged—in particular linden and horse chestnut trees—but that the aerial survey and photo interpretation method itself needed improvement and further study. As a consequence, funding was allocated for a collaboration between the Free University of Berlin and the Berlin Senate Department for Urban Develop-ment and the Environment for an accompanying terrestrial study of West Berlin's street trees. Through studying a sample of trees throughout the city on the ground, the researchers, under the guidance of landscape planner Ulrich Förster and geologist Michael Fietz from the Free University of Berlin, were able to establish a key for the interpretation of the aerial photographs. Continual monitoring and infrared photography of these trees on the ground could also complement the aerial survey, which would be flown only every five years. Furthermore, the researchers suggested that the accuracy of comprehensive citywide tree surveys could be improved by aerial surveys of smaller study areas whose results could then be extrapolated. One of the many challenges they were seeking to tackle was the length of time it took for comprehensive aerial surveys to be carried out and the resulting inaccuracy of their interpretation. Changing weather and light conditions, for example, affected photo quality and consequently rendered the interpretation of photos from different flights more difficult due to their compromised comparability.[54]

Michael Fietz had been working on the use of aerial infrared photography for street tree surveys for his 1981 thesis. He was tapping into an area of research that had first been established in the early 1970s by the young forestry scientist Hartmut Kenneweg at the University of Göttingen. Kenneweg had realized that aerial infrared photography, used for inventory and management in forestry, would provide a valuable tool in urban forestry as well. On the basis of Freiburg's 1971 aerial survey, he argued that infrared aerial photography could be used to assess urban trees and establish tree and green space inventories that could in turn help planners, citizen initiatives, and environmentalists guide urban development and protect trees and green open space. The analysis and interpretation of aerial infrared photographs allowed Kenneweg to complement the park area/inhabitant ratio, which had been common in the quantification of green open space since the Progressive Era, with a tree/inhabitant ratio for the various census tracts (figure 7.7). He maintained that the context and detail aerial photographs could provide—the comprehensive picture of the entire urban green space and canopy and the specific details of a single tree—would improve the design, planning, management, and maintenance of urban tree plantings and green space as a whole. Besides the assessment of a single tree's health, aerial photography also made possible an assessment of the tree's height and the photogrammetric measurement of its canopy. The latter data led Kenneweg to speculate on the use of aerial infrared photography in risk assessment. According to him, the photogrammetric measurement of the tree canopy in figure 7.8 would have shown that a falling tree limb was likely. This knowledge could have led to pruning the tree, thus preventing the damage that occurred to the parked car below and the potential effect on pedestrians.[55]

Michael Fietz applied much of Kenneweg's groundwork to the case of West Berlin. In his 1981 thesis, the young researcher used the 1979 aerial infrared survey to study how far street tree health, species, and age could be interpreted on the basis of aerial infrared photography. Fietz's study area was located in the Berlin district of Neukölln, where two-thirds of all street trees were reported as damaged or diseased (plate 14). Although the interpretation required significant training, Fietz agreed with Kenneweg and surmised that it ultimately would be possible to use infrared aerial photography to interpret street tree damage and its causes, an objective that he would pursue further in his dissertation, published a decade later in 1992. By that time, Fietz would also draw attention to the fact that it was not sufficient to compare aerial photographs of various years to track changes in tree health and canopy cover; it was im-

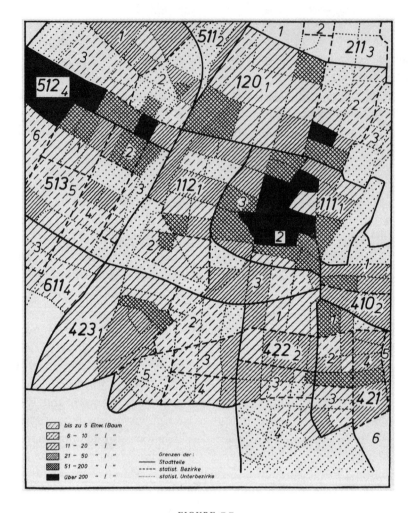

FIGURE 7.7.
Map showing the tree/inhabitant ratio for census tracts in Freiburg. (Hartmut
Kenneweg, "Objektive Kennziffern für die Grünplanung in Stadtgebieten aus
Infrarot-Farbluftbildern," *Landschaft und Stadt* 7, no. 1 [1975]: 35–43)

portant to factor in the climate and consequential phenological developments in the
respective years. In 1992 Fietz offered a thorough assessment of the potential of aerial
infrared photography for street tree management. Again using West Berlin as his case
study, he showed that it was possible to employ aerial surveys not only to provide comprehensive tree data for an entire city but to derive detailed information on single
trees as well. One of his contributions was to show how expertise and training in plant
science and aerial infrared photo interpretation made it possible to identify species,
numbers, and distribution.[56]

Although it was clear that aerial infrared photography and its interpretation was
a complicated and costly endeavor, it quickly moved from a means thought to help in
the assessment of street tree damage to a method that was hailed as facilitating the

FIGURE 7.8.
Photogrammetric measure-
ment of tree canopy reaching
over parking lot. (H. Kenneweg,
"Luftbildauswertung von
Stadtbaumbeständen—
Möglichkeiten und Grenzen,"
*Mitteilungen der Deutschen
Dendrologischen Gesellschaft*
71 [1979]: 159–92)

inventory, analysis, management, and planning of urban trees and urban green space
as a whole. Thus, West Berlin's 1979 aerial survey for the assessment of street tree
damage soon inspired the implementation of a systematic street tree inventory on
the ground that was supported by data gathered from aerial photographs. Trees were
marked and numbered on cadastral city maps and listed on index cards (figure 7.9).
Information on species, age, vitality, trunk circumference, canopy diameter, height,
root space, and any potential damage causes was listed on the card; the data could be
updated easily and fed into a computer system.[57]

Although West Berlin was the first West German city to experiment with a digi-
tal tree inventory, it followed earlier undertakings in Dresden that can be considered
forerunners of digitized inventories. In the years between 1972 and 1975, Siegfried
Sommer, at the time lecturer at Dresden's Technical University, had been assisted in
a citywide tree survey by his landscape architecture students. The city's objective had
been to establish a system that could be updated easily and that provided a basis for an
ecological-aesthetic tree concept plan. Besides documenting street trees on the city
map by applying a hachure for their life expectancy, Sommer recorded data related to

FIGURE 7.9.
An index card of West Berlin's street tree survey. (Ulrich Förster, "Vitalitätsbestimmung von Strassenbäumen. Ergebnisse der Color-Infrarot-Befliegung in Berlin [West]
1979," *Berliner geowissenschaftliche Abhandlungen*, series C3, vol. 3 [1984]: 43–51)

the street tree species, age and life expectancy, location, and formation/type of planting
(single tree, row, allée, mix of diverse species) on edge-notched cards. Punched holes
on the cards indicated number codes that corresponded to particular locations on the
city map. The mechanical card-sorting system facilitated the search for particular tree
data and eliminated the time-consuming job of refiling cards in any exact order (figure 7.10).[58]

In 1987 and the following years, high school students helped to survey and record street trees in East Berlin's inner-city district Pankow, and their data was fed into
computers used to list, count, and manage the trees. The Pankow district had been a
forerunner with regard to street tree surveys since the late 1960s, when its Department
of Gardens had developed a system of using index cards to register street trees. But
given the lack of labor, progress was slow. When the first round of work was finally
concluded by a student for a term paper in 1972, it revealed that the number of trees
in the district had to be adjusted downward. Not 17,000 trees, as had been thought,
but only 15,774 lined district streets. But the survey's upkeep was uncertain, and the
department's annual reports in later years failed to mention any updates.[59]

One of the advantages of the comprehensive and synoptic vision that aerial infrared surveys offered was that they facilitated the assessment of the spatial and temporal

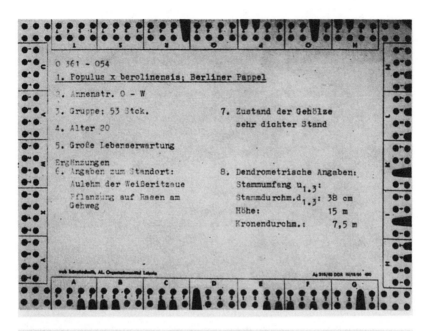

O 361 - 054

1. Populus x berolinensis; Berliner Pappel

2. Annenstr. O - W

3. Gruppe; 53 Stck.

4. Alter 20

5. Große Lebenserwartung

Ergänzungen
6. Angaben zum Standort:
   Aulehm der Weißeritzaue
   Pflanzung auf Rasen am
   Gehweg

7. Zustand der Gehölze
   sehr dichter Stand

8. Dendrometrische Angaben:
   Stammumfang $u_{1,3}$:
   Stammdurchm.$d_{1,3}$: 38 cm
   Höhe: 15 m
   Kronendurchm.: 7,5 m

FIGURE 7.10.
Documentation of Dresden's street trees on edge-notched cards and in city maps.
(S. Sommer, "Erfassung des Straßenbaumbestandes—Grundlage für Erhaltung
und Rekonstruktion," *Landschaftsarchitektur* 4, no. 4 [1975]: 113–14, 117)

relationship and development of tree damage and disease. At a time when remote sensing and geographical information systems were still in their infancy and suitably applied only on regional scales, aerial infrared photography—if undertaken at regular intervals—proved to be a practical means to track the dynamic of both urban development and the seasonal and life cycles of street trees. To make the information provided in aerial infrared photographs useful, it not only had to be interpreted but it consequently also had to be represented in a user-friendly format that could be updated easily. While it was an easy task to update tree lists on index cards and in computers, it was more complicated to keep updated tree maps, at the time still the only tool to visualize tree data in space. The cadastral survey maps at a 1:1,000 scale, which showed the location of individual tree trunks, were considered impractical because of their size. They were also unreliable as they were updated only rarely. On the occasion of the research undertaken in connection with the 1979 aerial infrared survey, Gottfried Borys, advised by Free University professor Bernd Meissner and the University of Applied Sciences (Technische Fachhochschule) professor Gerhard Pöhlmann, therefore developed a prototypical street tree map at a scale of 1:2,000 (plate 15). Produced as part of Borys's 1982 thesis before the widespread use of GIS, the street tree map was intended to provide a planning tool that could easily be adjusted and updated as trees and urban development changed over time. Borys achieved the easy adaptability of his map by selecting specific cartographic production methods and symbols. The map spatialized information gathered from aerial infrared photographs and terrestrial analysis available in the tables of Berlin's street tree inventory. It included information on tree species, tree age and vitality, trunk damage, soil type and permeability, and causes of damage. Borys's prototypical street tree map was never applied to street tree manage-

FIGURE 7.11.
Tree restoration measures performed on the horse chestnut trees along Kantstraße, 19 August 1986. (Photo by Ludwig Ehlers. Landesarchiv Berlin, F Rep. 290 [09], Nr. 0278629)

ment in West Berlin and, of course, it cannot compete with today's geo-information systems. But the cartographic symbols, colors, and method Borys devised for the map's production and printing did take into account time as a factor of change in the plant world and was a first response to the more rapid changes of urban plant life.[60]

Up in the air and in the domain of experts, aerial infrared photography was the urban government's response to grassroots street tree activism and a street tree situation that caused great distress among citizens, ecologists, and landscape architects alike. Although not the only official response to street tree disease and death, the comprehensive aerial view that West Berlin's first infrared surveys offered played a significant role in sparking the government's greater investment and engagement in street tree care and management. It promoted street tree surveys and street tree–planting campaigns that aspired to reach the prewar number of street trees in West Berlin. Care was taken to exchange polluted soils and aerate the soils street trees were rooted in (figure 7.11). Seeing the larger picture from above helped to identify, address, and ultimately remedy some of the problems on the ground. Both the comprehensive overview and the focal view would again become important in the street tree planting of a reunified Berlin.[61]

# 8
# Unity and Variety
## *Berlin's New Urban Forest*

I n 2001 the new German chancellery was inaugurated after four years of construction and a much longer process of deliberations about its location and architectural design. At first sight, the massive cube, soon nicknamed the washing machine, does not appear to have anything to do with trees. Instead, it forms the monumental yet transparent western anchor of the city's new symbolic east-west corridor—the Federal Strip (Band des Bundes)—consisting of a row of federal ministry buildings that bridge the river Spree north of the nineteenth-century Reichstag. But a close observer will marvel at the trees aloft at a height of fourteen meters on top of the columns that structure the space in front of the chancellery. In various ways trees played an important role in the building's design by architects Axel Schultes and Charlotte Frank.

Seeking to distance themselves from Berlin's nineteenth-century neoclassicism and Nazi monumentalism as well as from the low-lying, inconspicuous architecture that in West Germany had given expression to a new transparent democratic order in the postwar years, Schultes and Frank had sought inspiration in universal forms and ideas that transcend time and place. One of these ideas was the tree of life. Part of Schultes's repertoire, used already in past designs, the tree was co-opted again during the architects' final process of form finding for the chancellery. The two designs that Schultes and Frank presented in a 1996 news conference both employed a more or less abstract symbol of the tree of life. The designs gave expression to former chancellor Willy Brandt's exclamation on the occasion of Germany's reunification: "What belongs together must grow together." In environmentalist spirit, Schultes explained that the tree was also a warning against global destruction. But the tree symbolism appeared as a futile postmodern attempt to naturalize both an architecture that had grown to monumental proportions and a challenging political reunification process that would have winners and losers. Much ridiculed by architectural critics at the time and considered a rather curious faux pas, one design—entitled *Global Responsibility*—

featured the silhouette of a large, abstract tree on the building's eastern and western facades. As one journalist pointed out, the architects' superficial yet moralistic message broke with West Germany's postwar efforts to give expression in the architecture of its government buildings to the pluralism inherent in the idea of "democracy as client" (*Demokratie als Bauherr*). In the design that was ultimately realized, the tree had become a forest of columns, some carrying the chancellery's roof, others turned into oversized planters for real trees that now grow out of the columns. Although these elevated live trees may appear to many as surreal and as "architect's parsley"—superficial garnish—for Schultes they turned the chancellery into a garden. As he explained, either oblivious to Germany's Nazi legacy of rootedness or consciously seeking to overcome it, the elevated trees were meant to stress the importance of the reunited nation growing new roots, thus giving expression to the country's newfound self-confidence in a democratic future.[1]

But the elevated trees of the chancellery were not the only trees that played a substantial role in Germany's reunification, and tree symbolism did more than lie at the heart of parts of the new state architecture. It also played a role in the representation of the new reunified nation in other ways. Tree-planting activities abounded in the years after the fall of the wall. They became a practice through which engaged citizens and city officials sought to bridge the many discrepancies in the city of transition and transformation. Tree planting was a displacement activity, an activity that sought to heal the fissures of the German psyche as well as the physical fissures in the urban fabric, an activity that expressed the desire to make Berlin whole again and to overcome the different cultural traditions, frameworks, and collective memories in East and West Germany. It was an activity that defied the past and looked to the future. Trees, themselves in perpetual seasonal transformation throughout their life spans, could fill the space and time of uncertainty that characterized Berlin after the fall of the wall. The coalescence of an active citizenry in East and West Berlin that had already begun to lobby for trees in the 1970s, an environmental movement that had grown into a force in the 1980s, and the end of the Cold War fostered an increased interest in tree-planting activities in the 1990s. Celebratory of German reunification and reconstruction, and symbolic of peace and friendship on the one hand, and valuable as air-conditioners and carbon sink on the other, urban trees fulfilled multiple functions that called for action by the urban government and engaged citizens.

Among the first tree-planting efforts undertaken by interested individuals and members of various environmental groups from East and West Berlin was a planting campaign on Falkplatz, located along the former Berlin wall in Prenzlauer Berg. On 1 April 1990, members of the Green League (Grüne Liga), the Initiative against Spring Lethargy (Initiative gegen Frühjahrsmüdigkeit), the German Union for Environment and Nature Protection (Bund für Umwelt und Naturschutz Deutschland), and the Association for the Protection of Trees (Baumschutzgemeinschaft) cooperated in this planting effort with the district Department of Gardens, the VEB Kombinat Stadtwirtschaft, the tree nursery in Biesenthal, and members of the former East German border patrol (figure 8.1). Two other former border locations, Glienicker Bridge and

FIGURE 8.1.
Tree planting along the former border strip (in the location of today's Mauerpark),
1 April 1990. (Photo by Gerd Danigel)

the Brandenburg Gate, were the sites of the first Japanese cherry tree plantings sponsored by Japanese citizens and organized through the Japanese television corporation
Asahi. Numerous other cherry tree plantings followed. Many, like the 120 cherry trees
along Prenzlauer Berg's Schwedter Straße, were planted along or near the former
German-German border. Other tree species were also located along the former border strip, and citizens planted "green landmarks" (*grüne Merkpunkte*). Large quantities of polluted soil first had to be exchanged. For years, the GDR had not only applied herbicides and dioxides on the no-man's-land but had also used parts of it as a
dump for various toxins including heavy metals. After the fall of the wall, the former
border strip and adjacent areas like Falkplatz quickly became areas of speculation for
design ideas ranging from a north-south linear park to a motorway.[2]

In 1990 Ben Wargin's tree-planting activities gained renewed attention and mo-
mentum. He planted ginkgo trees on the former border strip at Potsdamer Platz and
along Friedrichstraße near the former border crossing. Celebrating the first anniversary of the fall of the wall, he installed *Parliament of Trees*, a monument to those who
had lost their lives attempting to cross the German-German border. It consisted of trees
planted along the former border strip bordering the Spree north of the Reichstag and
of parts of the wall, which Wargin painted with the numbers of East German refugees
who had died during escape attempts in every year of Germany's division. But the lo-

cation that attracted special attention to both Wargin's activism and planting activities by the city was Unter den Linden.[3]

## Unter den Linden

In 1993 Wargin positioned a forty-meter-long barge planted with five pine trees on Unter den Linden in front of Humboldt University (figure 8.2). The barge advertised his tree-planting projects and an underground installation in the nearby so-called Lindentunnel, a tunnel that had been opened in 1916 so that the tram could cross Unter den Linden underground. Unter den Linden and its tunnel became the anchor for Wargin's environmental initiatives in the 1990s. His installations above and below Unter den Linden in the years between the 1992 Rio and 1995 Berlin world climate conferences commented on environmental destruction and climate change, and they sought to encourage urban tree planting.[4]

Unter den Linden held an important symbolic meaning for the city. A day after the fall of the wall, the West German Federal Association of Garden, Landscape, and Sports Ground Construction (Bundesverband Garten-, Landschafts- und Sportplatzbau e.V.) declared its sponsorship of the planting of thirty silver lindens on Berlin's famous tree-lined boulevard. The trees were intended for the stretch between Otto-Grotewohl-Straße (today's Wilhelmstraße) and Pariser Platz near the former Berlin wall and its border strip.[5]

The project to replant this part of Unter den Linden was a symbolic act that signified rebirth and was intended to foster unity. The replanting of the boulevard had been emblematic of the city's rebirth once before, some forty-five years earlier after

FIGURE 8.2.
Ben Wargin's barge *Gräbendorf* loaded with five pine trees in front of the eastern wing of Humboldt University on Unter den Linden, Berlin, 1995. (Photo by Barbara Esch. Landesarchiv Berlin, F Rep. 290 [02] Nr. 0377109)

FIGURE 8.3.
Rubble removal on Unter den Linden near the intersection with Friedrichstraße,
Berlin, 1945. (Landesarchiv Berlin, F Rep. 290-06-06, Nr. 014)

the Second World War. The war had destroyed many trees and had left most others
along the boulevard scorched and charred (figure 8.3). Located near the Reich chancellery on Wilhelmstraße, Unter den Linden had been within the central target zone
of the air raids flown by the U.S. Air Force in the early months of 1945 and within the
last area of defense during the final battle of Berlin. Unter den Linden was the last
stronghold that Soviet soldiers had to conquer house by house, and the central promenade was used after the end of the battle as an open-air camp for wounded German
soldiers. To indicate the city's hope to rebuild a better future, only a few months after
the end of the war, the Berlin Magistrate had begun the boulevard's replanting with
silver lindens from the Späth tree nursery in Treptow (figure 8.4).

Over the centuries, Unter den Linden, or the Lindens, as Berliners affection-
ately came to call the tree-lined boulevard, had become a recognizable horticultural
landmark in the city center. The Lindens fostered unity and a collective identity
within a disparate citizenry. By the early eighteenth century—after the city had grown
to incorporate Dorotheenstadt and Friedrichstadt, which had developed north and
south of the linden allée in the late seventeenth century—the tree rows and the space
they enclosed had become a connecting element between these new parts of the city.
The Lindens again provided a space of collective identification for citizens of what

FIGURE 8.4.
Newly planted silver lindens and flexible seating on Unter den Linden near the
intersection with Friedrichstraße, 1949. (bpk Bildagentur/Kunstbibliothek Staatliche
Museen Berlin/Willy Römer/Art Resource, NY)

had before been independent towns when Berlin became Greater Berlin in 1920. Now
the Lindens stood at the heart of the new metropolis, providing citizens with an elegant tree-lined promenade that belonged to all. It had become the "axis of Berlin's
soul," as Swiss architect Robert Rittmeyer remarked in his entry for the 1925 design
competition for the boulevard's redesign. His colleague Max Heinrich described Unter
den Linden on the same occasion as the "backbone" and "main nerve" of the entire
capital.[6]

The Lindens' role in fostering a collective identity through the provision of a
beautiful, shady public space went hand in hand with its protection. Over the centuries, electors, kings, citizens, and urban governments repeatedly came to the defense
of the linden avenue. Whenever people saw the trees endangered or dying, ways were
found to protect or replace them. At the beginning of the eighteenth century, King
Frederick I issued a new patent especially for the conservation of the linden allée that
required all adjacent inhabitants to care for the trees in front of their homes, barred
citizens from fouling the area with debris and liquor, and prevented roaming pigs
from churning the earth and damaging the trees. In the nineteenth century, many
trees began to suffer and die from gas poisoning due to the introduction of gas lighting
in 1826. Others lost their vitality due to dust and the lowering of the water table resulting from construction projects including the Landwehrkanal. Consequently, many

trees had to be replaced. By the late nineteenth century, overcrowding and increasing traffic also threatened the trees. Now lined with banks, ministries, exclusive stores, cafés, and restaurants, Unter den Linden had become the most fashionable street in Berlin, attracting crowds, especially on the weekends. When some architects argued that the linden trees should be removed to make more space for the crowds and traffic, care was taken to relocate the two tree rows to the sidewalks rather than remove them entirely. And when the Ernst Wasmuth publishing house held a big competition for Unter den Linden's redesign in 1925, city planner Werner Hegemann considered the fate of the linden trees one of the most important questions, warranting several expert opinions including those of garden architects Paul Kache and Otto Werner. Nevertheless, two of the seven designs that were ultimately awarded prizes called for abolishing the linden trees. In other cases, designers eliminated two rows but retained the other two, for example, moving them to the sidewalks and opening the central promenade either for vehicular transportation or for an open-space design that maintained unobstructed views toward the Brandenburg Gate in the west and the monument to Frederick the Great in the east (figure 8.5).[7]

Some designs suggested that the linden canopies be boxed to create more space (figure 8.6), or that traffic be fed into underground tunnels to enable the central promenade to be maintained (figure 8.7). Only a few of the thirty-three designs submitted, however, tried to combine tree rows with new high-rise buildings. Most designs envisioning the latter were uninterested in the living matter that had given the avenue its name. There was one notable exception: the winning design entry by the young Dutch architect Cornelis van Eesteren entitled "Balance" (figure 8.8). Van Eesteren's balance was that between old and new. He suggested retaining the representative historic monuments at either end of the boulevard—the Brandenburg Gate and the monument to Frederick the Great—and the nearby palace plaza. In contrast, the commercial thoroughfare in between was to be redesigned and turned into a modern center with a high-rise building marking the intersection with Friedrichstraße. But despite the proposed makeover, van Eesteren kept the two double rows of linden trees. Seemingly unconsciously, therefore, the young designer also maintained the historic life cord between the monuments on either end of Unter den Linden.[8]

Heinrich Pape of the Biological Reich Institute for Agriculture and Forestry (Biologische Reichsanstalt für Land- und Forstwirtschaft) gave van Eesteren's design further uplift by stressing that it made most sense to maintain and care for the trees planted along the central promenade. In contrast, trees along the sidewalks would suffer from a lack of light as well as other hindrances. The Magistrate's building officer Roman Friedrich Heiligenthal finally deliberated that a "horticultural solution" for the design challenge was most appropriate, given the relatively little costs involved in such a solution and its easy and fast realization, in contrast to the many changes in architectural ideas that a long-term construction project would be subject to. Although none of the competition designs were realized, the discussions they triggered affirmed the recognition that the trees constituted a historic monument that was worth maintaining.[9]

Another redesign of the street cross-section *was* finally realized beginning in

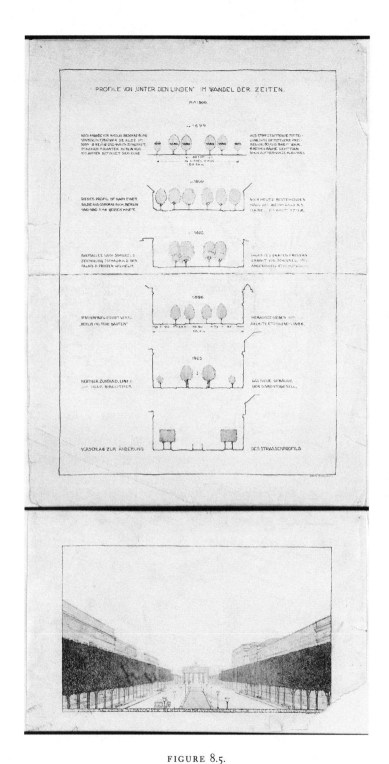

FIGURE 8.5.
Alexander Klein and Ernst Serck, "Monumentalstraße Unter den Linden," competi-
tion entry for the redesign of Unter den Linden, 1925, winner of the third prize.
The images show past and proposed street sections of Unter den Linden, and a
perspectival rendering of the design proposal looking toward the Brandenburg Gate.
(Architekturmuseum TU Berlin, Inv. Nr. 7972, 7974)

FIGURE 8.6.
A perspectival rendering of "Seelenachse," Robert Rittmeyer's competition entry for the redesign of Unter den Linden, 1925. (Architekturmuseum TU Berlin, Inv. Nr. 7909)

FIGURE 8.7.
N.N., "Neues Ortsgesetz," competition entry for the redesign of Unter den Linden, 1925. This sectional perspective shows the proposed tunnel for through traffic. (Architekturmuseum TU Berlin, Inv. Nr. 19384)

FIGURE 8.8.
Cornelis van Eesteren, "Gleichge-wicht," competition entry for the redesign of Unter den Linden, 1925. This bird's-eye perspective looks from Pariser Platz toward the east. (Collection Het Nieuwe Instituut/collection Van Eesteren-Fluck & Van Lohuizen Founda-tion, Amsterdam/archive: EEST, 8.66-2)

1934, a year after the Nazis seized power. The drives on either side of the central prom-enade were broadened to more easily accommodate increased traffic and Nazi pa-rades, and a new substructure was installed below the central promenade so that it could withstand heavy army vehicles. Although the promenade's width was decreased, care was taken to ensure that it could accommodate parades of twelve-person col-umns. Construction of the new north-south light rail line finally led the Magistrate to fell or move and replant all existing trees, but after citizen complaints they were hur-riedly replaced with young silver lindens on the occasion of the 1936 Olympics. The bleak and scrawny specimens that were planted only three months before the games' opening ceremony provoked no little mockery among Berlin citizens, who began call-ing their famous boulevard Barren Tree Allée (Kahlbaumallee). The young trees al-most disappeared between the tall posts with long vertical flags that the Nazis installed for this and other occasions, such as Berlin's seven-hundred-year celebration. For Mus-solini's 1937 state visit, additional steles mounted with the Nazi stylized Reich eagle and a swastika flanked the central promenade, overtowering the trees completely (figure 8.9). The Nazis turned Unter den Linden into their triumphal propaganda backbone—and not only the trees were deemed undesirable and disruptive. In December 1938, the regime prohibited Jews from using the northern sidewalk of the Lindens between the university and the Zeughaus in front of the Reichsehrenmal (Neue Wache), a prohibition added to existing access restrictions to parks, sports grounds, swimming pools, theaters, cinemas, concert and lecture venues, and museums.[10]

After the first replanting efforts immediately following World War II that lasted

FIGURE 8.9.
Unter den Linden's central promenade decorated with steles mounted with the
Nazi Reich eagle and swastika for Benito Mussolini's visit, 1937.
(bpk Bildagentur/Art Resource, NY)

until 1957, Unter den Linden, now located in East Berlin, was subject to several further makeovers during the city's division. In 1949 the Magistrate drafted the so-called
Lindenstatut, which, among other things, determined that six rows of lindens were to
line the boulevard between Staatsbibliothek and Pariser Platz. Although never signed
into law and although four tree rows would remain the maximum, the Lindenstatut
gave expression to the generally pervasive appreciation of the boulevard's trees. Several
entries in the 1957 West German urban design competition, which covered large parts
of East Berlin territory, envisioned a pedestrianized Unter den Linden, an idea that,
given its Western origin, never gained attention in East Berlin and was also quickly
rejected by the competition jury.[11]

Already by the late 1940s, Unter den Linden was recognized as part of East Berlin's
central east-west axis along which reconstruction was to occur in the Soviet-occupied
zone. But after the city's division, the boulevard quite literally became a dead end. The
Berlin wall and its deadly border zone behind the Brandenburg Gate turned Unter
den Linden into East Berlin's central parking space. Finally, to render the boulevard
attractive again on the occasion of the reconstruction of the city center, beginning in
1964 the central promenade obtained a new permeable surface and paving as well as
benches and chairs designed by artist blacksmith Fritz Kühn (figures 8.10 and 8.11).
Every effort was undertaken to achieve a unified and even-aged canopy, although that
endeavor soon proved futile. In the 1970s, tree grates and watering channels were installed; in the winter of 1978–79 the Magistrate finally prohibited de-icing salt on Unter
den Linden to save its damaged trees; and in 1980 a new storm water system was im-

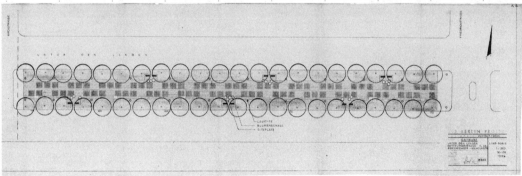

FIGURE 8.10.
Perspective and detail of plan for the redesign of the central promenade of Unter den
Linden, 1963. (Landesdenkmalamt Berlin, NGA, Mitte, 705, Unter den Linden)

plemented. Nevertheless, by 1988 only about half of the linden trees lining the boulevard were considered healthy. Throughout the decades of Berlin's division, dead and
dying trees had been replaced with various linden species, including silver linden,
littleleaf linden, largeleaf linden, and the cultivars and hybrids *Tilia x flaccida*, *Tilia x
vulgaris* 'Pallida,' and *Tilia x vulgaris*.[12]

The desolate condition of East Berlin's renowned representational boulevard,
also visible in the deterioration and destruction of its paving, caused the Institute for

FIGURE 8.11.
Unter den Linden's central promenade looking toward the Brandenburg Gate, 1967.
(Photo: agk-images, Paul Almasy/AKG252343)

Urban Design and Architecture to initiate studies for its improvement only a year before the fall of the wall. Landscape architects Stephan Strauss, Sigrid Eckmann, and
Petra Hennig recommended measures that could turn Unter den Linden back into a
"metropolitan boulevard with international flair" fit for concerts, exhibitions, demonstrations, fairs, convoys, marches, and processions. The landscape architects pointed
out that improving the boulevard's appearance and comfort would also involve its
comprehensive replanting to reestablish an even, unified tree canopy, work that was
anticipated to begin in 2001.[13]

When the wall fell in November 1989, the trees differed in health, shape, color,
and form, but at least citizens and visitors could still walk under the lindens. Due to
the continual replanting of trees along four to six rows since its 1647 inception by

Frederick William the Great Elector, the tree-lined boulevard is one of the oldest landmarks in Berlin's center. Thanks to its appropriation by the respective governments and citizenries throughout the ages, Unter den Linden both preceded and outlasted the Prussian winter palace, the city's architectural landmark from the eighteenth century to the Second World War. Whereas the socialist government considered the palace emblematic of royal power and therefore blew up its ruins after the war, Unter den Linden was replanted and subject to special care and treatment. The trees appeared apolitical and benign, and they were neutral enough eponyms for the boulevard to maintain its colloquial name, which had quickly replaced the original names Neustädtische Allee and Dorotheenstädtische Allee. Over time the Lindens had been sufficiently malleable to adjust to the agenda of whatever government was in power. The same inherent malleability of the trees' nature made them one of the longest-lasting elements of Berlin's urban design.

The trees of Unter den Linden were not alone among the city's lindens in tying East and West together again. Since the beginning of systematic street tree planting in Berlin in the nineteenth century, lindens had clearly dominated the street scene throughout the entire city. With German reunification, Berlin was to become an even greener linden city than ever before. In 1994 lindens therefore featured prominently in the plan for future street tree planting that the city had commissioned in the early 1990s. The plan's objective was to equalize the street tree canopy across the city and determine priorities and necessary improvements in ongoing street tree–planting practice. It suggested that lindens were to be planted along the east-west thoroughfares—for example, Karl-Marx-Allee. They were to remain the dominant species along the radial thoroughfares leading in and out of the city. In short, the major transportation routes in the inner city were to be lined with linden trees, even if concessions were made to maintain and plant some road sections and streets with London plane, maple, and oak. To achieve the necessary species variety in the city overall, the plan suggested that other species, like ash, pink hawthorn, birch, and poplar, could be planted as the leading species along inner district roads.[14]

## Berlin Air

Across the German-speaking states and countries since the nineteenth century, tree experts had applauded lindens for their resilience and aesthetic. While the genus proved popular in general, silver linden and Crimean linden stood out as the most desirable street tree species. Implicitly alluding to the standardized "normal tree" in scientific forestry, which had turned forests and their trees into quantifiable entities for timber production in the early nineteenth century, city gardener Carl Heicke even went so far as to describe the Crimean linden as the "normal street tree."[15]

Silver and Crimean lindens not only develop pleasant canopy shapes without pruning, they keep their leaves long into the autumn and resist air pollution. Furthermore, in the late summer, the sweet scent of the linden flowers complements these pleasant and practical attributes, even if they do not smell as strong as some of the native linden species like littleleaf linden. After their fresh green foliage appears throughout Berlin's city streets in May, at the end of the summer the linden trees dispense whiffs of sweet perfume into the air across the city. Not surprisingly, this was one of the trees' celebrated attributes noted by a journalist in Berlin's *Lokal Anzeiger* in July 1941, when many of the newly planted silver linden blossomed for the first time with, as the author attested, flowers larger and in stronger colors than those of other lindens in the city. Their flowers provided the city, which by the early twentieth century had accepted composer Paul Lincke's 1899 military march "Berliner Luft" (Berlin Air) as its unofficial anthem, with a special identifying quality. The name Unter den Linden therefore invoked both the spatial character of its luscious protective bright green canopy and the more ephemeral scent of sweet-smelling linden blossoms.[16]

Dried and used as an infusion, linden flowers had also been a household remedy for centuries, used to alleviate colds, fevers, headaches, and high blood pressure. During World War I, linden flowers were collected in cities at home and in the occupied territories. Besides the flowers' medicinal properties, during wartime linden infusions also substituted for tea, which could no longer be imported. In the spirit of "war botany," which sought to identify and test native plant species for their nutritional content as a contribution to the war effort and Germany's nutritional autarky, botanists had also entertained the idea that linden fruit and seeds could be used for producing oil. Some optimistically argued that the tree's oil could compete with the best olive oil and that it had the additional advantage of retaining its liquid state at low temperatures. Furthermore, lindens were considered "fat trees" because they transformed part of the starch in their wood and bark into fat during the winter. However, both oil production from fruit and seeds and fat extraction from wood and bark proved to be too cumbersome and inefficient to be used on a large scale. When pharmacologist Hermann Thoms used a linden tree from Berlin's Royal Botanical Garden to verify previous study results, it appeared that the tree produced much less fat than previous assessments had attested.[17]

In contrast, the collection of linden flowers turned out to be viable and could save the country from importing them. Yet, although it had been a common wartime practice elsewhere, in Berlin the Department of Gardens was reluctant to allow children and others to cut and collect the flowers out of fear that the trees would be damaged. It did concede, however, that its own laborers would collect them during their pruning work and sell them to the Society for the Cultivation and Use of Medicinal Plants (Gesellschaft für Anbau und Verwertung von Heilpflanzen), founded in 1920 by the retired Colonel Louis Ferdinand von Mathiessen in Berlin. Given their flowers' entrancing scent and medicinal and nutritional value, linden trees on occasion turned Berlin into an orchard that could be harvested.[18]

## The Silver Linden Case

In the late 1980s it was these same flowers that began to cause consternation, fear, concern, and confusion among scientists, nature preservationists, landscape architects, and city officials. Large quantities of dead bumblebees and honeybees were found below silver linden trees, causing a surge of emotional responses and disputes between scientists and landscape architects on the one side and environmentalists and nature preservationists on the other. In this "environmental crime thriller," which became a topic in professional, scientific, and popular media, environmentalists and nature preservationists accused the non-native silver linden and Crimean linden of being the culprits. Therefore, when the first linden trees on Unter den Linden were replanted in the early 1990s, there were disagreements and doubts about whether silver linden was the correct choice.[19]

Based on a 1975 study by the zoologist Günter Madel, who had claimed that the bees were intoxicated by mannose, a sugar apparently found only in the nectar of silver and Crimean linden, environmentalists argued that these non-native linden species were harmful to indigenous wildlife, particularly bees. Flavored with both environmentalist fundamentalism and botanical xenophobia, their demands included the felling of all silver and Crimean lindens and a prohibition on planting these species. Well aware that two of the most important street tree species were endangered by these radical arguments and demands, landscape architects and scientists in both East and West Germany sought to solve the mysterious silver linden case.[20]

Their activities began in 1987 and continued throughout the months and years of German reunification. In East Germany in 1989–90, after discussions had been held among ecologists, environmentalists, and landscape architects some years earlier, Humboldt University student Wolfgang Blenau surveyed and synthesized existing scholarship on silver lindens and the death of bumblebees under the direction of bee expert and professor Günter Pritsch. Indeed, reports of the phenomenon dated back to the 1920s, and studies seeking to find the cause of the bees' death had begun as early as the 1930s. But while the majority of scientists concluded that the fatalities were related to the linden flowers' nectar, in particular the sugar mannose, other theories pointed toward predator animals or simply the end of the bees' natural life cycle.[21]

In West Germany as well, research was undertaken in the early 1990s to figure out the role of linden trees in the mysterious bumblebee deaths. Sponsored by the Ministry of the Environment, Regional Planning, and Agriculture (Ministerium für Umwelt, Raumordnung und Landwirtschaft) of North Rhine-Westphalia, a group of scientists led by Bernhard Surholt at the Institute of Zoology of Münster University conducted an extensive study based on data collected in the summers of 1990, 1991, and 1992. The scientists found high death rates of bees under late-blossoming linden species, especially if these were isolated trees, but were unable to confirm what was causing the bees to die. The scientists were unable to verify that mannose played a role in the fatalities: the substance was not found in the flowers of the trees under which the most fatalities occurred, nor did captive bees die when fed with mannose. The researchers ultimately concluded that the insects starved to death because of an increased competition for fewer nectar sources at the end of the growing season. Pollinator populations grew during the summer when nectar sources were more effusive, but urbanizing areas often lacked late-flowering pollinator plants. To protect the bees, the scientists therefore argued, it would be beneficial to plant *more* silver and Crimean lindens rather than felling them, as demanded by many environmentalists.[22]

This was good news for landscape architects and urban governments opposing the overzealous environmentalists who wanted to ban non-natives from gardens, parks, and streets. Little more than a month had passed after the fall of the wall when Klaus-Dietrich Gandert, who had been on the case as chairman of the East German Central Expert Committee of Dendrology and Garden Architecture since 1987, shared his concerns regarding fundamentalist environmentalism in the silver linden case with his West German colleagues. While some East German experts had considered the environmentalist spirit an outgrowth of the capitalist world order, the quest to save some of the most resilient street tree species from eradication now united scientists and landscape architects in East and West Germany. But despite the now unhindered exchange of data and publications and several landscape architects' and researchers' efforts to lead evenhanded discussions about plant use, some environmentalists still could not be persuaded of the benefits of non-native lindens even ten years after the initial debates had erupted.[23]

On Unter den Linden between 1990 and 1994, seventeen silver linden and six common linden were planted. At the time, the city followed the advice of the Municipal Conference of Garden Directors, GALK, to plant silver linden. Sufficient proof of a causal effect between bee feeding on silver linden nectar and bee fatalities was still lacking, GALK had argued. But although the silver linden, hailed in 1959 as Berlin's quintessential street tree, had been found not guilty, species preferences had changed. In the following two decades, mostly common linden, its cultivar 'Pallida,' and the littleleaf linden cultivar 'Greenspire' were used to replace dead or decaying trees along Unter den Linden.[24]

Of all linden species, hybrids, and cultivars, 'Pallida' and 'Greenspire' were considered particularly amenable to the urban climate due to their canopy shape and relatively little care requirements compared to other lindens. The 'Pallida' cultivar, first propagated in the Späth nurseries, reportedly derived from a tree planted along Emperor William II's Siegesallee, a park boulevard built in Tiergarten around the turn of the century. Littleleaf linden 'Greenspire,' a much later breed from the Princeton nurseries in New Jersey, had originated from crossing the two unpatented littleleaf linden varieties 'Boston' and 'Euclid.' Nursery owner William Flemer III filed his patent in November 1960 (see figure I.2). Among the unique character traits of his new "carefree" linden tree were its straight and upright habit and growth, which did not require correction or staking and could resist wind damage; its attractive foliage, resistant to foliar aphids; and its symmetrical, oval canopy, which did not require frequent pruning. But although 'Greenspire' appeared to combine all character traits of the model street tree, in 2017 Berlin's Department of Gardens no longer considered any lindens carefree. Lindens in general were no longer a first choice for new plantings in the city due to their care requirements. The exception was Unter den Linden, where lindens were still used because of the boulevard's heritage status.[25]

# Unifying the Street Tree Canopy

Whereas the preoccupation with the trees of Unter den Linden was a way to symbolically stitch East and West Berlin together again in its urban center, the city's tree canopy at large also required attention and renewal. In 1993 the city counted around 390,000 street trees altogether, but in the late 1920s, the city had boasted circa 468,000 street trees. To maintain Berlin's status as the greenest German city work needed to be done, especially in the eastern districts, where liquid magnesium chloride had been applied to roads until 1990 and where thousands of gas leaks were damaging the trees. Although tying East and West Berlin together again after forty years of division was an unprecedented task, tackling urban planning and design comprehensively for the entire city was not completely new. Already in the early twentieth century, when Berlin consolidated with seven adjacent towns, the metropolis of Greater Berlin had elicited new plans that looked at the metropolitan region and included comprehensive green open-space plans. Parks and tree-lined streets were among the urban spaces that belonged to all and that could foster a new collective identity for a formerly dispersed citizenry of multiple cities.[26]

Reluctant to accept the reality of a divided city, in the immediate postwar years and up until the late 1950s, landscape designers had devised reconstruction plans for Berlin in its entirety, building on what had been developed in the prewar years. Reinhold Lingner advocated for a productive city landscape that enveloped a loose urban fabric. In contrast, his colleague Georg Pniower argued for a design based on Johann Heinrich von Thünen's nineteenth-century circular city diagram. Thünen's concentric land-use diagram consisted of belts of different productive land uses such as cattle ranching, crop rotation agriculture, husbandry, intensive agriculture, forestry, and market gardening. The most intense productive land use was carried out nearest to the city. Applying Thünen's concept to garden and landscape design, Pniower suggested that the aesthetic and psychological effects of urban green space should increase in intensity the nearer to the city center they were located. In terms of trees, this meant for Pniower that color and permanence was to be brought into the urban fabric by planting evergreens and pest- and pollution-resistant non-native species and cultivars. Urban vegetation demanded "the utmost finesse of horticultural technique and materials," and the choice of tree species had to respond not only to ecological and environmental requirements but to the citizens' social needs and aesthetic concerns.[27]

But even the most optimistic of planners finally had to accept realpolitik. Although in the 1950s West Berlin's new green open-space plan still included Berlin in its entirety (plate 16), it was clear by then that the city was divided. The 1957 capital city competition organized by the West German government and the Berlin Senate was a provocation to East Berlin. Disregarding the city's division, the call for entries invited European architects and urban designers to submit design proposals for a central area that covered more than a thousand hectares on largely East Berlin territory. The area under study stretched from the newly designed Hansaviertel in the west to Alexanderplatz in the east, and from Oranienburger Tor in the north to Mehringplatz in the south. A forced optimism underlay West Berlin's mayor Otto Suhr's declaration that the competition was to contribute to the country's reunification. The East German capital did not wait long to respond in style, initiating in 1958 its own Competition for the Socialist Redesign of the Capital Center of the German Democratic Republic—open only to architects from the Eastern Bloc. The hope of West Berlin's director of gardens Fritz Witte that a cooperative way forward could be found with like-minded East Berlin colleagues working on the greening of a dispersed urban fabric in their part of town now appeared in vain.[28]

When the political situation changed more than thirty years later, two urban governments with different administrative systems had to be consolidated, an undertaking that caused significant growing pains and personal hardships. At the time of German reunification in 1990, the street tree distribution throughout the city was very uneven, leading Michael Fietz to draw up a street tree density map for the city (figure 8.12). Not only did the street tree distribution vary significantly between East and West Berlin, which had around 140,000 and 252,000 street trees respectively, it also varied between districts. Analysis of aerial surveys showed that West Berlin's inner-city districts of Tiergarten, Wedding, and Schöneberg had the least street trees, with only between 6,000 to 12,000 each. Charlottenburg, Wilmersdorf, Steglitz, Tempelhof, and Neukölln each had between 20,000 and 30,000 street trees, whereas around 31,000 to 46,000 trees were standing along the streets of the peripheral and more loosely built-up districts of Zehlendorf and Reinickendorf respectively. More meaningful, however, were the numbers that were counted per street kilometer and per one hundred inhabitants. It was the wealthier districts—Charlottenburg, Wilmersdorf, Zehlendorf, Schöneberg, Steglitz, and Reinickendorf—that had more than 93 trees per street kilometer. With the exception of Schöneberg, these West Berlin districts also had more than 12 street trees per one hundred inhabitants, with a maximum of 33 trees per one hundred inhabitants in Zehlendorf. In contrast, the density of street trees in East Berlin's inner-city districts, particularly Mitte and Friedrichshain, was only half that of the Western inner-city areas. Whereas an average of around 87 trees per street kilometer were lining West Berlin's streets, only around 60 trees per kilometer were standing along East Berlin's streets. The dominant species were linden, maple, oak, London plane, horse chestnut, and black locust. Throughout the years of German division, West Berlin's district Departments of Gardens had planted an increasing number of oaks due to their resistance to pollution. Dutch elm disease, which had affected trees in Germany beginning in 1918, put an end to the use of elms as street trees in 1935. Nativist arguments that surged in the 1980s led to a temporary dwindling of the planting of the non-native black locust and London planes in West Berlin. But given their resistance to adverse urban conditions and ecologists' arguments for site specificity and against xenophobic emotions, these species have since then regained popularity.[29]

With the end of the Cold War, Berlin became a temporary laboratory for the study of past and potential future best management practices in street tree care. The comparisons made between street tree care and management in East and West Berlin in the early 1990s found that in general, despite the disproportionately lower number

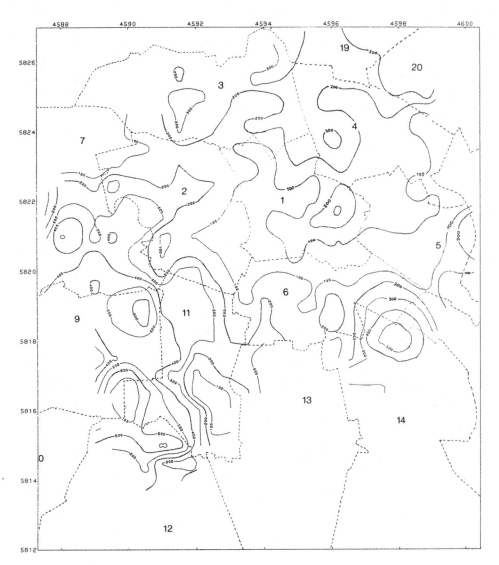

Abb. 3-1: Flächendichte der mittelalten und alten Straßenbäume in der Berliner Innenstadt für 1985 (Westteil) bzw. 1990 (Ostteil) als Isoliniendarstellung, Maßstab ca. 1:73.000; erhoben aus CIR-Luftbildern, berechnet mit dem Programm DIGIT/ISOLI (aus FIETZ et al. 1988; FIETZ, GLASER & MUNIER 1991, verändert). Dichte: Straßenbaumanzahl pro ½ km² (ca. 707 x 707 m) in Hunderterschritten; Baumaltersstufen: 15 - 40 und über 40 Jahre; Abbildungskoordinaten: Hoch- und Rechtswerte; Bezirksgrenzen: gestrichelte Linien; Bezirke: 1 = Mitte, 2 = Tiergarten, 3 = Wedding, 4 = Prenzlauer Berg, 5 = Friedrichshain, 6 = Kreuzberg, 7 = Charlottenburg, 9 = Wilmersdorf, 10 = Zehlendorf, 11 = Schöneberg, 12 = Steglitz, 13 = Tempelhof, 14 = Neukölln, 19 = Pankow, 20 = Weißensee.

FIGURE 8.12.

Map showing the density of old and middle-aged street trees in Berlin's center, based on 1985 data for the former West Berlin and 1990 data for the former East Berlin. (Michael Fietz, "Art- und schadensbedingtes Abbildungsverhalten von Berliner Straßenbäumen auf Color-Infrarot-Luftbildern," *Berliner geowissenschaftliche Abhandlungen*, series D 188, vol. 2 [1992]) of street trees in East Berlin, fewer trees had been felled for the widening of streets and there were more old, decaying, and dying trees. In West Berlin, many of these trees would have been felled already due to traffic safety requirements. In East Berlin, district Departments of Gardens complained about the inferior tree material they had had to work with and the consequent loss of trees during the phase when the plants take root (plate 17).

Already in the early 1980s, horticulture student Carola Wünsche, advised by East Berlin's director of plant protection Horst Kühn and Humboldt University professor Klaus-Dietrich Gandert, had criticized the low success rate of tree plantings in East Berlin in her thesis, "Care, Protection, and Increase of Street Trees in Berlin's District Prenzlauer Berg." Following the socialist ideal of a close link between academic research and its practical application, the thesis had challenged the VEB Kombinat Stadtwirtschaft to reach the international average of an 80 percent success rate in trees taking root.[30]

Gas leaks and the effects of de-icing salt, which in East Berlin had been used indiscriminately until the fall of the wall, especially along Schönhauser Allee, Prenzlauer Allee, and Greifswalder Straße—streets used by party officials who commuted between their homes north of the city and the city center—were causing trees to die in this part of the city, where the air quality, as a result of the use of coal, was clearly inferior as well. In West Berlin, where these issues were largely considered under control, the urine of an increasing number of dogs was causing concern instead.[31]

Street tree damage from gas had been a deficiency especially noted in Prenzlauer Berg in the 1980s. At the time, Prenzlauer Berg was undergoing the transition from illuminating gas to natural gas, leading to more than seven thousand gas leakages that caused the death of around two thousand street trees, among them all the mature trees along Hufelandstraße. To improve the situation, concrete planters with trees were positioned along Hufelandstraße and other affected streets (figure 8.13 and plate 18). After reunification it took a while to resolve the issue. The Berlin gasworks (GASAG), which had merged with the former East Berlin gas supplier (Erdgas AG) in 1993 and aspired to renew all leaking East Berlin gas pipes by 2001, refused to pay for damage that had been caused before German reunification.[32]

But gas leaks and battles about responsibilities for related tree death were an old story in Berlin. Landscape gardener and director of royal gardens Peter Joseph Lenné had already noted the damaging effects of illuminating gas in the middle of the nineteenth century. Once street trees on Unter den Linden had begun to die after the introduction of gas lighting in 1826, his colleagues began to study the phenomenon in the ensuing decades. After a first study by chemist Heinrich Poselger in 1869, the Berlin Magistrate and later the garden director Gustav Meyer initiated studies of the effects of illuminating gas on trees in Berlin's botanical garden. Studies subsequently continued at the tree nursery Späth. The scientists found that most trees whose root systems were exposed to varying quantities of gas over varying periods died either in the year of treatment or in the following year. They concluded that root exposure to large quantities of gas, and the exposure of roots to small gas quantities but over long

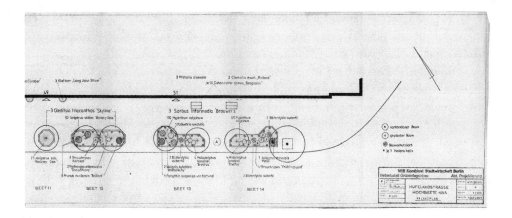

FIGURE 8.13.
Detail of planting plan for the containers on Hufelandstraße, fourth phase of
construction between Hans-Otto-Straße and Am Friedrichshain, August 1989.
(Landesdenkmalamt Berlin, NGA, Pankow, Prenzlauer Berg, 005,
Hochbeete Hufelandstraße, 4. BA)

time periods, were harmful to most trees. It was also found that gas was particularly
damaging during the vegetation period. Again and again, street trees suffering from
gas leaks had to be replaced, in 1882, for example, in Potsdamerstraße, in Landsberger
Allee, and along Luisenstädtischer Kanal.[33]

## Bau(m)boom

By 1995 many street trees were visibly suffering not only from soil and air pollution but
from the massive reconstruction efforts in the city, which had also caused the sinking
of the water table. On the occasion of the European Year of Nature Protection, the
grassroots Tree Protection Association (Baumschutzgemeinschaft Berlin e.V.) launched
a tree-protection campaign, the Bau(m)boom. A pun involving the words *Bauboom*—
construction boom—and *Baum*—tree—Bau(m)boom described the objective of up-
ping the ante in the city's tree planting and protection. Concerned about the speed
with which the city was changing and the lack of attention paid to urban trees, the association argued that "Berlin without trees is like a sprinter without air." A film, news
bulletins, and demonstrations against the felling of trees were to inform the public
and engage it in tree protection and care. In May, tree activists planted twenty tree pits
with wild roses, firethorn, mustard, and flax to demonstrate how tree pits could be both
embellished and prevented from drying out. The association suggested the institution
of tree-care contracts (Baumpatenschaften) administered by a coordinating body as
part of its city tree manifesto (Stadtbaummanifest). The manifesto also demanded the
establishment of a tree-protection fund to which both private and public developers
would have to contribute. This was to be used to compensate for tree damage when
the person responsible could not be identified or charged. The manifesto refuted the

use of tree static measurements for decisions about the felling of trees, and it promoted the establishment of district tree-preservation commissions consisting of members of citizen initiatives, interest groups, and associations.[34]

Although some of these ideas were reminiscent of the GDR's state-sponsored citizen initiatives, the lack of funds and the increasing domination of Berlin's urban development by private investors ultimately led the urban government to engage in public-private partnerships and implement citizen-sponsored tree-planting campaigns. Garden director Erhard Mahler had raised an alarm in the early 1990s. If the city was to balance its street tree density across all districts and maintain its green image, the 1993 budget of 8 million German marks was far too small. Mahler estimated that the removal and replacement of dead and dying trees alone would require 23 to 27 million marks annually. By the end of the 1990s, the Departments of Gardens were paying between 2,500 and 3,500 German marks for every new street tree, a cost that their budgets could not cover.[35]

Thus, whereas in the nineteenth century garden architects had considered it important that street tree planting and care be undertaken by the urban government, a century later, governmental responsibilities were consciously being eroded. To increase the number of street trees, in 1998 the Department of Gardens founded Lebensräume-Lebensbäume e.V., an association that was to elicit and manage one hundred tree-care contracts by the fall of 1998. But despite its success and its early prominent sponsors, such as Berlin's ice hockey club Eisbären, GASAG gasworks, and Dresdner Bank AG, the street tree loss could not be arrested. Ten years later *Der Tagesspiegel* reported that the city was losing around fifteen hundred urban trees annually.[36]

In 2012 the Berlin Senate Department for Urban Development and the Environment therefore undertook a concerted action to turn the situation around. Together with the district departments, it began the City Tree Campaign to raise private funds for the planting of trees. For every 500 euros that citizens donated for the cause, the city contributed around 1,200 euros, the amount required for the planting of one tree and its three-year care and maintenance. Although the city was still controlling tree planting and care, in 2017 its dependence on private funds was increasing. It was also outsourcing more and more of the necessary labor.[37]

This development had been set into motion already in the late nineteenth century when most street trees were still fully under municipal control. At the time, the planting of street trees contributed to Berlin's metamorphosis into an international modern metropolis partaking in the global economy. The systematic introduction of street trees in Berlin and some of its surrounding towns like Charlottenburg had begun in the second half of the nineteenth century, providing a new uniformity and cohesion among the various parts of the city and its neighboring towns, which would formally be consolidated first as an association of communes (*Zweckverband*) and finally as Greater Berlin in 1920. The establishment of a city garden director in 1870, a post filled by former Prussian court gardener Gustav Meyer until his death in 1877, was a decisive impulse for the establishment of new street tree plantings, parks, and squares. By 1876, Meyer's office was also in charge of Unter den Linden, formerly administered by the Prussian court. By 1888, Berlin counted 41,500 street trees that, following earlier plans drawn up by Peter Joseph Lenné, now lined in particular the ring roads and the broad radial avenues leading into and out of the city, where trees were also planted on medians. Many streets in new extension zones were planned with street tree planting in mind, and by 1909 seventy kilometers of streets were lined with trees. Yet, despite the increasingly systematic and standardized planting, care, and selection of street trees in early twentieth-century German cities, which was documented in a survey undertaken on the occasion of the 1904 horticultural exhibition in Düsseldorf, the accompanying report noted that differing soil and air quality as well as varying climates required a site-specific treatment and choice of trees.[38]

Since then, changes to our urban climates have changed trees' phenology, forcing nature to adapt. Climate change has also led to changes in the tree species that are considered viable choices for our cities today. Thus, with the exception of the trees of Unter den Linden, in the future few lindens may be left to scent Berlin's air.

# Epilogue
## Street Trees of the Future

When scientists in the United States began warning of climate change and a warming atmosphere in the 1950s, street tree planting was one of the antidotes put forward. In a widely noted paper presented at the 1958 National Conference on Air Pollution in Washington, DC, Chauncey D. Leake, president of the American Association for the Advancement of Science, warned of the "tremendous increase in the blanket of carbon dioxide" that would "inevitably tend to increase heat capture from the sun." He asked what we would do once the huge polar ice caps melted, leading to "gradual rise of our oceans, drowning out still further our shore lines." Tree planting was perhaps one possible countermeasure. "Maybe 10 trees planted for every automobile, with 100 for every truck, would help," Leake suggested. But regardless of the possible reduction of air pollution, he surmised, cities could "certainly benefit from such tree planting."[1]

For centuries street trees have been valued for both their aesthetic and climatological functions, but it has been especially the latter that have driven people to action again and again. Street trees began to take firm root in the urban environment in the nineteenth century once their climatological and hygienic functions in cities had been recognized during the public health movement. Trees did more than provide shade, bind dust, and cleanse the air. They also produced life-saving oxygen and contributed to ozone production, which in late nineteenth-century discussions about urban vegetation was still considered in its antiseptic rather than its toxic capacity. Ozone was discovered by chemist Christian Friedrich Schönbein in the 1840s before the advent of the germ theory; many believed that the ozone purportedly produced by trees was an effective protection against miasma and harmful gases, and that it could help to cure tuberculosis. Little did these tree promoters know that one hundred years later ozone near ground level would be understood as dangerously toxic for plants and animals alike, and that a diminishing ozone layer in the stratosphere would be threatening the global climate.[2]

Just as climate change has inspired tree-planting campaigns in cities across the world, it has also motivated studies to identify appropriate street trees of the future. Urban foresters, ecologists, and other scientists have informed us that climate change, bringing with it more frequent and stronger storm events, floods, and drought, will inevitably lead to new management and care regimens and new species selections in cities. The resulting consequences include changes in biodiversity and aesthetics. Ecologists anticipate trait shifts, meaning that species with character traits that function and look differently from our current common street tree species will come to dominate the urban landscape.[3]

What, then, will our cities look like in the future? Will a few selected species be used in cities around the globe? Given the urban heat island effect—the rise of temperatures in cities due to human activities and their modification of land surfaces— German tree experts suggested as early as the 1970s that cities in different climate and vegetation zones should exchange tree lists. After all, trees surviving New York City would also grow in Stuttgart, they argued. Will cities in the future still have site-specific tree selections? If so, what will be the future tree species distinguishing Berlin from Rome, New York from Los Angeles? What will be the species that will give Berlin its unmistakable atmosphere? Will Unter den Linden remain Unter den Linden? Or will it become Unter den Spanischen Eichen (Under the Spanish Oaks)? Or perhaps the name will continue to recall the boulevard's vegetal origins even though 'Green Vase' Japanese zelkova has replaced the lindens?[4]

Throughout history, various site conditions and climates in cities as well as the availability of trees have led to changes in the species planted over time. Species were tested for their adaptability to the urban environment, but their selection was also subject to aesthetic preferences and the human weakness for novelty, as American landscape gardener Andrew Jackson Downing complained in 1847. There is "a fashion in trees," he commented, that "sometimes has a sway no less rigorous than that of a Parisian *modiste*." Rooting for indigenous species with patriotic zeal, he decried the early nineteenth-century "Lombardy poplar epidemic" that had led to the use of this species in the case of "nine-tenths of all the ornamental planting of that period." In the 1870s, Lombardy poplars were among the trees planted in Washington, DC, under Governor Alexander R. Shephard. Like the Carolina poplars of the initial tree selection, Lombardy poplars grew quickly but their shallow roots damaged pavements and curbing. They were soon considered what landscape forester Wayne C. Holsworth later called a "weed tree." Like the initial silver and ash-leafed maples, which were short-lived and brittle, and attracted insects, the poplars were soon given up due to their damaging effects. Washington's first tree selection was replaced with American elms and lindens, oriental planes, ginkgos, sugar maples, and pin and red oaks.[5]

In mid-nineteenth-century New York City, the fashion shifted from ailanthus to elms, maples, and lindens, and by the early twentieth century to oriental planes. A hybrid between oriental plane and American sycamore, the London plane tree still led New York City's tree census in 2006. In Manhattan in the spring of 1942, 115 of a total 339 newly planted trees were London planes, their number surpassed only by the 124 silver lindens planted on Eighty-Sixth West and East Streets, West End Avenue, and Riverside Drive. Of a total of 1,149 street trees planted four years later, 690 — over 50 percent — were London planes.[6]

Studies to identify the most hardy and adaptable street tree species as well as planting success rates were undertaken in various cities in Germany and the United States beginning in the 1970s, but it took until the early 2000s for studies to focus on the effects of future climate change. In 2009 researchers at the Bavarian Office for Viniculture and Horticulture (Bayerische Landesanstalt für Weinbau und Gartenbau) began Project Urban Green 2021 (Projekt Stadtgrün 2021), a study that sought to understand which street trees should be planted in cities in the changing climatological conditions of the future. The project studies thirty tree species and cultivars that have been planted in three Bavarian cities with different climates, assessing the trees' adaptation and adaptability to extreme urban conditions. Many of the species and cultivars, like *Tilia mongolica*, *Malus tschonoskii*, *Acer rubrum* 'Somerset,' and the hybrid *Ulmus* 'Rebona,' hail from the Americas or Asia. Currently, they are not frequently planted on the country's city streets, but this may change. In fact, recommendations for German cities on the street tree list published regularly by the Municipal Conference of Garden Directors have changed over the last decades: species and cultivars like the sun-loving zelkova have been added, and others have been down- or upgraded. Whereas between 1978 and 1983 the silver linden was listed as suitable, by 1986 that verdict was changed to "suitable with reservations." 'Greenspire' littleleaf linden, in contrast, was first listed in 1978 as "suitable with reservations" but from 1983 onward was deemed "most suitable."[7]

In light of climate change and trees' importance for public health, social welfare, and justice, the question of who plants, cares for, and pays for trees in our cities remains as pertinent as ever. Although the responsibilities for street tree planting in New York City and Berlin have been different, in the early twentieth century most tree experts and urban reformers in the United States and Germany shared the conviction that street trees should be managed by municipal governments. A century later, cities in the two countries again shared some of their approaches to street tree planting. This time, however, planting activities reflected the erosion of governmental control due to an increasingly global neoliberal management regime.

In the United States in the early twentieth century, municipal ownership and management of street trees was unquestioningly preferred, as only "systematic, uniform, practical and economical development and treatment" could fend off the "devastating attacks of insects and fungus pests" and guarantee the "improved appearance of our streets in all parts of the city." In short, tree experts maintained that municipal government of street trees meant "a better sanitary condition for the people as a whole." Street trees were, as Ephraim Porter Felt and S. W. Bromley explained at the 1930 National Shade Tree Conference, an unalienable right. "The right to shade trees"

FIGURE E.1.
"The Appeal of the Innocents," created by the Newark Shade Tree Commission.
("Educating the People to Care for the Trees," *Park and Cemetery and Landscape
Gardening* 22, no. 12 [1913]: 302)

was, the tree experts believed, implicit in the equal opportunities for the pursuit of
happiness laid out in the Declaration of Independence.[8]

But because the street tree movement was a matter of public education and reform, city forester William Solotaroff reasoned as early as 1912, "In a democracy where
there is a strong individualism and . . . lack of co-ordination and interdependence between adjoining municipalities and states," progress would be slow. It would be slow
indeed, and as the New York City stories in this book show, it would always remain a
public-private affair that was heavily reliant on private funds and initiative. In other
U.S. cities, the situation was not dissimilar (figure E.1). For example, in the early twentieth century the Boston Department of Public Grounds enlisted citizens in the planting of street trees. Once a year it gave away one thousand trees grown in the municipal
nursery, encouraging people to plant them in their front yards and along sidewalks.
Through emulation, the city argued in 1906, "whole streets and neighborhoods are
today liberally planted that would not otherwise have had a tree." In Chicago, city
forester Jacob H. Prost initiated the "penny tree campaign," which sold trees to children in public schools for a penny each, encouraging them to plant their trees on the
sidewalk or in the yard in front of their home. Tying into Arbor Day activism, this
initiative had by 1914 resulted in the planting of 1 million trees within the city's bound-

aries. Such projects were necessary because the early city foresters only on rare occasions had an adequate budget and workforce.[9]

In New York City throughout the twentieth and into the twenty-first century, the city government and nonprofit groups again and again encouraged citizens to help the forestry forces by funding and caring for trees (plate 19). Not only in periods of crisis — during World War II and after hurricanes or particularly bad insect infestations — but even in years without extreme events it was impossible for the Department of Parks to maintain the city's street trees without the help of private contractors or to plant new trees without private funds. In 1956 272 climbers and pruners had to care for 536,000 street trees, leading the department to argue for an annual appropriation of $200,000 for the next five years to hire contractors to help with the tree work. The department also encouraged property owners to finance new street trees in front of their homes. Owners were required to apply to the department for a permit that would ensure the department's control over tree selection and location. Residents could then buy a tree for $75 from one of the suggested contractors, who would plant and care for the tree for one year, after which the department would take over. In the 1950s, street tree planting in Manhattan was largely determined by private interests and funding, particularly by philanthropist and health activist Mary Lasker. Complaining about the quality of flower plantings she had sponsored in other boroughs, in 1959 Lasker requested the planting of street trees along Park and Second Avenues in Midtown Manhattan instead.[10]

Observing the situation at the time, urbanist and housing expert Charles Abrams lamented that "efforts to make the city more interesting and natural in its setting have been left to the entrepreneur with little supplement by official action." Only during periods when special federal funds and grants were available — for example, Works Progress Administration funds after the world financial crisis in the 1930s and funds from the State Aid Post War Program in the 1940s — was the department able to complement its permanent workforce with temporary laborers paid with public funds. WPA laborers, for example, planted around four thousand trees throughout all boroughs in 1939. Two years later, they planted oriental plane trees to replace trees along various avenues and streets in Manhattan.[11]

In Berlin as well, job employment programs were used in the 1930s to increase the city's tree cover. The approximate synchronicity of two disastrous events — the financial crisis and the effects of Dutch elm disease — turned out to be a blessing in disguise. If in 1931 dead and affected elm trees still had to be felled and removed with the help of private property owners due to the lack of workers, by June 1933 the situation had changed. In the wake of the new law for the reduction of unemployment (Reichsgesetz zur Verminderung der Arbeitslosigkeit), the mayor's office announced the availability of new jobs for workers replacing dead street trees. As laid out in chapter 6, postwar job-creation programs provided much of the labor of replanting in both East and West Berlin. But the idea to systematically enlist citizens in street tree planting and care on a regular basis was unique to East Germany, where these activities were used in the creation of a "participatory dictatorship." After the fall of the wall, grassroots initiatives similarly argued for citizen participation in tree planting and care, but these suggestions were shot down by the Department of Gardens. Even if the city quickly conceded to indirect citizen participation, garnering private funds to finance tree planting in the reunited city, it was clear that supervision of planting and care needed to remain under municipal control. Already in the late nineteenth century, when many of Berlin's registered street trees had been planted and were cared for by private property owners, tree experts like Carl Hampel had demanded that property owners should be outright prohibited from planting street trees. Only government control could enable pleasant, ordered, and harmonious street tree plantings. Similar to early twentieth-century New York City practice, Hampel demanded that municipal governments should either plant and care for the trees themselves or issue planting permits that included strict orders on tree selection and location.[12]

Once established, throughout the twentieth century street trees' municipal management remained unquestioned except for the East German Participate! efforts during German division. Only after the fall of the wall, when other municipal services like water, gas, and electricity supply began to be privatized, did lack of funds and the increasing domination of Berlin's development by private investors lead the urban government to engage in public-private partnerships and implement citizen-sponsored tree-planting campaigns.

Not dissimilar to New York City's Million Trees campaign, which began in 2007 with the objective of planting 1 million trees by 2017, Berlin's more recent City Tree Campaign asked citizens to take over part of the costs of tree planting (plate 20). But in contrast to New York City, where citizens participated in tree surveys and where they were permitted to adopt and care for street and park trees, in Berlin, all street tree care remained with the city. In both cities, large business entities became involved in sponsoring street trees. Whereas in Berlin this development began after the fall of the wall and involved mostly local sponsors, New York City's public-private street tree history included multinational corporations like Toyota, a lead sponsor for Million Trees. Toyota's sponsorship was one of its local advertising initiatives undertaken across the world to highlight the corporation's green politics combating climate change. The history of multinational corporations' involvement in New York City's tree-planting campaigns reaches back to the late nineteenth century. Among the first trustees and members of the city's Tree Planting Association were notable industrialists, financiers, and business leaders including Edward Cooper, J. Pierpoint Morgan, W. Bayard Cutting, and William Collins Whitney. Their financial empires already spanned the world in the late nineteenth century, and they may even have been aware of urban trees' entanglement in far-reaching ecological networks. Although these men were concerned at the time about the aesthetics of their business headquarters and the local rather than the global climate, then as now street trees stood for the melding of nature, culture, science, technology, and the economy.

In 2007 New York City was for the first time able to put numbers to some of the more comprehensive values and benefits of its street trees with the help of various free software tools that had recently been introduced by scientists and foresters from the USDA Forest Service and the University of California–Davis. Known as i-Tree, these continually updated and expanded tools depend on satellite photography and geospatial data to enable the assessment of ecosystem services—what East German scholars in the 1960s called trees' "biological-technical" functions. These are carbon sequestration, pollution reduction, decrease of storm water runoff, and the reduction in energy consumption that are provided by single trees as well as entire urban forests. With the help of Stratum, one of the initial computer programs developed for the purpose, New York City's Department of Parks determined that its street trees provided an annual benefit of about $122 million. Although i-Tree's computer models and evaluations assess only trees' tangible benefits, leaving aspects like the aesthetics and cultural meanings of trees, their impact on mental health, and the reduction of crime unaccounted for, the computer software can provide numbers that have successfully been used for tree advocacy across the United States, not only in New York City.[13]

Like many other cities, at the beginning of the new millennium New York City and Berlin aspired to grow their tree canopy cover. The number of newly planted trees therefore had to be higher than the number of trees removed each year because of injuries, decline, and death. Concern about street tree mortality had been on the rise in the 1960s and 1970s, when increasing numbers of trees died from the effects of air and soil pollution, bad planting and care, lack of water, and vandalism. Trees were dying at an increasingly young age as well. In 1966 New York City's Department of Parks reported that the life expectancy of a street tree was forty to fifty years, as opposed to one hundred years for trees outside of cities. By the late 1970s, scientists had ascertained that the life expectancy of street trees in Washington was eighteen years and in Boston only ten. A 1991 survey of street trees in twenty cities, including New York City, established thirteen years as the average life span of an inner-city street tree. In 1995 the life expectancy of street trees in Manhattan was reported as seven years.[14]

In all respects, by the end of the twentieth century, street trees had lost a vital aspect attributed to them by American and German tree experts earlier in the century: trees' luscious, "thrifty"—that is, strong and healthy—canopies that unfolded their functions in the air space and occupied minimal real estate.[15]

By the end of the second millennium, street trees had become part and parcel of urban obsolescence and its economy of change. Yet, because of trees' dynamic nature, life cycles, and limited life spans, on the one hand, and their site-specificity, localism, and relative permanence on the other, they both embrace and resist change. Street trees exhibit the ambiguities of modernity. Their relatively slow growth resists speed and change, yet their inherent malleability is subject to rational management and the application of new technologies. They are appreciated for the durable physical spaces they create as well as for the temporary, ephemeral, and more intangible mental spaces and effects they produce. Trees' inbuilt a priori obsolescence—death— has saved them from becoming truly obsolescent in our cities and has enabled them to persist. Every tree death provides a literal and figurative space, an opening and a moment to decide whether and how the tree should be replaced. The opportunities this decision has offered societies and persons of subsequent generations has often led to the replanting of trees and therefore to their continuous presence in urban space, as in the prominent case of Unter den Linden. Indeed, as organisms that adapt to urban conditions, that develop over time in ways that are often unpredictable but are also the result of techno-scientific production, and that offer a material afterlife as wood, wood pulp, and compost, street trees are significant and long-standing modern elements in our urban environments. They are, as the histories of street trees in New York City and Berlin in this book show, relevant parts not only of our constructed urban environments but also of our urban social, cultural, and political lives.

# ABBREVIATIONS

## Archives and Collections

| | |
|---|---|
| BAB | Bundesarchiv Berlin-Lichterfelde |
| BC | Broadsides Collection, New-York Historical Society, New York |
| CA | Central Archives, American Museum of Natural History Archives, New York |
| CAP | Charles Abrams Papers (microfilm, 1974), Department of Manuscripts and University Archives, John M. Olin Library, Cornell University, Ithaca, NY |
| CH | Commissioner Heckscher Subject Files, New York City Municipal Archives |
| CHR | Commissioner Hoving Reference Files, New York City Municipal Archives |
| DPR | Department of Parks and Recreation, General Files, New York City Municipal Archives |
| GFK | George Frederick Kunz Papers, Huntington Library, San Marino, CA |
| IRS | Leibniz-Institut für Raumbezogene Sozialforschung, Erkner |
| LAB | Landesarchiv Berlin |
| LDB | Landesdenkmalamt Berlin |
| NGA | Archiv der Naturschutz- und Grünflächenämter, Landesdenkmalamt Berlin |
| NGABP | Depot des Naturschutz- und Grünflächenamtes Berlin-Pankow |
| NL Funcke | Nachlass Walter Funcke, Staatsbibliothek Berlin |
| NL Gandert | Nachlass Klaus-Dietrich Gandert, Studienarchiv Umweltgeschichte, Institut für Umweltgeschichte und Regionalentwicklung e.V., Hochschule Neubrandenburg |
| NL Greiner | Nachlass Johannes Greiner, Leibniz-Institut für Raumbezogene Sozialforschung, Erkner |
| NL Pniower | Nachlass Georg B. Pniower, Universitätsarchiv Dresden |
| NYCHAC | New York City Housing Authority Collection, La Guardia and Wagner Archives, New York |
| NYCPDAR | New York City Parks Department Annual Reports, https://www.nycgovparks.org/news/reports/archive#ar |
| NYCPPR | New York City Parks Press Releases, https://www.nycgovparks.org/news/reports/archive#pr |
| ORR | Office of the Messrs. Rockefeller Records, Business Interests, Series C, 1886–1961, Investments: Rockefeller Center, Inc., Rockefeller Archive Center, Sleepy Hollow, NY |
| PCR | Parks Council Records, Avery Architectural & Fine Arts Library, Drawings and Archives Department, Columbia University, New York |
| PT | Papiertiger, Archiv und Bibliothek der sozialen Bewegungen, Berlin |
| RFWC | Robert F. Wagner Collection, La Guardia and Wagner Archives, New York |
| RMR | Robert Moses Records, New York City Municipal Archives |
| STUG | Studienarchiv Umweltgeschichte, Institut für Umweltgeschichte und Regionalentwicklung e.V., Hochschule Neubrandenburg |

UDL             Wettbewerb zur Umgestaltung der Straße Unter den Linden, Berlin, 1925,
                Architekturmuseum, Technische Universität Berlin
WLPRPR          Women's League for the Protection of Riverside Park Records, New-York
                Historical Society, New York

## Newspapers

BZ              *Berliner Zeitung*
CDT             *Chicago Daily Tribune*
ND              *Neues Deutschland*
NYAN            *New York Amsterdam News*
NYDN            *New York Daily News*
NYDT            *New-York Daily Times*
NYT             *New York Times*
NZ              *Neue Zeit*
TS              *Der Tagesspiegel*
WS              *Washington Star*

# NOTES

## Introduction

1. G. A. Schulze, "Welches ist der Zweck der Strassenbäume im Innern der Grossstadt und wie erfüllen sie denselben?" *Sammlung gemeinnütziger Original-Vorträge und Abhandlungen auf dem Gebiete des Gartenbaues*, 2nd ser., no. 4 (1881): 5–13 (13); F. L. Mulford, *Street Trees*, U.S. Department of Agriculture Bulletin no. 816 (Washington, DC: Government Printing Office, 1920), 12; C. B. Whitnall, "Environmental Influence of City and Regional Planning," *Parks & Recreation* 11, no. 4 (1928): 240–46; Alfred MacDonald, "City Forestry in Connection with a Park System," *Parks & Recreation* 7, no. 2 (1923): 154–59 (155); Frank S. Santamour Jr., "Breeding and Selecting Better Trees for Metropolitan Landscapes," in *Better Trees for Metropolitan Landscapes, Symposium Proceedings, USDA Forest Service General Technical Report NE-22* (Darby, PA: Forest Service, 1976): 1–8 (1); Charles A. Stewart, "Developing an Urban Forestry Program: The Role of the Consultant," in *Proceedings of the National Urban Forestry Conference, November 13–16, 1978, Washington DC*, vol. 2 (Syracuse: State University of New York, 1978): 714–19 (718).
2. L. D. Cox, "Design with Respect to Street Tree Planting," *Eighth National Shade Tree Conference, Proceedings of Annual Meeting* (1932): 42–48 (42–43); Dieter Hennebo, "Städtische Baumpflanzungen in früher Zeit," in *Bäume in der Stadt*, ed. Franz Hermann Meyer (Stuttgart: Ulmer, 1977), 11–44 (44).
3. Joanna Dean, "Seeing Trees, Thinking Forests: Urban Forestry at the University of Toronto in the 1960s," in *Method and Meaning in Canadian Environmental History*, ed. Alan MacEachern and William J. Turkel (Toronto: Nelson Education, 2009), 236–53; George R. Cook, "Report of the General Superintendent of Parks," in *Second Annual Report of the Board of Park Commissioners of the City of Cambridge* (Cambridge, MA: Harvard Printing, 1894), 71–72; Cecil C. Konijnendijk, Robert M. Ricard, Andy Kenney, and Thomas B. Randrup, "Defining Urban Forestry—A Comparative Perspective of North America and Europe," *Urban Forestry & Urban Greening* 4 (2006): 93–103.
4. Dean, "Seeing Trees"; Gordon King, "The Role of Education in Urban Forestry-Arboriculture," in *Proceedings of the National Urban Forestry Conference, November 13–16, 1978*, 681–85; G. A. Schulze, "Welches ist der Zweck der Strassenbäume im Innern der Grossstadt und wie erfüllen sie denselben?" *Sammlung gemeinnütziger Original-Vorträge und Abhandlungen auf dem Gebiete des Gartenbaues*, 2nd ser., no. 5 (1881): 5–16; Verein Deutscher Gartenkünstler, *Allgemeine Regeln für die Anpflanzung und Unterhaltung von Bäumen in Städten* (Berlin: Gebrüder Bornträger, 1901); Elbert Peets, "Street Trees in the Built-up Districts of Large Cities," *Landscape Architecture* 6 (October 1915): 15–31 (15); G. A. Schulze, "Welches ist der Zweck der Strassenbäume im Innern der Grossstadt und wie erfüllen sie denselben?" *Sammlung gemeinnütziger Original-Vorträge und Abhandlungen auf dem Gebiete des Gartenbaues*, 2nd ser., no. 6 (1881): 11–15 (15).
5. William Penn Mott Jr., "A Growing Street Tree Problem," *Parks & Recreation* 33, no. 7 (1950):

244–45; S. R. Bassett, J. E. Kuntz, and G. L. Worf, "Maple Decline in Urban Wisconsin I: Rootlet Mortality Associated with Maple Decline," *Wisconsin Urban Forester* 3, no. 2 (1980): 1–3; Arthur H. Westing, "Sugar Maple Decline: An Evaluation," *Economic Botany* 20, no. 2 (1966): 196–212; Peter Kiermeier, "Entwicklung von neuen Strassenbaumsorten in den USA," *Das Gartenamt* 30, no. 2 (1981): 92–106; David Karnosky, "Testing the Air Pollution Tolerances of Shade Tree Cultivars," *Journal of Arboriculture* 4, no. 5 (1978): 107–10; David F. Karnosky, "Chamber and Field Evaluations of Air Pollution Tolerances of Urban Trees," *Journal of Arboriculture* 7, no. 4 (1981): 99–105; David F. Karnosky, "Screening Urban Trees for Air Pollution Tolerance," *Journal of Arboriculture* 5, no. 7 (1979): 159; Santamour, "Breeding and Selecting Better Trees for Metropolitan Landscapes"; Frank S. Santamour Jr., "The Selection and Breeding of Pest-Resistant Landscape Trees," *Journal of Arboriculture* 3, no. 8 (1977): 146–52; Gary L. Koller, "New Trees for Urban Landscapes," *Arnoldia* 38, no. 5 (1978): 157–72; Norman A. Richards, "Modeling Survival and Consequent Replacement Needs in a Street Tree Population," *Journal of Arboriculture* 5, no. 11 (1979): 251–55; Henry D. Gerholt, "Landscape Trees from Other Countries," *Journal of Arboriculture* 5, no. 7 (1979): 156.

6.  C. S. Harrison in *Independent Farmer*, cited as "Intelligence in Trees," *Tree Talk* 2, no. 3 (February 1915): 6; A. F. Bartlett, "Trees Commit Suicide," *Tree Talk* (1935): 5–9; Orville W. Spicer, "Do Trees 'Commit Suicide'?" *Tree Talk* 9, no. 1 (1933): 31; Massachusetts Forestry Association, *Practical Suggestions for Tree Wardens* (Boston: Massachusetts Forestry Association, 1900), 7.

7.  "Preserving Leaves for Compost," 23 October 1935, NYCPPR; "Save the Leaves," *Parks & Recreation* 21, no. 3 (1937): 112–13; Gustav Rohde, "Kompostierung der Stadtabfälle," *Wissenschaftliche Zeitschrift der Humboldt-Universität zu Berlin, mathematisch-naturwissenschaftliche Reihe* 4, no. 5 (1954–55): 417–33. For the argument of the city as a locus of production of natural resources, see Sonja Dümpelmann, "'Tree Doctor' vs. 'Tree Butcher': Material Practices and Politics of Arboriculture in Chicago," in *Landscript 05: Material Culture*, ed. Jane Hutton (Berlin: Jovis, 2017), 90–113.

8.  Mrs. Miriam Cooper to Department of Parks, 4 May 1944, DPR, Queens, 1944, folder 20; Francis Cormier, 13 January 1941, DPR, Queens, 1941, folder 27; Erle Cocke Jr., "Factors for a Stronger America," in *Thirty-Fourth National Shade Tree Conference, Proceedings of Annual Meeting* (Wooster, OH: Collier, 1958): 3–11 (10).

9.  Lists in DPR, Administration, 1943, folder 28.

10. Gustav Allinger, *Das lebendige Grün in Bauentwürfen: Bäume und Bauten* (Berlin: Institut für Allgemeine Bautechnik, 1946); Heinrich Friedrich Wiepking, *Umgang mit Bäumen* (Munich, Basel, and Vienna: BVL, 1963).

11. B. Pflanzer, "Straßenbepflanzung," *Mitteilungen der deutschen Dendrologischen Gesellschaft* 35 (1925): 124–26; Max Bromme, "Sind Alleen und Einzelbäume an der Straße in ihrer Beziehung zum Verkehr, zur Landschaft und zum Ortsbild künftig lebensberechtigt?" *Mitteilungen der deutschen Dendrologischen Gesellschaft* 49 (1937): 165–73 (173).

12. Henry Lawrence, *City Trees* (Charlottesville and London: University of Virginia Press, 2006), 287; "A Waste of Wealth," *NYT*, 5 August 1888, 4.

13. Peets, "Street Trees in Built-up Districts," 15; Laurie D. Cox, *A Street Tree System for New York City, Borough of Manhattan: Report to Honorable Cabot Ward, Commissioner of Parks, Boroughs of Manhattan and Richmond, New York City* (Syracuse: Syracuse University, 1916), 56–57; Frederic Shonnard, *Street Forestry: Report on Selection, Planting, Cultivation and Care of Street Shade Trees* (Yonkers: Meadows Bros., 1903), 19–21. Many early manuals suggested the German method of combating certain types of insects by preventing them from climbing up tree trunks through attaching paper bands covered with a sticky insect lime—*Raupenleim*— or with a similar compound called Dendrolene invented by E. L. Nason of Brunswick, New Jersey. For the use of Raupenleim in American arboriculture, see *Annual Report of the Tree Planting and Fountain Society of Brooklyn, N.Y., December 1895* (Brooklyn: Eagle Book and Job Printing Department, 1896), 37; Bernhard Eduard Fernow, *The Care of Trees in Lawn,*

*Street and Park* (New York: Henry Holt, 1910), 145; Elbert Peets, *Practical Tree Repair* (New York: McBride, Nast, 1913), 66. For a report on the subterranean irrigation of street trees that began in German cities in the 1870s, see "Subterranean Irrigation of Street Trees at Dresden," *Park and Cemetery and Landscape Gardening* 10, no. 10 (1900): 228–30. For the use of the American tree mover in Berlin and Germany, see Herbert Wichmann, "Tree-Mover," *Das Gartenamt* 19, no. 2 (1970): 72–74; A. Weil, "Der 'Tree-Mover' ein Verpflanzgerät für grosse Bäume," *Baum-Zeitung* 3, no. 2 (1969): 31–32; "Baumverpflanzen kostet 600 Mark," *TS*, 25 September 1971, 8; "200 Bäume ziehen um," *TS*, 30 November 1971, 9; *TS*, 23 May 1975, 8; Wolfang Reckling, "Erfassung des Straßenbaumbestandes Berlins, Hauptstadt der DDR; Maßnahmen zur Pflege und Erhaltung des Straßenbaumbestandes" (Diplomarbeit [master's thesis], Humboldt University, Berlin, 1974), 61–62.

14.  For a comparative analysis of New York City and Berlin, see R. Heiligenthal, "New York und Berlin," *Der Neubau* 8, no. 11 (1926): 121–31. John Y. Culyer, "Tree Planting in Berlin and New York," *NYT*, 5 March 1908, 6.

15.  "Berlin as a Model for American Cities," *Park and Cemetery and Landscape Gardening* 20, no. 8 (1910): 1; *Bulletin of the Tree Planting Association of New York City*, September 1913, 2; copy of pupil essay in WLPRPR, box 7, folder 6; Robert J. Caldwell to Helen C. Kerr, 13 May 1929, WLPRPR, box 7, folder 12. For Berlin as model tree city, see "Straßenbäume in Berlin," *Das Gartenamt* 24, no. 6 (1975): 350.

16.  Peets, "Street Trees in Built-up Districts," 18; Frederick W. Kelsey, "Tree Planting in Streets," *NYT*, 20 November 1898, 15; "New York Is Being Surveyed for Planting of Trees," *NYT*, 4 January 1914, SM6; Malcolm Howard Dill, "The Progress of Systematic Street Tree Planting in American Cities," *American City* 34, no. 3 (1926): 300–305 (304); Stephen Smith, "Trees in the City," *NYT*, 6 July 1903, 6; "Trees in Paris Streets," *Park and Cemetery and Landscape Gardening* 12, no. 5 (1902): 320–21; "A Report on the Street Trees of Hartford, Conn.," *Park and Cemetery and Landscape Gardening* 13, no. 3 (1903): 39–40; E. Böttcher, "Statistisches über die Entwickelung der öffentlichen Park-, Garten- und Baumanlagen in den Weltstädten," *Die Gartenkunst* 3, no. 10 (1901): 201–4.

17.  For Trueman and Clifford Lanham, see "Lanham, 67 Today, Has Major Role in Beautifying Capital," *WS*, 31 March 1943, B-9. For the role of Paris and Washington, DC, as model street tree cities, see William Solotaroff, *Shade-Trees in Towns and Cities* (New York: John Wiley and Sons, 1911), 235–36; William F. Fox, *Tree Planting on Streets and Highways* (Albany: J. B. Lyon, 1903), 6–7; C. Lanham, *The Tree System of Washington* (Washington, DC: CF Judd and Detweiler, 1926), 23; "Beautifying the City," *WS*, 8 May 1889; *Bulletin of the Tree Planting Association of New York City*, September 1913, 2; J. H. Prost, *Trees and Lawns for the Streets*, Special Park Commission pamphlet no. 6 (April 1914), 1; J. H. Prost, *Street Tree Planting for Illinois*, Illinois Outdoor Improvement Association pamphlet no. 2 (April 1910), 6; Henry M. Hyde, "Paris Outshines Cities of World in Tree Culture," *CDT*, 25 April 1913, 1; "Capital's Trees Finest in World," *Washington Herald*, 18 March 1914, 6; Victoria Faber Stevenson, "Capital Appears Green to Airmen," *WS*, 6 November 1927, 3.

18.  "The Trees of Washington," newspaper article, 1886, folder "Trees A-D-1971," Washington, DC, Public Library, the Washingtoniana Collection. On the one-upmanship compared to Berlin's Unter den Linden, see "Beautifying the City," *WS*, 18 May 1889, 8, and "Parks and Street Trees of Washington, DC," *Park and Cemetery and Landscape Gardening* 17, no. 2 (1907): 32–34 (34). "Shade Trees of Washington," *CDT*, 20 January 1884, 12. *Annual Report of the Tree Planting Association of New York City, 1905* (New York: Gilliss, 1906), 18.

19.  John Y. Culyer, "Spare the Trees," *NYT*, 27 September 1909, 8; Newark Shade Tree Commission, cited in *Bulletin of the Tree Planting Association of New York City*, September 1913, 2.

20.  For a history of the American chestnut and chestnut blight, see Susan Freinkel, *American Chestnut: The Life, Death, and Rebirth of a Perfect Tree* (Berkeley, Los Angeles, and London: University of California Press, 2007); Department of Parks, City of New York, *Annual Report for the Year 1913* (New York: M. B. Brown, 1914), 256–59, 274–75. For lumbering in Forest Park,

see Department of Parks, City of New York, *Annual Report for the Year 1912* (New York: J. J. Little and Ives, 1913), 269–70, 354–59 (appendix 4); Department of Parks, City of New York, Borough of the Bronx, *Annual Report for the Year 1919* (New York: M. B. Brown, 1920), 25; Department of Parks, City of New York, Borough of the Bronx, *Annual Report for the Year 1922* (New York: Herald Square, 1923), 32; "Central Park Utilizes Dead Trees," *American Forestry* 27, no. 1 (1921): 38; "Gives Fuel to the Poor: Mayor's Women's Committee Uses Wood Gathered in Parks," *NYT*, 17 February 1918, 8; "Poor May Get Free Wood Saturday," *NYT*, 4 February 1920, 6.

21. 16 September 1944; 23 October 1944; 18 September 1945, NYCPPR. Also see the documents in DPR, Brooklyn, 1944, folder 76; press release, 3 October 1944, DPR, Administration, 1944, folder 2; memorandum by borough director Warren C. Donnelly, "Storm Damage," 15 September 1944, DPR, Queens, 1944, folder 16.

22. "763. Polizeiverordnung zur Bekämpfung der Ulmenkrankheit," *Amtsblatt für den Landespolizeibezirk Berlin*, 19 September 1931, 217; Joseph Treffert to Magistrat Berlin, Abteilung Gartenbau, 22 January 1930, LAB, A Rep. 007, Nr. 68.

23. "The Life of the Modern Tree," *Parks & Recreation* 18, no. 7 (1935): 267–69. For Boston's "tear-down week," see Massachusetts Forestry Association, *Tear-Down Week*, pamphlet published in 1917 [?]. *Verordnungsblatt der Stadt Berlin* 1, no. 9, 10 October 1945, 115.

24. For the damage to trees by rock salt water, see *Annual Report of the Tree Planting and Fountain Society of Brooklyn, N.Y., December 1896* (Brooklyn: Eagle Job and Printing Department, 1897), 57. "Prevention of Nuisances by Dogs or Other Animals in Public Places, Section 227 . . . Adopted by the Board of Health, November 4, 1918," in *Municipal Ordinances, Rules, and Regulations Pertaining to Public Health, 1917–1919*, supplement 40 to the Public Health Reports, ed. Jason Waterman and William Fowler (Washington, DC: Government Printing Office, 1921), 21; [unreadable name] to Robert Moses, 7 November 1955, DPR, Manhattan, 1955, folder 37; David Schweizer to William H. Latham, 4 June 1940, DPR, Manhattan, 1940, folder 8.

25. For trees and dogs in Berlin, see *Gemeinde-Blatt der Haupt- und Residenzstadt Berlin*, 21 July 1912, 320; "Der ungepflasterte Sandstreifen . . . ist der richtige Platz für den Hund," *Neuköllner Tageblatt*, 15 July 1937, n.p.; "Zweieinhalb Straßenbäume je Hund," *TS*, 9 June 1957, 11; photo in *TS*, 7 February 1989, 9; bk, "Wunsch nach Bürgerhilfe beim Schutz der Straßenbäume," *TS*, 10 August 1989, 12; Hartmut Balder, "Hundeurin als Schadagens an Bäumen," *Das Gartenamt* 39, no. 11 (1990): 736–38.

26. For early tree damage through human urine, see "Die Baumnoth in großen Städten," *Berlin und seine Entwicklung, Städtisches Jahrbuch* 3 (1869): 96–106; C. Heicke, *Die Baumpflanzungen in Straßen der Städte* (Neudamm: J. Neumann, 1896), 7.

27. For the bee complaint, see June 1966, CHR, box 3, folder 7. For caterpillar complaints, see Joseph Posniack to Robert Moses, 26 June 1944, DPR, Brooklyn, 1944, folder 79; Mrs. L. Moore to Robert Moses, 29 June 1944, DPR, Brooklyn, 1944, folder 79; Mrs. Samuel Parver to Robert Moses, July 1948, DPR, Brooklyn, 1948, folder 4; Joseph Hyman to Department of Parks, 21 June 1949, DPR, Brooklyn, 1949, folder 26. For roosting starlings, see Mr. and Mrs. I. Weingold and Neighbors to William O'Dwyer, 18 September 1947; William H. Latham to Fairfield Osborn, 1 October 1947, DPR, Brooklyn, 1947, folder 48. "Protecting Bird Life in German Woodlands," *Park and Cemetery and Landscape Gardening* 20, no. 2 (1910): 255–56, xi; "The Birds and Outdoor Improvement," *Park and Cemetery and Landscape Gardening* 20, no. 12 (1911): 1. For bird protection in Berlin, see Georg Hengstenberg to garden director, 10 October 1928; gez. Hilsheimer, 9 February 1929; gez. Hilsheimer, 22 June 1929; Treptow district Department of Gardens, gez. Harrich, 28 October 1929, LAB, A Rep. 036-08, Nr. 5; Dr. Wagner, 29 July 1930, LAB, A Rep. 036-08, Nr. 11; LAB, A Rep. 007, Nr. 218. "Stirbt Berlins Vogelwelt aus?" *Berliner Lokal-Anzeiger*, evening edition, 16 August 1929, supplement.

28. For the effects of London plane trees, see Joseph Stübben, *Hygiene des Städtebaus* (Jena: Gustav Fischer, 1896), 415; St. Olbrich, "Allee- und Straßenbäume und ihre Verwendung,"

*Mitteilungen der deutschen Dendrologischen Gesellschaft* 17 (1906): 108–18 (114–15); Alfred Hoffmann, "Der Straßenbaum in der Großstadt unter besonderer Berücksichtigung der Berliner Verhältnisse" (doctoral diss., Humboldt University, Berlin, 1954), 55; Dr. Hoffmann, "Bäume in Stadtstrassen," *Das Gartenamt* 5, no. 5 (1956): 87–91; Harri Günther, "Über das Verhalten von Gehölzen unter großstädtischen Bedingungen, untersucht an einigen Gehölz-arten in Berlin" (doctoral diss., Humboldt University, Berlin, 1959), 68.

29. Bezirksrat to Luzie Martin, 9 February 1948, LAB, A Rep. 044-08, Nr. 274.

30. Lucke to Gartenverwaltung, 17 May 1929, LAB, A Rep. 044-08, Nr. 275; Langer, 4 November 1936, LAB, A Rep. 036-08, Nr. 126; Pertl, "Behinderung der Straßenübersichtlichkeit durch Gehölze," 6 July 1939, LAB, A Rep. 036-08, Nr. 11; Polizeipräsident to Gartenbauamt Neukölln, "Strassenbeleuchtung in der Sonnenallee," 20 August 1949; Amt für Tiefbau to Gartenamt Neukölln, "Strassenbeleuchtung," 12 September 1949, LAB, A Rep. 044-08, Nr. 275.

31. As New York was one of the biggest and densest cities in the United States, its example could help convince those skeptics who argued that trees could not survive in cities due to air pollution. See Robert C. Weinberg to David Schweizer, 15 March 1944; Robert C. Weinberg to Francis Cormier, 20 March 1944, DPR, Administration, 1944, folder 50; Santiago Sanjuan to Department of Parks and Recreation, 4 October 1965, DPR, Administration, 1965, folder 5.

# Chapter 1. Tree Doctor vs. Tree Butcher

1. "Shade Tree Dear to Them," *NYT*, 28 December 1897, 11.

2. Frederick Leland Rhodes, *Beginnings of Telephony* (New York and London: Harper and Brothers, 1929), 196–206; "How to Contend with the Tree Butchers," *Park and Cemetery and Landscape Gardening* 17, no. 4 (1907): 1.

3. "Tree Butchers," *NYT*, 17 June 1904, 8; "A Shade-Tree War in Jersey City," *NYT*, 21 July 1881, 8. "An Act in Relation to Telegraph and Electric Light Companies in Cities of This State," in *Laws of the State of New York Passed at the One Hundred and Seventh Session of the Legislature*, vol. 1. (Albany: Banks and Brothers, 1884), 647.

4. *Arboriculture* was used in early nineteenth-century French and English horticultural and agricultural treatises to describe the planting, propagation, care, and maintenance of trees in the context of rural estate management and horticulture. See, for example, John Claudius Loudon, *An Encyclopaedia of Gardening: Comprising the Theory and Practice of Horticulture, Floriculture, Arboriculture and Landscape-Gardening Including All the Latest Improvements* (London: Longman, Hurst, Rees, Orme, and Brown, 1822); Alphonse Du Breuil, *Cours élémentaire théorique e pratique d'arboriculture* (Paris: Langlois et Leclercq et V. Masson, 1846); H. I. Mabille, *Le propriétaire paysagiste; ou, Manuel d'horticulture, d'arboriculture fruitière et forestière . . .* (Paris: Librarie de L'Agriculture, Andre Sagnier, Editeur, 1869).

5. For the use of the word *straighten*, see memorandum by David Schweizer, "Straightening of Trees in Streets by Forestry Forces," 14 November 1944, DPR, Administration, 1944, folder 51.

6. "Tree-Planting," *New-York Daily Tribune*, 24 October 1845, 2; "Setting out Shade Trees," *New-York Evening Post for the Country*, 19 November 1822, n.p.; "Our Shade-Trees: Great Dearth in the City," *NYT*, 19 August 1873, 2; *Bulletin of the Tree Planting Association of New York City*, September 1913, 5.

7. Charles Thaddeus Terry, "Extracts from the Annual Report," in *Annual Report of the Tree Planting Association of New York City, 1905* (New York: Gilliss, 1906), 14–16 (15); American Association for the Planting and Preservation of City Trees, *Arbor Day Message to Boys and Girls* (Brooklyn, 1911); Tree Planting Association of New York City, *A Brief History of the Organization and the Annual Report for 1897* (New York: Tree Planting Association, n.d.), 4; Tree Planting Association of New York City, *Information Bulletin* (1910); Stephen Smith, "President's Report," in *Annual Report of the Tree Planting Association of New York City* (New York: Tree Planting Association, 1912), 9–20 (16).

8. Physician George M. Beard attributed the medical phenomenon of nervousness to "modern

civilization, which is distinguished from the ancient by these five characteristics: steampower, the periodical press, the telegraph, the sciences, and the mental activity of women." See his *American Nervousness: Its Causes and Consequences* (New York: G. P. Putnam's Sons, 1881), vi. For Pinchot, see *Annual Report of the Tree Planting Association of New York City*, 1905, 1.

9.  Stephen Smith, "Vegetation a Remedy for the Summer Heat of Cities," *Appleton's Popular Science Monthly*, 1 February 1899, 433–50; "Shade-Trees as Disinfectants," *NYT*, 7 April 1873, 4; William F. Fox, *Tree Planting on Streets and Highways* (Albany: J. B. Lyon, 1903), 6; William Solotaroff, *Shade-Trees in Towns and Cities* (New York: J. Wiley and Sons, 1911), 4; J. H. Prost, *Trees and Lawns for the Streets*, Special Park Commission pamphlet no. 6 (April 1914), 2; Emil Mische, "Street Trees," *Park and Cemetery and Landscape Gardening* 10, no. 11 (1901): 252–53.

10. Asa Gray, *First Lessons in Botany and Vegetable Physiology* (New York: Ivison and Phinney, 1857), 54; "Shade-Trees as Disinfectants"; C. B. M., "Trees in Our Streets," *NYT*, 9 April 1899, 18. For the attribution to Peirce, see "New York City Is Being Surveyed for Planting of Trees," *NYT*, 4 January 1914, SM6; *Bulletin of the Tree Planting Association of New York City*, September 1913, 1; *Annual Report of the Tree Planting Association of New York City*, 1912 (New York: Tree Planting Association, 1912), 14–15.

11. Stephen Smith, "Trees in the City," *NYT*, 6 July 1903, 6; John Y. Culyer, "The Boulevard Trees," *NYT*, 7 October 1900, 18.

12. "Trees in City Streets," *NYT*, 7 January 1900, 3; William Solotaroff, "Municipal Control of Planting and Care of Shade Trees," *Park and Cemetery and Landscape Gardening* 16, no. 11 (1907): 219; A. T. Erwin, "Municipal Control of Street Trees," *Park and Cemetery and Landscape Gardening* 19, no. 9 (1909): 153–54; "Brooklyn Wants Shade Trees," *NYT*, 17 May 1882, 3; *Annual Report of the Tree Planting and Fountain Society of Brooklyn, N.Y., December 1896* (Brooklyn: Eagle Job and Printing Department, 1897), 4. For the question of ownership and responsibility of street trees, also see "Some Court Decisions on the Ownership of Shade Trees," *Park and Cemetery and Landscape Gardening* 16, no. 12 (1906): 248–49.

13. Cornelius B. Mitchell, "Report of the Chairman of the Executive Committee," in *Annual Report of the Tree Planting Association of New York City, 1901–1902* (New York: Tree Planting Association, 1902), 10–11.

14. Datus C. Smith, "Shade Trees in the City," *NYT*, 15 March 1903, 27; "Tree Planting," *NYT*, 1 March 1903, 6.

15. Cornelius B. Mitchell, "Report of the President, February 14th, 1905," in *Annual Report of the Tree Planting Association of New York City*, 1905, 9–12 (11); "Trees in New Delancey Street," in *Annual Report of the Tree Planting Association of New York City*, 1905, 22–23; Cornelius B. Mitchell, "Report of the President," in *Annual Report of the Tree Planting Association of New York City*, 1906 (New York: Tree Planting Association, 1906), 7–8 (7); Jno. Y. Cuyler, "The New Delancey Street: Report of Committee," in *Annual Report of the Tree Planting Association of New York City*, 1906, 14; Department of Parks, City of New York, *Annual Report for the Year 1913* (New York: M. B. Brown, 1913); Department of Parks, City of New York, *Annual Report for the Year 1914* (New York: J. J. Little and Ives, 1915).

16. A Cape Cod native, Francis had graduated in 1910 with a BS from Massachusetts Agricultural College. Before joining the faculty at Syracuse, where he would eventually become professor of forest recreation and landscape extension and teach until 1937, he had worked for the Minneapolis Park System and as landscape engineer of a large nursery in the same city. "Henry Francis," *Empire Forester* (1936): 18.

17. See Henry R. Francis, *Report on the Street Trees of the City of New York*, issued by the Tree Planting Association of the City of New York (Syracuse: New York State College of Forestry at Syracuse University, 1914), 27.

18. Francis, *Report on the Street Trees of the City of New York*, 17–18, 20–22, 24.

19. Cox, a Canadian with an American high school diploma and a 1908 BS degree in landscape architecture from Harvard University, had worked for the city planning departments in Boulder, Colorado, and Detroit, and as landscape architect in the park department in Los

Angeles. He taught at the College of Forestry, where he eventually became a professor of landscape engineering, until 1946. "Laurie D. Cox," *Empire Forester* (1935), 13–14; "Laurie D. Cox," *Empire Forester* (1936), 15.

20. Laurie D. Cox, *A Street Tree System for New York City, Borough of Manhattan: Report to Honorable Cabot Ward, Commissioner of Parks, Boroughs of Manhattan and Richmond, New York City* (Syracuse: Syracuse University, 1916), 35; "Studying New York City Trees," *Park and Cemetery and Landscape Gardening* 25, no. 7 (1915): 1.

21. Solotaroff, *Shade Trees*, 231–57; Wayne C. Holsworth, "City Forestry as Park Work," *Park and Cemetery and Landscape Gardening* 32, no. 9 (1922): 222–24.

22. Laurie D. Cox, "A Street Tree System for New York City," *Bulletin of the New York State College of Forestry at Syracuse University* 16, no. 8 (March 1916).

23. Many early city foresters, like Chicago's Jacob H. Prost and Cleveland's George Rettig, were trained as landscape gardeners or landscape architects. City foresters' connection to city planning, in turn, became manifest through appointments and memberships. William Solotaroff, secretary and superintendent of the Shade Tree Commission of East Orange, New Jersey, was appointed to the Advisory Board of the *American City*, and the chairman of the Shade Tree Commission of Dallas was a member of the City Plan Institute of America. Similarly, Buffalo's city forester was a member of the City Planning Board. But city foresters also created their own venues. In 1924, they began to meet at national and regional Annual Shade Tree Conferences, which began to be formally organized as the National Shade Tree Conference in 1928. Wayne C. Holsworth, "Street Tree Planting in Its Relation to City Planning," *Parks & Recreation* 7 (July–August 1924): 650–54 (651); Wayne C. Holsworth, "Street Tree Planting in Its Relation to City Planning" (master's thesis, Harvard University, School of Landscape Architecture, 1921), 71, 89; proceedings of the first gathering were published as "First Shade Tree Conference," *Tree Talk* 6, no. 3 (1924): 6–8, 26–29; C. C. Lawrence, "The Third Annual Shade Tree Convention," *Tree Talk* 8, no. 1 (1926): 24–26. For the Shade Tree Conferences, also see Richard J. Campana, *Arboriculture: History and Development in North America* (East Lansing: Michigan State University Press, 1999), 145–80; Laurie D. Cox, *Street Tree System*, 17.

24. See Laurie D. Cox, "The Department of Landscape Engineering," *Empire Forester* (1917): 45–46 (45); "New York State College of Forestry," *Park and Cemetery and Landscape Gardening* 22, no. 11 (1913): 277; Nelson C. Brown, "Possibilities of Municipal Forestry in New York," *Bulletin of the New York State College of Forestry at Syracuse University* 14, no. 2(d) (1914): 13; George R. Armstrong and Marvin W. Kranz, *Forestry College: Essays on the Growth and Development of New York State's College of Forestry, 1911–1961* (Buffalo: Wm. J. Keller, 1961), 180, 200.

25. L. D. Cox, "Design with Respect to Street Tree Planting," *Eighth National Shade Tree Conference, Proceedings of Annual Meeting* (1932): 42–48 (42–43); Armstrong and Kranz, *Forestry College*, 208–10, 349. The first private arboricultural service companies, like the Davey Tree Expert Company in Ohio and the F. A. Bartlett Tree Expert Company in Connecticut, had sought to compensate for the lack of training opportunities by founding training institutes for their own employees in 1909 and 1923 respectively.

26. B. E. Fernow, "A Forest Policy for Massachusetts," *Forestry Quarterly* 2, no. 2 (1904): 49–76; "City Forester, or Gardener?" *Park and Cemetery and Landscape Gardening* 13, no. 12 (1904): n.p.; Solotaroff, *Shade-Trees*, 245; William Solotaroff, "The City's Duty to Its Trees," *American City* 4, no. 3 (1911): 131–34; William Solotaroff, "The City's Duty to Its Trees (Concluded)," *American City* 4, no. 4 (1911): 166–68; *Forestry Quarterly* 2, no. 1 (November 1903): 33; Elbert Peets, *Practical Tree Repair* (New York: McBride, Nast, 1913), i–ii; Malcolm Howard Dill, "The Progress of Systematic Street Tree Planting in American Cities," *American City* 34, no. 3 (1926): 300–305; A. T. Erwin, "Forester or Tree Warden," *Park and Cemetery and Landscape Gardening* 19, no. 9 (1909): 152, viii–ix.

27. Wayne C. Holsworth, "Street Tree Planting in Its Relation to City Planning" (*Parks &*

*Recreation*), 1; Holsworth, "Street Tree Planting in Its Relation to City Planning" (master's thesis), 2; O. W. Spicer, "Shade Tree Legislation," *Twelfth National Shade Tree Conference, Proceedings of Annual Meeting* (Boston, 1936), 66–74.

28. A. F. Bartlett called his tree workers "dendricians" to indicate that they were trained experts. Gilbert S. Jones, "What's a 'Dendrician'?" *Tree Talk* (1935): 29; Peets, *Tree Repair*, i–ii; 233–38.

29. *Annual Report of the Tree Planting and Fountain Society of Brooklyn, N.Y., December 1895* (Brooklyn: Eagle Book and Job Printing Department, 1896), 46–47, 66–67; *Annual Report of the Tree Planting and Fountain Society, 1896*, 12; *Annual Report of the Tree Planting and Fountain Society of Brooklyn, N.Y., December 1897* (Brooklyn: Eagle Book and Job Printing Department, 1898), 11; J. H. Prost, *Street Tree Planting for Illinois*, Illinois Outdoor Improvement Association pamphlet no. 2 (April 1910), 7; City of Dallas, Forestry Department, *The Ordering and Planting of Shade Trees*, bulletin no. 1 (October 1917), 1; Fox, *Tree Planting*, 23; H. A. Surface, "Tree Fakers in Pennsylvania," *Tree Talk* 2, no. 3 (February 1915): 14; Bernhard E. Fernow, *The Care of Trees in Lawn, Street and Park* (New York: Henry Holt, 1910), 123; "Tree Butchery," *Tree Talk* 3, no. 2 (November 15): 37; "Quack Tree Surgery," *Park and Cemetery and Landscape Gardening* 23, no. 2 (1913): 1; "Operations of the Quack Tree Doctor," *Park and Cemetery and Landscape Gardening* 25, no. 6 (1915): 165.

30. "Street Trees in the Bronx," *NYT*, 12 June 1914, 12; Tree Lover, "Bronx Street Trees: Suffer from Careless Pruning by the Park Department," *NYT*, 16 June 1914, 8; Thomas W. Whittle, "Bronx Trees: Not Butchered, but Saved If Possible, Mr. Whittle Says," *NYT*, 24 June 1914, 10.

31. *Annual Report of the Tree Planting and Fountain Society, 1896*, 18; Solotaroff, *Shade-Trees*, xi, cited in Jacob H. Prost, "Shade Trees in Cities," *CDT*, 15 April 1911, 10. *Annual Report of the Tree Planting and Fountain Society, 1895*, 30–31; Holsworth, "Street Tree Planting in Its Relation to City Planning" (*Parks & Recreation*), 25.

32. Fox, *Tree Planting*, 18, and footnote on p. 18; G. E. Stone, "The Clogging of Drain Tile by Roots," *Torreya* 11, no. 3 (1911): 51–54; "Detroit Report on City Tree Planting," *Park and Cemetery and Landscape Gardening* 20, no. 3 (1910): 273–76, xii (275); A. T. Hastings, "The Carolina Poplar as a City Shade Tree," *Park and Cemetery and Landscape Gardening* 23, no. 11 (1914): 208–10; A. T. Hastings, "The Carolina Poplar as a City Shade Tree," *Park and Cemetery and Landscape Gardening* 23, no. 12 (1914): 227–28.

33. Florence Finch Kelly, "To a Little Ailanthus Tree," *NYT*, 12 August 1925, 20.

34. D. J. Browne, "On the Choice of Trees and Shrubs for Cities and Rural Towns," in *Assembly, Report of the New York Department of Agriculture*, no. 150 (Albany, 1846): 376–404 (389); Andrew Jackson Downing, *A Treatise on the Theory and Practice of Landscape Gardening* (New York and London: Wiley and Putnam; Boston: C. C. Little, 1841), 206–8; Andrew Jackson Downing, "Trees in Towns and Villages," *Horticulturalist* 1, no. 9 (1847): 393–97 (397). "The Ailanthus, or Tree of Heaven," *NYDT*, 27 December 1852, 3; "Deferred Nuisances," *NYDT*, 1 August 1855, 4. "A Popular Cry," *NYDT*, 5 July 1855, 2; "The Poisonous Ailanthus," *NYT*, 30 June 1859, 1; "Our Upas Trees," *NYDT*, 9 July 1855, 6; "Our Upas Trees," *NYDT*, 2 July 1855, 4. For the public perception of the tree being poisonous, also see "Ailanthus Poisoning," *NYT*, 22 July 1877, 10.

35. "Our Upas Trees," *NYDT*, 9 July 1855, 6; "Our Upas Trees," *NYDT*, 2 July 1855, 4; "The Awful Ailanthus," *NYT*, 4 July 1859, 8; Andrew Jackson Downing, "Shade Trees in Cities," *Horticulturalist*, 1 August 1852, 345–49 (345–46); "A Friend of the Ailanthus," *NYDT*, 10 July 1855, 4; "A Plea for the Ailanthus," *NYDT*, 12 July 1855, 8; "The Ailanthus Tree," *NYDT*, July 25, 1855, 6; "The Ailanthus," *NYDT*, 18 July 1855, 2; "Julia Pleads for the Ailanthus," *NYDT*, 5 July 1855, 2; *Annual Report of the Tree Planting and Fountain Society of Brooklyn, N.Y., December 1898* (Brooklyn: Eagle Book and Job Printing Department, 1899), 25; "Men Armed with Wires Brushing Trees in the Parks," *NYT*, 10 August 1902, 32.

36. Department of Parks, City of New York, *Annual Report for the Year 1913*, 49–50; Cox, *Street Tree System*, 51–53; "The Ailanthus—A Word for the Defense," *NYT*, 7 July 1859, 3. After the survey initiated by Cox in 1927, the borough of Manhattan's Department of Parks undertook

another street tree census. See Department of Parks, City of New York, Borough of Manhattan, *Annual Report for the Year 1927* (New York: I Smigel, 1928), 17–18. For a disagreement within the Department of Parks and Recreation in 1944, see memorandum by David Schweizer, 30 June 1944; Francis Cormier to Mrs. James Devlin, 22 June 1944; Mrs. James Devlin to Robert Moses, 19 June 1944, DPR, Administration, 1944, folder 48. While senior landscape architect Cormier pledged for the tree's positive character traits, arguing that the ailanthus was not so "lowly" after all, park engineer Schweizer was against the tree's use given its susceptibility to insects and wilt disease, and its propensity to disseminate. T. B. Richards, "A Plea for Ailanthus," *NYT*, 26 July 1934, 18; Arthur Guiterman, "Favoring the Plane Tree," *NYT*, 1 August 1934, 16; Lawrence A. Tassi, "The Hardy Ailanthus," *NYT*, 30 July 1934, 12. Memorandum by David Schweizer, "Ailanthus Trees," 8 September 1947; and memorandum by David Schweizer, "Removal of Ailanthus Trees," 11 September 1947, DPR, Administration, 1947, folder 33. Leonard Weinles to Commissioner Hoving, 10 August 1966; and Carl J. Schiff to Leonard Weinles, 18 August 1966, DPR, Administration, 1966, folder 17.

37. Emil T. Mische, "Street Trees—I," *Park and Cemetery and Landscape Gardening* 11, no. 3 (1901): 43–44; Cox, *A Street Tree System*, 20.

# Chapter 2. Street Tree Aesthetics

1. Department of Parks, City of New York, *Annual Report for the Year 1871* (New York: Department of Parks, n.d.), 79. Elbert Peets, "Street Trees in the Built-up Districts of Large Cities," *Landscape Architecture* 6 (October 1915): 15–31 (16); J. H. Prost, *Street Tree Planting for Illinois*, Illinois Outdoor Improvement Association pamphlet no. 2 (April 1910), 6; J. H. Prost, *Trees and Lawns for the Streets*, Special Park Commission pamphlet no. 6 (April 1914), 1.

2. Peets, "Street Trees in Built-up Districts," 16; William F. Fox, *Tree Planting on Streets and Highways* (Albany: J. B. Lyon, 1903), 26.

3. Sid J. Hare, "Avoiding Monotony in City Street Planting," *Park and Cemetery and Landscape Gardening* 17, no. 4 (1907): 101; Phelps Wyman, "Street Tree Planting—Its Relation to City Planning," *Parks & Recreation* 19, no. 5 (1931): 223–25 (224).

4. Laurie D. Cox, *A Street Tree System for New York City, Borough of Manhattan: Report to Honorable Cabot Ward, Commissioner of Parks, Boroughs of Manhattan and Richmond, New York City* (Syracuse: Syracuse University, 1916), 35; memorandum by horticulturalist C. J. Schiff, "Drought Damage," 15 September 1965; C. J. Schiff to S. M. White, "Tree Pruning," 16 November 1965, DPR, Administration, 1965, folder 6.

5. Henry R. Francis, "Suggestions for Proper Procedure in Systematic Street Tree Planting for Towns and Cities of New York," *Bulletin of the New York State College of Forestry at Syracuse University* 15, no. 4 (Syracuse: University of Syracuse, 1915): 20, 23; Fox, *Tree Planting*, 8–9; Peets, "Street Trees in Built-up Districts," 31; Prost, *Trees and Lawns*; Wayne C. Holsworth, "Street Tree Planting in Its Relation to City Planning," *Parks & Recreation* 7 (January–February 1924): 1–32 (5).

6. Fox, *Tree Planting*, 26; L. D. Cox, "Design with Respect to Street Tree Planting," *Eighth National Shade Tree Conference, Proceedings of Annual Meeting* (1932): 42–48 (46).

7. Fox, *Tree Planting*, 25. According to Charles W. Eliot II, the painter worked in two dimensions, the sculptor in three, and the architect in four dimensions "because he moves around inside the structures which he builds." In contrast, people who "trained to work with living materials have to think of a fifth dimension—life, for our compositions change with every season and are always growing." Charles W. Eliot, "Parks and Shade Trees in City Plans," paper presented to the western chapter of the National Shade Tree Conference, May 1945, published in *Twenty-First National Shade Tree Conference 1945*, ed. Paul E. Tilford (Wooster, OH: Collier, 1946), 61–63, also printed as "Local Needs for Parks and Shade Trees" in *Parks & Recreation* 28, no. 4 (1945): 198–200; Francis, "Systematic Street Tree Planting," 34; Adolphe Chargueraud, *Les arbres de la ville de Paris* (Paris: J. Rothschild, 1896), 265–66. C. Heicke,

*Baumpflanzungen in Straßen der Städte* (Neudamm: Verlag von J. Neumann, 1896), 77–78; E. Petzold, *Anpflanzung und Behandlung von Alleebäumen* (Berlin: Verlag von Wiegandt, Hempel und Parey, 1878), 7–10.

8. Wayne C. Holsworth, "Street Tree Planting in Its Relation to City Planning" (master's thesis, Harvard University, School of Landscape Architecture, 1921), 35.

9. Fox, *Tree Planting*, 45.

10. "Trees on Fifth Avenue," *NYT*, 7 March 1939, 17.

11. Francis, "Systematic Street Tree Planting," 31; Holsworth, "Street Tree Planting in Its Relation to City Planning" (master's thesis), 30.

12. Massachusetts Forestry Association, *Practical Suggestions for Tree Wardens* (Boston, 1900), 7. For the critique of the unsymmetrical shapes of silver maple and Carolina poplar, see "Tree Planting," *NYT*, 1 March 1903, 6. Charles Sprague Sargent, introduction to A. Des Cars, *A Treatise on Pruning Forest and Ornamental Trees*, trans. from the 7th French ed., 3rd ed. (Boston: Massachusetts Society for the Promotion of Agriculture, 1894), 19–27; Fox, *Tree Planting*, 24, 107–9; Bernhard E. Fernow, *The Care of Trees in Lawn, Street, and Park* (New York: Henry Holt, 1910), 105–6; William Solotaroff, *Shade-Trees in Towns and Cities* (New York: John Wiley and Sons, 1911), 128–29; J. Wesseln, "Vorwort," in A. Des Cars, *Das Aufästen der Bäume* (Cologne: Verlag der M. Dumont-Schauberg'schen Buchhandlung, 1868), v–viii.

13. Des Cars, *Treatise on Pruning*, 21; Chargueraud, *Les arbres*, 107–9; Jules Nanot, *Guide de l'ingénieur pour l'établissement et l'entretien des plantations d'alignement d'arbres fruitiers, forestiers et d'ornement sur les routes, boulevards et avenues* (Paris: Librairie centrale d'agriculture et de jardinage), 220–22 (Nanot's treatise and reference to the dendroscope were translated into German by Ludwig Beissner in *Der Straßen-Gärtner* [Berlin: Parey, 1887]); Fernow, *Care of Trees*, 105; Wayne C. Holsworth, "Science of Tree Pruning," *Parks & Recreation* 8, no. 3 (1925): 228–31; Massachusetts Forestry Association, *Practical Suggestions*, 7; *Annual Report of the Tree Planting Association of New York City, 1906* (New York: Tree Planting Association, 1906), 25; *Annual Report of the Tree Planting and Fountain Society of Brooklyn, N.Y., December 1895* (Brooklyn: Eagle Book and Job Printing Department, 1896), 7–11.

14. "The Science of Doctoring Trees," *NYT*, 31 July 1910, X1; George E. Stone, "Repairing Defective Trees with Cement Filling," *Park and Cemetery and Landscape Gardening* 16, no. 11 (1907): 223–24; "Treatment of Trees," *CDT*, 25 March 1876, 10.

15. Department of Parks, City of New York, Boroughs of Manhattan and Richmond, *Annual Report for the Year 1912* (New York: J. J. Little and Ives, 1912), 91–92; "City's Oldest Tree Honored at Inwood," *NYT*, 31 October 1912, 22; Arthur E. Scott, "Where Hudson Stood in New York," *New York Herald*, 31 December 1922, 3; Department of Parks, City of New York, Borough of Manhattan, *Annual Report for the Year 1930* (New York: Beacon, 1930), 13–14; Sherman Graft, "Taming Inwood Hill," *NYT*, 10 April 1938, 159.

16. James A. G. Davey to Mrs. J. G. William Greeff, 25 January 1929, WLPRPR, box 7, folder 13; box 6, folders 6, 11; "Flower Show Ends with Allied Day," *NYT*, 22 March 1920, 26.

17. Solotaroff, *Shade-Trees*, 220–25; "The Care of Trees," *American City* 4, no. 2 (1911): 92. For NuwuD and the Bartlett Company's cavity-filling methods, see "Tree Surgery to Date," *Tree Talk* 5, no. 1 (Spring 1923): 7–9; F. A. Bartlett, "The Development of Tree Surgery," *Tree Talk* (1935): 26–28; W. H. Rankin, "Pathological Work of the Bartlett Tree Research Laboratories," *Tree Talk* (1935): 64–73 (66–68); J. Franklin Collins, "Tree Surgery," *Tree Talk* (1935): 99–100. Paul R. Davey, "The Use of Concrete Fillings for Tree Cavities," *NYT*, 9 July 1939, 42; Wilbur H. Seubert, "Ounce of Prevention Theory Is Effective in Tree Surgery," *NYT*, 2 April 1939, 56. For cavity fillings, also see Richard J. Campana, *Arboriculture: History and Development in North America* (East Lansing: Michigan State University Press, 1999), 335–51.

18. George E. Stone, "Street Shade Trees and Their Troubles," *Park and Cemetery and Landscape Gardening* 17, no. 5 (1907): 130–31 (131). For cement as modern material, see Adrian Forty, *Concrete and Culture* (London: Reaktion, 2012), 14.

19. John Davey, *The Tree Doctor* (New York, Chicago, and Akron, OH: Saalfield, 1904), introduction, n.p.

20. Holsworth, "Street Tree Planting in Its Relation to City Planning" (master's thesis), 10.

21. *The Report of the Department of Parks to August 1934*, n.p. NYCPDAR; 28 October 1934, NYCPPR; DPR, Administration, 1935, folder 009.

22. Memorandum by Queens director of parks James J. Mallen, 17 June 1941, DPR, Queens, 1941, folder 29; General Superintendent to Mr. Varga, 21 December 1939, DPR, Administration, 1939, folder 46. On the street tree situation in New York City in 1934, also see Frederick W. Kelsey, "Parks, Park Ways, and Street Planting," *Parks & Recreation* 18, no. 4 (1934): 122–23.

23. DPR, Administration, 1935, folder 009; "Street Trees," *New Yorker*, 23 March 1935, 12–13 (13); Robert Moses, 10 September 1935, NYCPPR; Robert Moses to Irving V. A. Huie, 12 January 1942, RMR, box 107866, folder 80: Works Projects Administration, 1942; "Report of Mrs. Arthur Hays Sulzberger," in *Seventh Annual Report of the Park Association of New York City* (New York: Park Association of New York City, 1935), 3; "Park Department to Plant Street Trees," 15 March 1935, NYCPPR; 17 January 1936, NYCPPR.

24. All WPA personnel earning a minimum of $130 per month were required to purchase their own uniform. Suggestions to outfit workers sponsored by the Temporary Emergency Relief Administration included khaki shirts and trousers as well as armbands or badges that could easily be handed out daily and thereby also provided a tool to monitor workers reporting for work. At first, many Department of Parks employees failed to wear or pay for their uniforms, causing tension within the department as well as supervision and enforcement of the dress code. See DPR, Administration, 1935, folders 51–55. For Arbor Day 1935, see J. Y. Rippin to E. P. King, "Arbor Day Planting," 2 May 1935, DPR, Administration, 1935, folder 009. *The Report of the Department of Parks to August 1934*, 18–20, NYCPDAR. Robert Thompson, *Safety for Tree Workers, Tree Preservation Bulletin* no. 8 (Washington, DC: Government Printing Office, 1937), 5.

25. 7 May 1937, NYCPPR; "Twelfth Annual Report of the President Mrs. Arthur Hays Sulzberger," in *The President's Report for 1939* (New York: Park Association of New York City, 1939), 8; "Controlling the Dutch Elm Disease," 3 August 1934, NYCPPR.

26. For the development and use of chainsaws and spraying technology, see Campana, *Arboriculture*, 365–70, 379–91; Department of Parks, City of New York, *Annual Report for the Year 1915* (New York: Clarence S. Nathan, n.d.), 198; William H. Latham to Mr. H. W. DelVecchio, 7 May 1949, DPR, Brooklyn, 1949, folder 26; press release, 28 July 1958, DPR, Administration, 1958, folder 18; Robert Moses to Mayor Robert F. Wagner, 28 July 1958, DPR, Administration, 1958, folder 19; memorandum by G. L. Quigley, "Trees," 2 August 1957, DPR, Administration, 1957, folder 22. In 1941, park commissioner Robert Moses was confronted with the death of several Scotties from arsenic and lead poisoning due to tree-spraying activities in Washington Square. Rather than question the use of the insecticides in the first place, signs were posted around the sites of operation advising people to keep pets away. See communication between John S. Bradley and Park Commission in October 1941, DPR, Manhattan, 1941, folder 004; 30 April 1944, NYCPPR.

27. Memorandum by S. M. White, "Tree Spraying," 22 July 1959, DPR, Administration, 1959, folder 18; commissioner Newbold Morris to Dr. David Lehr, 3 May 1961, DPR, Manhattan, 1961, folder 32; executive officer Stuart Constable to Mr. Emil J. Ruckert, 24 April 1957, DPR, Manhattan, 1957, folder 25. For a history of toxicology, DDT, and other chemicals used as insecticides in the U.S., see Frederick Rowe Davis, *Banned: A History of Pesticides and the Science of Toxicology* (New Haven and London: Yale University Press, 2014).

## Chapter 3. Tree Ladies

1. See Cornelius B. Mitchell, "Report of the Vice-President," in Tree Planting Association of New York City, *A Brief History of the Organization and the Annual Report for 1897* (New York:

Tree Planting Association, n.d.), 29–30 (30); *Annual Report of the Tree Planting Association of New York City, 1906* (New York: Tree Planting Association, 1906), 17; Cornelius B. Mitchell, "President's Report," in *Annual Report of the Tree Planting Association of New York City, 1909* (New York: Tree Planting Association, 1909), 6–8 (8); Joseph L. Delafield, "Secretary's Report," in *Annual Report of the Tree Planting Association of New York City, 1909*, 11–12 (12); Cornelius B. Mitchell, "Report of the Chairman of the Executive Committee," in *Annual Report of the Tree Planting Association of New York City, 1901–1902* (New York: Tree Planting Association, 1902), 10–11 (10).

2. Mary Cynthia Dickerson, *Trees and Forestry* (New York: Museum of Natural History, 1910).

3. George E. Waring, *Village Improvements and Farm Villages* (Boston: James R. Osgood, 1877), 16; Mary Ritter Beard, *Women's Work in Municipalities* (New York: D. Appleton, 1915), 307–8.

4. Lydia Adams-Williams, "Conservation-Woman's Work," *Forestry and Irrigation* (14 June 1908): 350–51 (351); E. G. Routzahn, "The Tree Planting Movement," *Chautauquan: A Weekly Newsmagazine* 41, no. 4 (1905): 337; "Women's Clubs and Forestry," *Park and Cemetery and Landscape Gardening* 15, no. 2 (1905): 1. For women in the conservation movement and forest preservation, see Carolyn Merchant, *Earthcare* (New York: Routledge, 1996), 109–36.

5. Beard, *Women's Work in Municipalities*, 323; Suzanne M. Spencer-Wood, "Turn of the Century Women's Organizations, Urban Design, and the Origin of the American Playground Movement," *Landscape Journal* 13, no. 2 (1994): 124–37.

6. Sallie D. Richards, "No River-Bank Airport," *NYT*, 4 December 1930, 24; Sallie D. Richards in a draft letter to the editor of the *NYT*, 29 June 1931, WLPRPR, box 8, folder 11.

7. On the role of the Chicago Women's Club in the promotion of forestry, see Sonja Dümpelmann, "Designing the 'Shapely City': Women, Trees, and the City," *Journal of Landscape Architecture*, no. 2 (2015): 6–17.

8. For tree-planting campaigns by the Municipal Art Society, which took place repeatedly throughout the twentieth century beginning in the 1910s, see Nathalie Dana, "The Municipal Art Society: Seventy-Five Years of Service to New York"; *Street Trees Make a City More Livable*, brochure by the Municipal Art Society of New York, DPR, Administration, 1949, folder 020; "Campaign Begun to Beautify City Streets," *NYT*, 21 March 1947, 23; "City of Tree-Lined Streets Will Be Goal of Year-Long Drive by Municipal Art Society," *NYT*, 20 November 1947, 31. "Prendergast Says Riverside Is Secure," *NYT*, 22 January 1917, 11; Frances Peters, "Suffragists Not to Blame. Deny Harming Garden of Women's Municipal League," *NYT*, 19 May 1913; Emily Legutko, *The First 85 Years: A History of the City Gardens Club of New York City, 1918 to 2003* (New York: City Gardens Club, 2004), 8, 10; Constitution of the City Gardens Club of New York, 1923, CA, 1923, 130, City Gardens Club. "Memory Trees All Over: War Has Given a Great Impulse to Tree Planting," *NYT*, 24 March 1920, 25; *Minutes of the Park Board of the Department of Parks of the City of New York, for the Year Ending December 31, 1923* (New York: M. B. Brown, 1924), 73.

9. Annual report of the Women's League for the Protection of Riverside Park for 1926–27, WLPRPR, F 128.64.R58 W67 1924/1926; WLPRPR, F 128.64.R58 W678 1926, Women's League for the Protection of Riverside Park Year Book, 1930–31.

10. See documents in WLPRPR, box 7, folder 2.

11. William T. Hornaday, *Wild Life Conservation in Theory and Practice* (New Haven: Yale University Press, 1914), 5, 22–23.

12. Department of Parks, City of New York, *Report for the Year Ending December 31, 1900* (New York, 1901), 11; Department of Parks, City of New York, *Report for the Year Ending December 31, 1901* (New York: Mail and Express, 1902), 17, 22–24; Department of Parks, City of New York, *Report for the Year Ending December 31, 1902* (New York: Martin B. Brown, 1903), 45–47; L. O. Howard, *The Leopard Moth: A Dangerous Imported Insect, Enemy of Shade Trees* (Washington, DC: U.S. Dept. of Agriculture, 1916), 3–4; Southwick, cited in James W. Chapman, *The Leopard Moth and Other Insects Injurious to Shade Trees in the Vicinity of*

*Boston* (Cambridge, MA: Bussey Institution, Harvard University, 1911), 7–8; Edmund B. Southwick, "Report," in *Report 1901*, 22–24.

13. A. R. Grote, "Worms on Shade Trees," *NYT*, 16 June 1883, 5. "Insect Pests in the Park," *NYT*, 21 February 1884, 3; "Insects in the Parks," *NYT*, 12 April 1885, 4; Department of Parks, the City of New York, *Annual Report for the Year 1908* (New York: Martin B. Brown, 1909), 52.

14. *Annual Report of the Tree Planting and Fountain Society of Brooklyn, N.Y., December 1895* (Brooklyn: Eagle Book and Job Printing Department, 1896), 13–14, 75; *Annual Report of the Tree Planting and Fountain Society of Brooklyn, N.Y., December 1897* (Brooklyn: Eagle Book and Job Printing Department, 1898), 86.

15. Howard, *The Leopard Moth*, 3; "Birds as Health Officers," *Tree Talk* 3, no. 3 (February 1916): 73; Tree Planting Association of New York City, *A Brief History of the Organization and the Annual Report for 1897*, 9; John Y. Cuyler, "Protect the Songbirds," *NYT*, 22 April 1898, 6.

16. U.S. Department of Agriculture, Division of Biological Survey, circular no. 17: *Bird Day in the Schools* (Washington, DC, 1896), 4.

17. "A Bird Day for School," *NYT*, 20 September 1896, SM15; "To Protect the Wild Birds," *NYT*, 12 November 1897, 2; J. Sterling Morton to C. A. Babcock, cited in Charles A. Babcock, *Bird Day: How to Prepare for It* (New York, Boston, and Chicago: Silver, Burdett, 1901), 10.

18. Babcock, *Bird Day*, 18, 41; Frank M. Chapman, "Birds and Bonnets," *Forest and Stream* 26, no. 5 (1886): 84; Frank M. Chapman, *Autobiography of a Bird-Lover* (New York and London: D. Appleton-Century, 1933), 38–39; "New York Fashions," *Harper's Bazaar*, 4 December 1897, 1003. For Chapman, the millinery trade, women, and the foundation and work of the Audubon Society, see Frank Graham Jr. with Carl W. Buchheister, *The Audubon Ark* (New York: Knopf, 1990), 25; Felton Gibbons and Deborah Strom, *Neighbors to the Birds* (New York and London: Norton, 1988), 114–25, 175–89; Robin W. Doughty, *Feather Fashions and Bird Preservation* (Berkeley, Los Angeles, and London: University of California Press, 1975), 16. For a feminist bent on women, the millinery trade, and bird protection, see Jennifer Price, *Flight Maps* (New York: Basic Books, 1999), 57–109.

19. For the New York Bird and Tree Club's position on the proposed construction of a ferry terminal along Riverside Drive, see WLPRPR, box 6, folder 18. New York Bird and Tree Club, New York City, *Report of Committee-Memorial Fruit Trees for France* (1921); "Miss Anna Jones Dies at Park Avenue Hotel," *NYT*, 26 January 1925, 17; "Trees for France a Christmas Plan," *NYT*, 24 November 1918, 43; "Fruit Trees for France, The New York Bird and Tree Club," *NYT*, 22 January 1919, 10; "Replanting Trees in France," *NYT*, 6 April 1919, 44. Mayor of Nancy to George Frederick Kunz, 2 June 1920, GFK, box 2 (49); French Consulate, NYC, to George Frederick Kunz, 6 October 1919, GFK, box 2 (52); "Reports of Affiliated State Societies and of Bird Clubs," *Bird-Lore* 23, no. 6 (1921): 359–98 (375).

20. BC, SY1919 no.14; "Scout Tree Patrol Guards Riverside Memorial Grove," *NYT*, 14 February 1926, X22; scoutmaster Sereno Stetson to the president, 24 February 1928, WLPRPR, box 8, folder 11.

21. *Summary of the Eighteenth Annual Report of the Department of Parks for the Borough of Brooklyn, Comprising the Fifty-Fifth Annual Report of the Old City (Now Borough) of Brooklyn for the Year 1915* (New York City: O'Connell, 1916), 28; Edwin A. Osborne, "City Aids Birds to Stay in Zones of New Housing," *NYT*, 17 May, 1925, XX13; "Five Bird Havens Planned in Parks," *NYT*, 29 June 1935, 17; *The Report of the Department of Parks to August 1934* (New York, n.d.), 31; "Report of the Committee on Bird Protection and Propagation," *Parks & Recreation* 1, no. 4 (1918): 45–47; E. F. A. Reinisch, "Attracting Birds by Furnishing Them Nesting Sites," *Parks & Recreation* 1, no. 3 (1918): 52; Robert J. Terry, "A Bird Sanctuary for St. Louis," *Parks & Recreation* 2, no. 2 (1919): 29–43; L. P. Jensen, "Parks as Bird Protectorates," *Parks & Recreation* 4, no. 1 (1920): 16–18; Arthur Hawthorne Carhart, "The Bird Refuge Park," *Parks & Recreation* 10, no. 5 (1927): 483–84.

22. "50 Bird Houses Placed along Riverside Drive," *NYT*, 29 June 1928, 16; WLPRPR, box 6,

folder 6, especially letters by the president to Miss Rochester, 20 April 1928, and Mr. Boylan, associate superintendent, Board of Education, 16 May 1928; Department of Parks, City of New York, Borough of Manhattan, *Annual Report for the Year 1929* (New York: F. Hubner, 1930), 17; Department of Parks, City of New York, Borough of Manhattan, *Annual Report for the Year 1930* (New York: Beacon, 1931), 11; WLPRPR, box 7, folders 14, 16; box 8, folders 17, 19; "Prizes for Pupils Who Aid Birds," *Sun*, 11 April 1929, 31; Mrs. John Klapperton Kerr to Walter R. Herrick, 15 July 1927, and related correspondence, WLPRPR, box 6, folder 3; A. F. Burgess to Mrs. William R. Luce, 19 April 1930, WLPRPR, box 8, folder 19. The construction of birdhouses by schoolchildren was common in cities throughout the nation at the time. See, for example, Neltje Blanchan, *How to Attract the Birds and Other Talks about Bird Neighbors* (New York: Doubleday, Page, 1903), 1–18, 165–204; "Washington Schools Vote for the Oak as the National Tree," *American Forest* 27, no. 327 (March 1921): 169–72; also see the bird-related articles in the early issues of *Tree Talk*, the journal issued by the Bartlett Tree Expert Company in Stamford, Connecticut. Mrs. James Edwin Morris, "Bird Protection as a Business Proposition," *Tree Talk* 2, no. 3 (February 1915): 21; Paul G. Howes, "Bird Houses," *Tree Talk* 2, no. 4 (May 1915): 23; "Appliances for Bringing Back the Birds," *Tree Talk* 3, no. 1 (August 1915): 25; "Bringing Back the Birds," *Tree Talk* 3, no. 2 (November 1915): 43–44; "A National Bird Sanctuary," *Tree Talk* 3, no. 4 (May 1916): 109; B. S. Bowdish, "Birds and Trees," *Tree Talk* 4, no. 2 (Winter 1916): 43–48; "Houses for the Birds," *Tree Talk* 4 (Spring 1917): 86–87; "A Plea for Our Native Birds," *Tree Talk* 5, no. 3 (1923): 1–2. On the history of nature study in the United States, see Sally Gregory Kohlstedt, *Teaching Children Science* (Chicago and London: University of Chicago Press, 2010).

23. "Homes for Park Birds: Girl Scouts Solve One Pressing Housing Problem," *NYT*, 3 October 1920, 141; John D. Haney, "How Some Children Helped Birds," *Animal News* 1, no. 2 (April 1912): 5–6 (5); Paul B. Riis, "Nesting Boxes," *Parks & Recreation* 3, no. 3 (1920): 38–41; *American Bird House Journal* 5, no. 1 (January 1921), 9, 13.

24. See CA, 1268, 1928; and the brochures in the American Museum of Natural History Archives, Department Records, Department of Education. On the role of the American Museum of Natural History in the nature study movement, see Kohlstedt, *Teaching Children Science*, 64–66, 206.

25. Henry R. McCartney, "Why I Killed My Cat," in *Second Report of the Meriden Bird Club*, 1912 (Boston: Poole, 1912), 35–36 (36); Ernest Harold Baynes, "Meriden—the Bird Village," *Animal News*, 1, no. 5 (July 1912): 5–6; the first, second, and third reports of the Meriden Bird Club, 1911, 1912, and 1916 (Boston: Poole); Frederic H. Kennard, "A List of Trees, Shrubs, Vines, and Herbaceous Plants Native to New England, Bearing Fruit or Seeds Attractive to Birds," *Bird-Lore* 14, no. 4 (July–August 1912): 201–3.

26. See "Berlepsch Bird Houses," *Second Report of the Meriden Bird Club*, 1912, 17–21; Frederic H. Kennard, "My Experience with von Berlepsch Nesting-boxes," *Bird-Lore* 14, no. 1 (January–February 1912), 1–12; Frederic H. Kennard, "A Rustic Food-House," *Bird-Lore* 14, no. 6 (November–December 1912): 337–38; Baynes, "Meriden—the Bird Village."

27. Edward Howe Forbush, *Bird Houses and Nesting Boxes*, the Commonwealth of Massachusetts State Board of Agriculture circular no. 47 (Boston: Wright and Potter, 1915), 6; Edward Howe Forbush, "Nesting-Boxes," *Bird-Lore* 7, no. 1 (January–February 1905): 5–10. On nesting boxes in urban environments, also see Paul B. Riis, "Nesting Boxes," *Parks & Recreation* 3, no. 3 (1920): 38–41.

28. *Third Report of the Meriden Bird Club*, 1916, 11–38; "President Watches Daughter in Play," *NYT*, 13 September 1913, 11; "A Masque to Aid Birds," *NYT*, 2 September 1913, 2.

29. On Coffey, see Thaisa Way, *Unbounded Practice* (Charlottesville: University of Virginia Press, 2009), 142, 254. Coffin was a generation older than Coffey; her interest in trees was still largely confined to designs for private house and estate gardens. She encouraged her readership to "think as the architect does" and use trees as architectural elements in garden designs. Marian

Cruger Coffin, *Trees and Shrubs for Landscape Effects* (New York: Charles Scribner's Sons, 1940), 1; Coffin, "Gardening with Trees," *Garden and Home Builder* 44, no. 2 (October 1926): 128–30.

30. "6 Women Seek Jobs as Tree Climbers," *NYT*, 13 October 1939, 22. For the Department of Parks' original request to add ninety-four climbers, pruners, and gardeners to its inadequate workforce, see memorandum "1939–1940 Budget Request for the Department of Parks," 9 February 1939; Robert Moses to Fiorello La Guardia, 24 May 1939, NYCPPR. The Department of Parks employed five arboriculturists and 154 climbers and pruners at the time. City of New York Civil Service Commission, *Civil Service Bulletin* (June 1940): 254, 272; "Climber and Pruner," *Civil Service Bulletin* (October 1939): 381–82, 384–85.

31. John Davey, *The Tree Doctor* (New York, Chicago, and Akron, OH: Saalfield, 1904), introduction, 29; Christine Sadler, "No Puddlers? Women Tree-Climbers Sooth [*sic*] Outraged Judge," *Washington Post*, 12 November 1939, 3. Another fact noted at the conference was that census figures listed two hundred female woodchoppers. For Davey's first female tree worker hire, see Richard J. Campana, *Arboriculture: History and Development in North America* (East Lansing: Michigan State University Press, 1999), 110.

32. See Campana, *Arboriculture*, 206; and the membership and participant lists in the *Proceedings of the National Shade Tree Conferences* in these years.

33. *Bulletin of the Tree Planting Association of New York City* (September 1913), 3.

34. "New York City Is Being Surveyed for Planting of Trees," *NYT*, 4 January 1914, SM6.

35. C. B. Mitchell, "Trees on Fifth Avenue: Difficulties in the Way," *NYT*, 10 May 1897, 7. Tree Planting Association of New York City, *A Brief History of the Organization and the Annual Report for 1897*, 25–26; Department of Parks, City of New York, *Annual Report for the Year 1909* (New York: M. B. Browne, 1909), 40; "A Glimpse at Fifth Avenue's Dwindling Shade Line," *NYT*, 13 March 1910, SM6; Department of Parks, City of New York, *Annual Report for the Year 1913* (New York: M. B. Brown, 1913), 47, 50; Department of Parks, City of New York, Boroughs of Manhattan and Richmond, *Annual Report for the Year 1912* (New York: J. J. Little and Ives, 1912), 98; Henry R. Francis, *Report on the Street Trees of the City of New York*, issued by the Tree Planting Association of the City of New York (Syracuse: New York State College of Forestry at Syracuse University, 1914), 16.

36. Charles R. Lamb, "Report on Conditions in the Borough of Manhattan," in *Annual Report of the Tree Planting Association of New York City, 1912* (New York: Tree Planting Association, 1912), 21–23 (21); "Fighting to Beautify Fifth Avenue with Trees," *NYT*, 30 April 1911, SM3; Laurie D. Cox, *A Street Tree System for New York City, Borough of Manhattan: Report to Honorable Cabot Ward, Commissioner of Parks, Boroughs of Manhattan and Richmond, New York City* (Syracuse: Syracuse University, 1916), 35, 50–51.

37. F. L. Mulford, *Street Trees*, U.S. Department of Agriculture bulletin no. 816, 19 January 1920 (Washington, DC: Government Printing Office, 1920), 3; Ada Rainey, "Shade Trees for City Streets," *Craftsman* 24, no. 6 (1913): 611–21 (611–12); "The Value of a Tree," *Tree Talk* 3, no. 2 (November 1915): 55. For early court decisions regarding street trees, also see "Important Court Decisions Protecting Street Trees," *Park and Cemetery and Landscape Gardening* 16, no. 6 (1906): 118–19.

38. J. H. Prost, *Trees and Lawns for the Streets*, Special Park Commission pamphlet no. 6 (April 1914), 2; William F. Fox, *Tree Planting on Streets and Highways* (Albany: J. B. Lyon, 1903), 26; correspondence between Robert Moses, T. J. McInerney, Walter J. Salmon, and Harry Brandt, 18 July 1956, 19 July 1956, 27 July 1956, and 3 August 1956, NYCPPR; L. D. Cox, "Design with Respect to Street Tree Planting," *Eighth National Shade Tree Conference, Proceedings of Annual Meeting* (1932): 42–48 (45). Malcolm Howard Dill, "The Progress of Systematic Street Tree Planting in American Cities," *American City* 34, no. 3 (1926): 300–305. None of Dill's surveyed cities had undertaken efforts to plant street trees in business districts.

39. "Report of the President Mrs. Arthur Hays Sulzberger," in Park Association of New York City,

*The President's Report for 1937 and 1938* (New York: Park Association of New York City, 1938), 11–12; "Elms," *New Yorker*, 11 March 1939, 18–19 (19).

40. "Gotham Sees a Tree and Gasps," *Seattle Daily Times*, 10 March 1939. For the inadequacy of the workforce to deal with the tasks at hand, see memorandum "1939–1940 Budget Request for the Department of Parks." For the Rockefeller tree plantings, see press release, 5 March 1939, NYCPPR; "Park Department Regulations for Street Tree Planting," 1 October 1939, NYCPPR; William L. Latham to Charles E. Hall, 30 January 1940, DPR, Manhattan, 1940, folder 10. For tree-planting proposals in the wake of the Fifth Avenue elm tree plantings, see documents in ORR, box 99, folder 746. For the Fifth Avenue Association's opinion on tree planting on the avenue, see William J. Pedrick to Mrs. Arthur Hays Sulzberger, 14 June 1945, DPR, Manhattan, 1945, folder 35, and a copy of the same letter in PCR, box 1, folder 71; "Opposes Trees on 5th Avenue," *New York Sun*, 25 July 1945. "City Tree Planting Begins Wednesday," *NYT*, 5 March 1939, 6; "Rockefeller City Plants Huge Elm," *NYT*, 10 March 1939, 25; "Another Tree in Fifth Avenue," *NYT*, 12 March 1939, 59.

41. "St. Patrick's Gets a Gift of 12 Trees," *NYT*, 21 August 1939, 11; "Third Elm Planted at Cathedral Front," *NYT*, 20 October 1939, 12. The tree gift was also an opportunity to experiment with a new chemical, "Dowax," produced by the Dow Chemical Company for application on trees to reduce evaporation and thereby aid transplanted trees in setting root. See F. F. Rockwell, "Round about the Garden: Little Bulbs Indoors, Rubber Coats for Trees," *NYT*, 19 November 1939, D10; "Drive on to Add 10,000 Trees to City Streets," *New York Herald Tribune*, 27 September 1942, C2; memorandum by John F. Walsh, "Rubber Coats for Trees," 28 November 1939, DPR, Administration, 1939, folder 46; memorandum by A. V. Grande, "Street Tree Planting," 3 January 1941; press release for 19 May 1941, DPR, Administration, 1941, folder 8.

42. Philip LeBoutillier to Iphigene Ochs Sulzberger, cited in report by the Connecticut Tree Protective Association, DPR, Manhattan, 1946, folder 033.

43. Report by the Connecticut Tree Protective Association, DPR, Manhattan, 1946, folder 033; "Released for AMS, Monday, June 11, 1945," DPR, Administration, 1945, folder 42; collection of letters by the professionals who were surveyed in DPR, Manhattan, 1945, folder 34, esp. Gilmore D. Clarke to Cameron Clark, 21 May 1945, and A. F. Brinckerhoff to Cameron Clark, 15 May 1945. "LeBoutillier Scores Nathan in Tree Row," *NYT*, 21 August 1945, 16.

44. Frank Sullivan, "Friendly Advice to a Recalcitrant Mercer," *New Yorker*, 2 June 1945, 18. For public reactions for and against trees on Fifth Avenue, see "A Tree May Grow in Brooklyn, but How about 5th Ave.?" *New York World-Telegram*, 23 August 1945; Albert A. Volk, "Letters to the Times: Trees on Streets Is Question," *NYT*, 30 June 1945, 16.

45. See the display ad "Best's Stand on 5th Avenue Trees," *NYT*, 22 June 1945, 7; "Trees in Business Streets: Esthetic or Practical or Both?" *American City* 60, no. 7 (1945): 90.

46. "Curran Takes Up Cudgels for Tree Planting," *NYT*, 9 June 1945, 15; "Poem Made by P. LeB. to Hit Tree Publicity," *NYT*, 13 June 1945, 25. The public submitted poetry as well. See, for example, a poem by Mark Werthmann, 22 June 1945, PCR, box 1, folder 71.

47. Report by the Connecticut Tree Protective Association, DPR, Manhattan, 1946, folder 033; "5th Ave. Tree Row Jumps State Line," *NYT*, 10 February 1946, 11.

48. "Third Elm Planted at Cathedral Front: Fifth Avenue Rock Hampers Work," *NYT*, 20 October 1939, 12; memorandum by David Schweizer, 16 August 1940, DPR, Manhattan, 1940, folder 10; "7 of 8 Fifth Ave. Elms Thrive," *NYT*, 22 April 1940, 20; "Center Gets New Tree," *NYT*, 23 April 1948, 18; "New Trees for Fifth Ave.," *NYT*, 21 April 1948, 31; "4 U.S. Elms 'Reforest' Rockefeller Center, Replace English Type . . . ," *NYT*, 9 April 1949, 19; 19 May 1941, NYCPPR. In 1946, landscape architects Francis Cormier and Gilmore D. Clarke had suggested to park engineer William H. Latham that the elms should be replaced with London plane trees or honey locust trees. See memorandum by Francis Cormier, "Rockefeller Center Trees on Fifth Ave.," 7 October 1946, and Gilmore D. Clarke to William H. Latham, 4 October 1946, DPR, Manhattan, 1946, folder 22; Victor Borella to George S. Avery, 27 August 1951, ORR,

box 99, folder 746; correspondence in DPR, Manhattan, 1946, folder 22; "Honey Locusts, Starting at 50-Foot Height, Are Replacing Rockefeller Center Elms," *NYT*, 4 October 1951, 35.

49. Beard, *Women's Work in Municipalities*, 307–8.

## Chapter 4. Planting Civil Rights

1. Nellie Burchardt, *What Are We Going to Do, Michael?* (New York: Franklin Watts, 1973); Harold X. Connolly, *A Ghetto Grows in Brooklyn* (New York: New York University Press, 1977), 140–41.

2. The term *plant-in* was used in one of the Magnolia Tree Earth Center's brochures to describe tree-planting activities in Brooklyn's Tompkins Park in 1971.

3. Albert Kramer, "Honor Hattie Carthan; Saved Tree from Ax," *NYDN*, 11 February 1979, 49, B1; "Magnolia Tree Benefit," *NYAN*, 9 November 1974, D20; "For a 'Tree Lady,' a City's 'Thank You,'" *NYT*, 20 May 1975, L42. In the 1950s, Glenn H. Beyer, professor of housing and design at Cornell University, defined a "slum" as a housing area that had reached "the point of deterioration and obsolescence, where the dwelling units have passed any possible state of rehabilitation in order to provide decent living quarters": Glenn H. Beyer, *Housing: A Factual Analysis* (New York: Macmillan, 1958), 226.

4. Pratt Institute, Planning Department, *Stuyvesant Heights: A Good Neighborhood in Need of Help: A Study Prepared for Church Community Services* (Brooklyn: Community Education Program, Planning Department, Pratt Institute, 1965), 26; 16 May 1967, NYCPPR; "Praise for Spruce Up Project," *NYAN*, 2 December 1967, 25; "Over $4,000 Collected by Bed-Stuy Block Assns," *NYAN*, 9 December 1967, 25; Hattie Carthan, "Our Pleasure," *NYAN*, 5 November 1966, 8; ". . . A Poem as Lovely as a Tree," 28 October 1969, NYCPPR; Mrs. Thomas E. Jones to Commissioner Hoving, 5 March 1966, DPR, Brooklyn, 1966, folder 42.

5. *Annual Report of the Tree Planting and Fountain Society of Brooklyn, N.Y., December 1896* (Brooklyn: Eagle Job and Printing Department, 1897), 18, 30–32; *Annual Report of the Tree Planting and Fountain Society of Brooklyn, N.Y., December 1895* (Brooklyn: Eagle Book and Job Printing Department, 1896), 44; *Tree Planting and Fountain Society of Brooklyn 1890* (n.p., n.d.); commissioner to Miss Odile R. Follenius in Brooklyn, 20 November 1944, DPR, Brooklyn, 1944, folder 76.

6. For the Neighborhood Tree Corps, see Hattie Carthan, "Magnolia Tree Earth Center of Bedford-Stuyvesant," *Street: Magazine of the Environment*, no. 8 (December 1972), 6–8, 35 (35); E. A., "Tree Center Starts Services," *Phoenix*, 23 August 1979, 9, C4. See also Robert Fleming, "Botanic Garden Fete Will Raise Seed Money," newspaper article marked as "NYDN, 17 June 1987, KSI 14.3," in folder "Magnolia Tree Earth Center of Bedford-Stuyvesant, Inc.," Brooklyn Collection, Brooklyn Public Library; Bryant Mason, "They Put a Shine on Big Apple," *NYDN*, 18 December 1975, ML7; Kenneth P. Nolan, "Bed-Stuy Children Learn How to Nurture Trees," *NYT*, 9 January 1972, A16.

7. Kim Lem, "Youths Working to Save Trees," *NYT*, 17 August 1975, 77; "Magnolia Sets Pace in Street Trees Survey," *NYAN*, 13 August 1975, B3; George Todd, "New Director: Loves Challenge," *NYAN*, 4 October 1980, 11.

8. For the trash problem and rat infestation in Bedford-Stuyvesant and related discriminatory practices in the Department of Sanitation, see Brian Purnell, *Fighting Jim Crow in the County of Kings* (Lexington: University of Kentucky Press, 2013), 129–69; Malcolm McLaughlin, "The Pied Piper of the Ghetto: Lyndon Johnson, Environmental Justice, and the Politics of Rat Control," *Journal of Urban History* 37, no. 4 (2011): 541–61.

9. "Gala Magnolia Tree Festival On," *NYAN*, 18 July 1970, 23; Edward Ranzal, "Hearing Divided on How to Save Tree," *NYT*, 4 March 1970, 39; Peter Coutros, "Brooklyn's Magnolia Stands in Limbo," *NYDN*, 30 January 1970, 38; "Noted Poet Contributes," *News*, 10 August 1969; Deirdre Carmody, "Wreckers, Spare That Tree," *NYT*, 29 March 1969, 37; Fred Ferbetti, "Urban Conservation: A One-Woman Effort," *NYT*, 8 July 1982, C1, C6; George Todd, "Boost

Magnolia Tree Project," NYAN, 6 November 1976, B2; Joe Flaherty, "A Magnolia Tree Grows in Brooklyn," *Village Voice*, 17 September 1970, 13; "Tree Declared Landmark," NYAN, 30 May 1970, 23; Edward C. Burks, "Brooklyn's Magnolia Tree Nears Landmark Status," NYT, 4 February 1970, 34; "Youngsters Act to Save Tree Growing in Brooklyn," NYT, 15 January 1970, 37. Also see documents pertaining to the preservation of the magnolia tree in PCR, box 11, folder 19.

10. See "Milton Mollen Donation, Speeches: Tree Planting Ceremony, W 88th Street, 11/24/64," RFWC, box 060194, folder 3,

11. See "The Windowbox Program: A Summary of the Park Association's Self-Improvement Program for Declining Neighborhoods in New York City," PCR, box 11, folder 21; Moira Johnston, preliminary draft "Proposal for a Program of Neighborhood Improvement in the Bedford Stuyvesant Section of Brooklyn," 16 January 1965, PCR, box 11, folder 25. For the original Philadelphia program, see Louise Bush-Brown, *Garden Blocks for Urban America* (New York: Charles Scribner's Sons, 1969).

12. Jeff Offerman and Mrs. Sedlin, "Report and Analysis of the Window Box Program," January 1967, p. 8 and appendix A, PCR, box F, folder 1; Donald Neuwirth and Prof. C. S. Asher, "A Study of the Park Association of New York's Neighborhood Improvement Project," 19 December 1966, p. 16, PCR, box F, folder 1; "Neighborhood Improvement Program," PCR, box 11, folder 21.

13. Between 1963 and 1969 the number of participants increased from a thousand to forty-six hundred. Peggy Mann, "Miracle of the Flower Boxes," *Reader's Digest* 103, no. 7 (1973): 106–10 (110); C. A. Lewis, "Public Housing Gardens—Landscapes for the Soul," in *Yearbook of Agriculture 1972*, ed. U.S. Department of Agriculture (Washington, DC: U.S. Government Printing Office, 1972), 277–81; Charles A. Lewis, "Garden and Home: Can You Grow Flowers on Avenue D?" NYT, 28 March 1971, G1; Ira S. Robbins, "Tenants' Gardens in Public Housing," in *Small Urban Spaces*, ed. Whitney North Seymour Jr. (New York: New York University Press, 1969), 139–47. See also documents in NYCHAC, box 0071C1, folder 01 (1970–84) and folder 02 (1964–72).

14. 4th Annual Flower Garden Competition Manual 1966 and 5th Annual Flower Garden Competition Manual 1967, issued by the New York City Housing Authority, NYCHAC, box 0071B8; "flower garden competition manuals"; "Flower and Vegetable Garden Competition, April 1977–September 1977," NYCHAC, box 0092E3, folder 22; Robbins, "Tenants' Gardens in Public Housing," 145. In 1969, Mrs. Vincent Astor also made a private donation of $1,000 to support the attendance of Housing Authority tenants at the International Flower Show, held at the Coliseum in 1970 where the Housing Authority was exhibiting its Tenant Flower Garden Competition. See NYCHAC, box 0071C1, folder 02 (1964–72).

15. For Lady Bird Johnson's tree-planting campaign in Washington, DC, see Lewis L. Gould, *Lady Bird Johnson and the Environment* (Lawrence: University Press of Kansas, 1988), 88, 91–93, 104. For the Special Award of Merit, see Richard J. Campana, *Arboriculture: History and Development in North America* (East Lansing: Michigan State University Press, 1999), 179.

16. "Statement of August Heckscher, Administrator of Parks, Recreation and Cultural Affairs and Commissioner of Parks at the Board of Estimate and City Council Finance Committee Hearings on the 1968–69 Capital Budget on Tuesday, February 13, 1968," NYCPPR; Murray Schumach, "Shades of Kilmer to Beautify City: Public and Private Funds to Support Tree Planting," NYT, 5 March 1967, 62. For Heckscher's politics, also see "Heckscher Emphasizes Neighborhood Street and Park Improvements," 26 October 1967, NYCPPR; "Upon Receipt Heckscher Asks Plan Commission to Restore $9 Million to Budget," 15 December 1967, NYCPPR. For earlier attempts to increase the budget for street trees, see "A Tree Is a Tree, but Not in City," NYT, 21 August 1966, 66. "On the Street Where You Live," 1 October 1969, NYCPPR. In 1958–59, Lasker gave $32,472.10 for street tree–planting projects in Manhattan. These included street trees along Forty-Second Street from Second Avenue to Times Square;

Fifth Avenue from Thirty-Eighth Street to Fifty-Seventh Street; Madison Avenue from Forty-Second Street to Fifty-Seventh Street; Lexington Avenue from Forty-Second Street to Fifty-Seventh Street; First Avenue from Forty-Second Street to Fifty-Seventh Street; Fifty-Seventh Street between Sutton Place and Avenue of the Americas to Second Avenue. See memorandum by John A. Mulcahy, 31 October 1958, DPR, Manhattan, 1958, folder 21; memorandum by Stuart Constable, 5 June 1959, DPR, Manhattan, 1959, folder 21. Executive officer to Hon. Louis A. Cioffi, 28 August 1959; Mrs. Albert D. Lasker to Stuart Constable, "Location and Estimate of Cost of Proposed Tree Planting, Manhattan," 23 July 1959; Stuart Constable to John A. Mulcahy, 11 September 1959, DPR, Manhattan, 1959, folder 23. Assistant executive director to Mrs. Albert D. Lasker, 18 November 1959; assistant executive officer to Hon. Lawrence E. Gerosa, 30 November 1959; assistant executive officer to Mrs. Albert D. Lasker, 1 December 1959, DPR, Manhattan, 1959, folder 24. "Proposed Tree Planting"; "5,000 Street Tree Project"; Alexander Victor to J. H. Mulcahy, 6 April 1961; Newbold Morris to J. H. Mulcahy, 29 January 1962, DPR, Administration, 1962, folder 3. Executive officer to June Fauldo, 2 December 1963, DPR, Administration, 1963, folder 20. "Communication from the Mayor—Transmitting Approval of Amendment of 1964–1965 Capital Budget, as Requested by the Commissioner of Parks, to Add $675,000 for Street Tree Planting, Various Boroughs," and "Report of the Committee on Finance in Favor of Approving and Adopting the Amendment of 1964–1965 Capital Budget . . . ," in *Proceedings of the Council of the City of New York* (from 6 January to 29 June 1965), vol. 1 (New York: The Council, 1965), 42–43, 245–46.

17. Newbold Morris to J. H. Mulcahy, 29 January 1962, DPR, Administration, 1962, folder 3; memorandum by S. M. White, "Street Tree Planting Program for East Harlem," 8 August 1960, DPR, Administration, 1960, folder 18; memorandum by William Kirk, "Additional Tree-Planting in East Harlem," 11 July 1962, and commissioner to Hon. Robert Low, 7 August 1962, DPR, Manhattan, 1962, folder 20.

18. See the survey conducted by John F. Kraft, Inc., "Attitudes by Harlem Residents toward Housing, Rehabilitation and Urban Renewal, August 1966," published in United States Senate, 89th Cong., 2nd Sess., Committee on Government Operations, *Federal Role in Urban Affairs, Hearings before the Subcommittee on Executive Reorganization, August 31; September 1, 1966,* part 6 (Washington, DC: U.S. Government Printing Office, 1966), 1409–23. The study was referenced widely in the literature of the time. See, for example, Whitney North Seymour Jr., "An Introduction to Small Urban Spaces," in Seymour, *Small Urban Spaces,* 1–10 (7–8); and Charles E. Little, "The Double Standard of Open Space," in *Environmental Quality and Social Justice in Urban America,* ed. James Noel Smith (Washington, DC: Conservation Foundation, 1974), 73–84. Fred Powledge, "Haryou Project Is Found Lacking: Panel Calls Summer Plan Victim of Inefficiency," *NYT,* 18 January 1966, 22; Alphonso Pinkney and Roger R. Woock, *Poverty and Politics in Harlem: Report on Project Uplift 1965* (New Haven: College and University Press, 1970), 135–36; Sidney H. Alexander Jr. to Newbold Morris, 3 August 1965, DPR, Administration, 1965, folder 5. On HARYOU-ACT, the embezzlement and misappropriation of funds, and the power play involved, see Irwin Unger, *The Best of Intentions: The Triumphs and Failures of the Great Society under Kennedy, Johnson, and Nixon* (New York and London: Doubleday, 1996), 148–97. On Bouza, see John Kostouros, "Trees, Crime, and Tony Bouza," *American Forests* 95, nos. 9–10 (1989): 41–44; Anthony V. Bouza, "Trees and Crime Prevention," in *Proceedings of the Fourth Urban Forestry Conference, St. Louis, Missouri, October 15–19, 1989,* ed. Phillip D. Rodbell (Washington, DC: American Forestry Association, 1989), 31–32.

19. *Annual Report of the Tree Planting and Fountain Society,* 1895, 16, 27–28. See A. A. Low and Lewis Collins, "Circulars (No. 11)," in *Annual Report of the Tree Planting and Fountain Society,* 1896, 29–30; *Annual Report of the Tree Planting and Fountain Society of Brooklyn, N.Y., December, 1898* (Brooklyn: Eagle Book and Job Printing, 1899), 11–12; Charles Thaddeus Terry, "Prosecutions for Injuries to Trees," in *Annual Report of the Tree Planting Association of New York City,* 1906 (New York: Tree Planting Association, n.d.), 15–16.

20. "Bill Would Force Planting of Trees," *NYT*, 17 June 1959, 37; Thomas W. Ennis, "Bill Seeks Trees for City Streets," *NYT*, 17 May 1959, R1; *Proceedings of the Council of the City of New York from January 5th, 1959 to June 25th, 1959*, vol. 1 (New York: The Council, 1959), 2312–13.

21. Charles Abrams, "Planting Trees Favored: Bill to Authorize Area Planting by Boroughs Viewed as Solution," *NYT*, 1 August 1959, 16; Charles Abrams, "More Trees Advocated: Bill Favored to Stimulate Planting, Either by Owners or by City," *NYT*, 22 October 1959, 36; "Memorandum in Support of Bill to Amend the City Charter to Authorize the Designation of 'Tree Areas' by the City Planning Commission," DPR, Administration, 1958, folder 16; press release from the office of Charles Abrams, for a.m. papers, October 21 and thereafter, CAP.

22. Robert Moses to Stanley H. Lowell, 2 October 1958, DPR, Administration, 1959, folder 18; Robert Moses to Joseph T. Sharkey, 12 November 1959; memorandum by Robert Moses, 22 October 1959, "New York Times Clipping with Letter from Charles Abrams," DPR, Administration, 1959, folder 18. Memorandum by Meyer Scheps, "Proposed Local Law for Designation of Tree Areas"; memorandum by William S. Lebwohl, 26 September 1958, "Proposed Local Law for Designation of Tree Areas," DPR, Administration, 1958, folder 16. "Forestry Personnel," 20 August 1959, DPR, Administration, 1959, folder 15.

23. For Abrams and the battle of Washington Square Park, see Robert Fishman, "Revolt of the Urbs: Robert Moses and His Critics," in *Robert Moses and the Modern City*, ed. Hilary Ballon and Kenneth T. Jackson (New York and London: Norton, 2007), 122–29.

24. Joseph C. DiCarlo to Charles Abrams, 29 December 1959; Mrs. Jerome L. Strauss to Charles Abrams, 2 June 1959; "Bob" to Charles Abrams, 4 June 1959; Charles Abrams to members of the Temporary Committee for Tree Planting, 29 October 1959; Charles Abrams, "Memorandum in Support of a Bill to Amend the City Charter to Authorize the Planting of Trees by Borough Presidents," CAP.

25. Manhattan borough president Hulan Jack did not support the bill. Both Jack and Stanley M. Isaacs, minority leader of the City Council, doubted "the constitutionality and wisdom" of the bill. See Millard L. Midonick to Charles Abrams, 1 June 1959, CAP; Stephen J. Healy to Robert Moses, 18 June 1947; William H. Latham to Stephen Healy, 23 June 1947, DPR, Brooklyn, 1947, folder 51. David Katzoff to Robert Moses, 23 March 1948; William H. Latham to David Katzoff, 30 March 1948, DPR, Brooklyn, 1948, folder 1. Jesse Kashner, "Planting of Trees Queried," *NYT*, 2 November 1959, 30; Rose Lane to Mayor Fiorello La Guardia, 19 March 1940; Rose Lane to Commissioner Lewis Valentine, 8 June 1938; memorandum by Richard C. Jenkins, 28 March 1940, DPR, Brooklyn, 1940, folder 44. Various New York City departments' responses and memoranda related to the case of Rose Lane are in the same folder.

26. Kenneth B. Clark, *Dark Ghetto: Dilemmas of Social Power* (New York, Evanston, and London: Harper and Row, 1965), 32–33; Isidor Chein, "The Environment as a Determinant of Behavior," *Journal of Social Psychology* 39 (1954): 115–27; Gerald D. Suttles's study of the Addams area on the Near West Side of Chicago, *The Social Order of the Slum* (Chicago and London: University of Chicago Press, 1968), 73–93; Marc Fried and Peggy Gleicher, "Some Sources of Residential Satisfaction in an Urban Slum," *Journal of the American Institute of Planners* 27 (1961): 305–15 (312).

27. Oscar Newman, *Architectural Design for Crime Prevention* (Washington, DC: U.S. Government Printing Office, 1973), xv. For territoriality in 1960s social studies, see Stanford M. Lyman and Marvin B. Scott, "Territoriality: A Neglected Sociological Dimension," *Social Problems* 15, no. 2 (1967): 236–49; Robert Sommer, *Personal Space: The Behavioral Basis of Design* (Englewood Cliffs, NJ: Prentice-Hall, 1969). See also "flower garden competition manuals," box 0071B8, NYCHAC.

28. Jane Jacobs, *The Death and Life of Great American Cities* (New York: Random House, 1961). For the "Sidewalks" conference and interest in pedestrianism in New York City, see documents in PCR, box 2, folder 41; brochure "Neighborhood Street Festivals New York City," CH, 1969, box 2, folder 35.

29. For the Open Space Action Committee and Institute, see materials in PCR, box 9, folder 7. Little, "The Double Standard of Open Space," 75, 77, 83.

30. Oscar Newman, *Defensible Space* (New York: Macmillan, 1972); Oscar Newman, *Design Guidelines for Creating Defensible Space* (Washington, DC: U.S. Government Printing Office, 1976), 101–25.

31. Stephen Kaplan, "A Model of Person-Environment Compatibility," *Environment and Behavior* 15, no. 3 (1983): 311–32.

32. Geoffrey H. Donovan and Jeffrey P. Prestemon, "The Effect of Trees on Crime in Portland, Oregon," *Environment and Behavior* 44, no. 1 (2012): 3–30; Kevin L. Kalmbach and J. James Kielbaso, "Resident Attitudes toward Selected Characteristics of Street Tree Plantings," *Journal of Arboriculture* 5, no. 6 (June 1979): 124–29; Rachel Kaplan and Stephen Kaplan, *The Experience of Nature* (Cambridge: Cambridge University Press, 1989), 99–103, 161; Dale A. Getz, Alexander Karow, and J. James Kielbaso, "Inner City Preferences for Trees and Urban Forestry Programs," *Journal of Arboriculture* 8, no. 10 (October 1982): 258–63; Janet Frey Talbot and Rachel Kaplan, "Needs and Fears: The Response to Trees and Nature in the Inner City," *Journal of Arboriculture* 10, no. 8 (August 1984): 222–28; Roger S. Ulrich, "The Role of Trees in Human Well-Being and Health," in Rodbell, *Proceedings of the Fourth Urban Forestry Conference*, 25; Herbert W. Schroeder, "Psychological Value of Urban Trees: Measurement, Meaning, and Imagination," in *Proceedings of the Third National Urban Forestry Conference, Orlando, Florida, December 7–11, 1986*, comp. Ali F. Phillips and Deborah J. Gangloff (Washington, DC: American Forestry Association, 1987), 55–60.

33. Joan M. Welch, "An Assessment of Socio-Economic and Land-Use Histories That Influence Urban Forest Structure: Boston's Neighborhoods of Roxbury and North Dorchester" (PhD diss., Boston University, 1991). One of the few studies predating Welch's and dealing with urban forest cover and socioeconomics is Gary L. Talarchek's study of New Orleans, "The Urban Forest of New Orleans: An Exploratory Analysis of Relationships," *Urban Geography* 11, no. 1 (1990): 65–86. For notes on the imbalanced distribution of trees in American cities, also see Donald Appleyard, "Urban Trees and Forests, What Do They Mean?" (working paper no. 303, Institute of Urban and Regional Development, University of California, Berkeley, 1978), 17–19.

34. Dana W. Cole, "Oakland Urban Forestry Experiment: A Cooperative Approach," *Journal of Forestry* 77, no. 7 (1979): 417–19; Richard G. Ames, "The Sociology of Urban Tree Planting," *Journal of Arboriculture* 6, no. 5 (1980): 120–23; Fred Sklar and Richard G. Ames, "Staying Alive: Street Tree Survival in the Inner-City," *Journal of Urban Affairs* 7, no. 1 (1985): 55–65.

35. Maureen E. Austin and Rachel Kaplan, "Identity, Involvement, and Expertise in the Inner City: Some Benefits of Tree-Planting Projects," in *Identity and Natural Environment: The Psychological Significance of Nature*, ed. Susan D. Clayton and Susan Opotow (Cambridge, MA: MIT Press, 2003), 205–25; Dana R. Fisher, Erika S. Svendsen, and James J. T. Connolly, *Urban Environmental Stewardship and Civic Engagement* (London and New York: Routledge, 2015), 113. For a first critical assessment of MillionTreeNYC's political ecology, see Lindsay K. Campbell, "Constructing New York City's Urban Forest," in *Urban Forests, Trees, and Greenspace: A Political Ecology Perspective*, ed. L. Anders Sandberg, Adrina Bardekjian, and Sadia Butt (London and New York: Routledge, 2015), 242–60.

36. See *Annual Report of the Tree Planting and Fountain Society of Brooklyn, 1896*, 18.

37. "Shade-Trees as Disinfectants," *NYT*, 7 April 1873, 4.

38. "Tree Planting," *NYT*, 1 March 1903, 6; E. G. Routzahn, "The Tree Planting Movement," *Chautauquan: A Weekly Newsmagazine* 41, no. 4 (1905): 337; "Trees and the City Streets," *Outlook* 76, no. 14 (1904): 766–67.

39. Routzahn, "Tree Planting Movement."

40. See "Magistrate Gives Two Boys Some Homework for Damaging Tree That Grew in Brooklyn," *NYT*, 17 May 1944, 21.

## Chapter 5. Burning Trees

1. J. Sch., "Gingko-Baum und Sumpfzypresse," *BZ*, 24 January 1946, 4. For war damage to trees in general, also see "Kontrolle von Baumschäden," *Das Gartenamt* 3, no. 2 (1954); "Stadtbäume — Wohltat und Gefahr," *Das Gartenamt* 16, no. 12 (1967): 544–45. For a 1944 report on the condition of street trees, see LAB, A Rep. 037-08, Nr. 383; Max Taut, Betrachtungen zum Aufbau Berlins," *Der Bauhelfer* 1, no. 2 (1946): 1–9 (1).

2. "Zehn Jahre lang Sprengsätze an Straßenbäumen," *TS*, 26 July 1955, 8; "Sprengsätze in Kladower Bäumen," *NZ*, 28 July 1955, 6.

3. Blauhardt to Charlottenburg district mayor, 12 March 1945, and response, "Erfassung der Ebereschenfrüchte," LAB, A Rep. 037-08, Nr. 383. "Schutz der Wildobstgehölze," *Verord-nungsblatt der Stadt Berlin* 2, no. 20, 13 May 1946, 155. For the use of mountain ash in city streets and along roads, see L. Fintelmann, *Über Baumpflanzungen in den Städten* (Breslau: Kern's, 1877), 17, 48–49; E. Petzold, *Anpflanzung und Behandlung von Alleebäumen* (Berlin: Wiegandt, Hempel und Parey, 1878), 60–61; Karl Hampel, *Stadtbäume* (Berlin: Parey, 1893), 4, 12; Verein Deutscher Gartenkünstler, *Allgemeine Regeln für die Anpflanzung und Unter-haltung von Bäumen in Städten* (Berlin: Gebrüder Bornträger, 1901); Georg Thiem, *Der Alleebaum* (Stuttgart: Ulmer, 1906), 3; Max Koernicke, "Zur Kenntnis der mährischen 'süßen' Eberesche, einer wertvollen, natürlichen Vitamin-C-Spenderin," *Die deutsche Heilpflanze* 9, no. 6 (1943): 49–54; B. Hörmann, *Unsere natürlichen Vitamin-C-Spender* (Munich: Pflanzen-werke, 1941); Horst Müller, *Die Edeleberesche* (Berlin: Deutscher Bauernverlag, 1956), 26–28; L. Späth Baumschule, ed., *Späth-Buch 1720–1920* (Berlin: Rudolf Mosse, 1920), 76–77, 214–15, 259–60; L. Späth Baumschule, ed., *Späth-Buch 1720–1930* (Berlin: Berlin-Baumschulenweg, 1930), 121, 292–94, 351. Mountain ash was appreciated for its fruit and leaves already during World War I. While medicinal properties were attributed to the fruit, which was also used in compotes, the leaves made for "German tea." See "Gehölz-Produkte," *Mitteilungen der Deutschen Dendrologischen Gesellschaft* 27 (1918): 249–58; note in *Mitteilungen der Deutschen Dendrologischen Gesellschaft* 25 (1916): 248; H. Roß, "Pflanzt Holzgewächse welche Arznei-drogen liefern," *Heil- und Gewürz-Pflanzen* 1, nos. 10–11 (1917–18): 270–77. In the years of World War II, native medicinal and spice plant sources were tested for their vitamin C con-tent, and studies determined how plants should best be dried and prepared so that their nutri-tional value was maintained. See Max Löhner, "Vitamin C in Heil- und Gewürzpflanzen," *Heil- und Gewürz-Pflanzen* 20, no. 2 (1941): 17–28. Further research was undertaken to deter-mine the usefulness of native trees and other plants for tea production. See Edgar Cordes, "Untersuchungen einheimischer Pflanzen auf ihre Eignung als Hausteepflanzen," *Heil- und Gewürz-Pflanzen* 20, nos. 3–4 (1941): 49–90. On early twentieth-century German vitamin research and its role in Nazi politics, see Heiko Stoff, "Vitaminisierung und Vitaminbestim-mung. Ernährungsphysiologische Forschung im Nationalsozialismus," *Dresdener Beiträge zur Geschichte der Technikwissenschaften* 32 (2008): 59–93; Petra Werner, *Vitamine als Mythos* (Berlin: Akademie, 1998), 11–21.

4. The fences were erected along Bieselheider Weg and Rosamundeweg in Frohnau. "Achtung-Zonengrenze," *TS*, 9 September 1951, 4; "Noch mehr Grenzzäune," *TS*, 6 March 1953, 4.

5. See photographs by Gert Schütz and Horst Siegmann, LAB, F Rep. 290 (02), Nr. 0088907 and 0077049.

6. See documents in LAB, B Rep. 004, Nr. 3644.

7. L. H., "Östliches Sprühgift schädigt West-Berliner Pflanzen," *TS*, 16 June 1966, 8. On the browntail moths, see documents in LDB, NGA, Friedrichshain Kreuzberg, no. 299; "Goldafterraupen bedrohen Grün," *BZ*, 27 September 1970, 8; L. H., "Raupen auf Weddinger Bäumen," *TS*, 11 May 1971, 8.

8. August Endell, "Frühlingsbäume," in *Zauberland des Sichtbaren* (Berlin-Westend: Verlag der Gartenschönheit, 1928), 18–26.

9. Guidelines for the replacement of street trees were instituted on 28 July 1945. See LAB, C

Rep. 109, Nr. 174. "Linden und Laternen," *BZ*, 27 September 1945, 2; L. R., "Spaziergang unter östlichen Linden," *TS*, 16 April 1950, first supplement; Heinz Dannemann, "'Unter den Linden in Berlin,'—alte Allee in neuer Gestalt," *Garten und Landschaft* 68, no. 5 (1958): 121. For the 1949 "Statut zum Schutze der Strasse 'Unter den Linden' in Bezug auf einheitliche Gestaltung des Straßenbildes, ihres Einflußraumes und der städtebaulichen Einheit," see LAB, C Rep. 109, Nr. 99.

10. For Lingner's plans, see "Besprechung mit den Leitern der Bezirks-Grünplanungsämter und Planungsämter am 6.11.46," LAB, C Rep., Nr. 1046, 1945–49; Reinhold Lingner, "Aufgaben und Ziele der Grünpflanzung in Berlin," *Berliner Gärtnerbörse* 1, no. 1 (1947): 4; Reinhold Lingner, "Aufgaben und Ziele der Grünplanung," *Der Bauhelfer* 2, no. 4 (1947): 5–11. For plans to turn the destroyed urban areas into parkland and create "green breakthroughs" (*Gründurchbrüche*), see Paul Vogler and Gustav Hassenpflug, *Gesundheitswesen in der Baugestaltung* (Berlin: Dr. Werner Saenger, 1948), 76–82.

11. "Energieeinsparung," 23 August 1943, LAB, A Rep. 037-08, Nr. 383.

12. Pertl to Charlottenburg district mayor, 20 July 1937; Pertl to Charlottenburg district mayor, 26 November 1940; Lindig to Berlin mayor, "Pflege der Neugestaltungs-Straßenbäume," 17 April 1943; Plath to Charlottenburg district mayor, "Prachtstraßenbäume," 2 February 1944; Bach-mann to Charlottenburg district mayor, "Pflege der Bäume für die künftigen Prachtstrassen," 5 August 1944, LAB, A Rep. 037-08, Nr. 383. At the Buch tree nursery in 1938, four thousand street trees were cultivated for Speer's boulevards. Nursery land that was initially to be ex-changed for land areas elsewhere was kept for nursery purposes, revealing the trees' priority in administrative planning. See documents in "Baumschule Buch," NGABP, especially the note by supervising master gardener Seelig, 18 February 1938. Hermann Mattern, "Eine Aufstel-lung von Sträuchern und Gehölzen für die Reichshauptstadt Berlin, zusammengestellt im Auftrage des General-Bauinspektors Prof. A. Speer," 1944, NL Greiner, R16, no. 37.

13. For the 1946 Magistrate's directive to use all parks, gardens, and market gardens for the culti-vation of vegetables—an order that the district Departments of Gardens sought to amend the following year, arguing that the production of fruit and street trees as well as trees for shelter-belts was essential to Berliners' well-being—see *Verordnungsblatt der Stadt Berlin*, 2, no. 2, 14 January 1946, 7–8; Hardtke to Magistrate of Greater Berlin, 21 October 1947, LAB, C Rep. 109, Nr. 175. Some street trees were planted in 1946–47, but overall, preference was given to urban farming. LAB, C Rep. 109, Nr. 174; report of meeting of directors of district Departments of Gardens, 7 May 1947, LAB, C Rep. 109, Nr. 1046; Gf., "Grünanlagen bitten um Schonung," *BZ*, 10 June 1947, 6; "Bäume ertrinken unter Schutthügeln," *BZ*, 4 March 1948, 4. For the cultivation of vegetables in parks, gardens, and nurseries during the war years, see Kurt Pöthig, "Die Kriegsmaßnahmen der städtischen Garten-, Friedhofs-, und Kleingartenverwaltung Berlins," *Die Gartenkunst* 55, no. 6 (1942): 1–2. In 1947, Lingner sought to initiate a campaign for the protection of trees, plants, and parks in general. Lingner, "Zur Pressekonferenz am 9. Juni 1947. Schutz der Grünanlagen!" NL Greiner, 16A, no. 3.

14. "Bericht des Amtes für Tiefbau über 2 Jahre Leistungen nach Kriegsende. Berichtszeit vom 2.5.1945 bis 31.3.1947," p. 4, LAB, C Rep. 109, Nr. 453. Newly planted street and park trees and tree poles were often stolen and used as firewood and to construct fences. See NZ, 22 April 1947, 3. For the cutting of trees protected under the 1922 Berlin tree ordinance, see "Sitzung der Gartenamtsleiter am 3.3.1948"; "Sitzung der Gartenamtsleiter am 2.3.1949," LAB, C Rep. 109, Nr. 1046, 1945–49. E. T., "Für ein Stück Holz in den Tod," NZ, 5 March 1947, 3; "Mehr sprengen, weniger einreißen," *BZ*, 3 January 1948, 4. Willy Alverdes, "Der grosse Tiergarten," *Garten und Landschaft* 62, no. 9 (1952): 9–19 (10). Otto Mai, "Sechs Millionen Sämlinge im Grunewald," *TS*, 27 September 1950, 4; Jockel Turm, "Aufforstung vor dem Abschluß," *TS*, 23 April 1952, 4. "Tag des Baumes," *TS*, 16 April 1953, 4; Documents in LAB, C Rep. 109, Nr. 174.

15. Volker Koop, *Tagebuch der Berliner Blockade* (Berlin: Bouvier, 1998), 45; DENA, "350 000 Kubikmeter Holz sollen geschlagen werden," *TS*, 10 October 1948, n.p.; "Um den Holzein-schlag: Magistrat will neuen Plan vorschlagen," *TS*, 8 October 1948, n.p.; "Protokoll über die

(86.) Ordentliche Sitzung der Stadtverordnetenversammlung von Groß-Berlin, Donnerstag, 21. Oktober 1948"; "Drucksache Nr. 142 für die (86.) Ordentliche Sitzung der Stadtverordnetenversammlung von Groß-Berlin am 21. Oktober 1948," LAB, C Rep. 001, Nr. 1. Directive no. USMG/107, directive no. MGBS/102, and directive no. GMFB/104 of 6 October 1948, regarding "Holzschlage-Programm in den Westsektoren von Berlin" in *Verordnungsblatt für Groß-Berlin* [West] 4, part 1, no. 46 (30 October 1948), 448, 450, 451.

16. "Amtlicher Stenographischer Bericht über die 86. Ordentliche Sitzung der Stadtverordnetenversammlung von Groß-Berlin am Donnerstag, den 21. Oktober 1948 im Studentenhaus, Hardenbergstraße," LAB, C Rep. 001, Nr. 1.

17. See "Drucksache"; "Amtlicher Stenographischer Bericht"; "Holzeinschlag in Etappen," *TS*, 14 October 1948, n.p.

18. "Geringerer Holzeinschlag," *TS*, 30 October 1948, n.p.; "Holzeinschlag hat begonnen," *BZ*, 19 October 1948, 6; "Holzzuteilung in Westberlin," *TS*, 30 November 1948; Koop, *Tagebuch*, 117; "Hilfe für Frierende amtlich verboten," *BZ*, 9 March 1949, 6.

19. "Westsektoren ohne Holz," *ND*, 3 September 1948, 4; "Blick auf Berlin," *NZ*, 3 September 1948, 3; "Aus den Bezirksparlamenten," *ND*, 16 September 1948, 4; "Spandau fällt seine Bäume," *BZ*, 17 September 1948, 6; "Magistrat gesteht seine Unfähigkeit," *BZ*, 5 October 1948, 6. On the politics of the German press in the postwar years, see Holger Impekoven and Victoria Plank, *Feigenblätter: Studien zur Presselenkung in Drittem Reich und DDR* (Münster: Scriptorum, 2004); Anke Fiedler and Michael Meyen, eds., *Fiktionen für das Volk: DDR-Zeitungen als PR-Instrument* (Berlin: Lit, 2011).

20. O. T., "Hände weg von unseren Bäumen in Berlin!" *ND*, 8 October 1948, 4. Also see O. T., "Schluss! Keinen Axthieb mehr weiter!" *ND*, 10 October 1948, 6; "Zum Schaden Berlins," *BZ*, 8 October 1948, 1; M. N., "Der Berliner Wald kann gerettet werden," *ND*, 10 October 1948, 1.

21. See the photographs published in association with O. T., "Hände weg." "Waldschlächter oder Waldverwalter," *ND*, 13 October 1948, 1; "Holzeinschlag muß verhindert werden!" *ND*, 28 October 1948, 6; *ND*, 22 October 1948, 4; "Zum Schaden Berlins," *BZ*, 8 October 1948, 1; "Proteste gegen die Abholzung," *BZ*, 13 October 1948, 6.

22. "Aus dem Grunewald wurde ein Ödland," *BZ*, 2 February 1949, 6. Ironically, a discussion about felling every second street tree arose in East Berlin in 1954 when citizens began to complain about street trees that were hindering light penetration into first- and second-floor apartments. See "Streit um die Strassenbäume," *BZ*, 4 November 1954, 6.

23. "Winternotprogramm in Spandau," *TS*, 16 September 1948, n.p.; "Keine Holzfäller im Botanischen Garten," *TS*, 13 October 1948, n.p.

24. For Tiegs's biography, see H. Melchior, "Ernst Tiegs: Ein Leben für die Deutsche Botanische Gesellschaft," *Berichte der Deutschen Botanischen Gesellschaft* 80, no. 4 (1967): 255–64; Sch., "Bäume erhalten das 'Kleinklima,'" *TS*, 10 October 1948, n.p.; "Zunächst geringer Holzeinschlag"; "Der ganze Baumbestand gefährdet," *TS*, 13 October 1948, n.p.

25. "Die Linde am Großen Stern. Berlin kommt wieder: Die ersten Anpflanzungen im Tiergarten," *Telegraf*, 18 March 1949, 4; Wo—, "Der schönste Fleck im Stadtinnern," *TS*, 18 March 1949, n.p.; Fritz Witte, "Grüne Brücke nach Berlin," *Garten und Landschaft* 61, no. 2 (1951): 1–4; "Westdeutsche Spenden für Tiergarten," *TS*, 26 May 1951, 4; "77 000 Bäume aus Niedersachsen," *TS*, 12 November 1950, 4; *TS*, 4 November 1950, 4; "Wie einst unter Buchen und Erlen," *TS*, 28 January 1951, 4. On the postwar development and planting of Tiergarten, also see Willy Alverdes, "Der grosse Tiergarten," *Garten und Landschaft* 62, no. 9 (1952): 9–19; W. Alverdes, "Berliner Denkmal," *Das Gartenamt* 1, no. 12 (1952): 11–12; Charlotte Stephan, "Der neue Tiergarten ist ganz anders," *Garten und Landschaft* 62, no. 9 (1952): 20; Willy Alverdes, "Die Neubepflanzung des grossen Tiergartens zu Berlin," *Garten und Landschaft* 65, no. 2 (1955): 3–5.

26. Wo—, "Der schönste Fleck im Stadtinnern"; "Reuter pflanzt Brennholz," *BZ*, 18 March 1949, 2. For an early account of the Tiergarten's replanting efforts, which were sponsored largely by gifts from cities in the Western zones of Germany, see Witte, "Grüne Brücke."

27. Lingner, "Wiederaufbau der Wälder und Parks," 10 July 1946, LAB, C Rep. 109, Nr. 174; "Aufforstung unseres Waldes," *BZ*, 5 May 1949, 1; "Neue Wälder werden grünen," *ND*, 5 May 1949, 2; "Der deutsche Wald im Zweijahresplan," *BZ*, 21 September 1948, 5; Lingner, "Aufgaben und Ziele der Grünplanung."

28. For forestry and the Great Stalin Plan for the Transformation of Nature, see Stephen Brain, *Song of the Forest: Russian Forestry and Stalinist Environmentalism, 1905–1953* (Pittsburgh: University of Pittsburgh Press, 2011).

29. R. Lingner, "Landschaftsplanung in Stadt und Land," lecture on 9 September 1949 in Leipzig at a meeting of the economic advisors to the Association of Farmers' Mutual Aid (Vereinigung der gegenseitigen Bauernhilfe [VdgB]), NL Greiner, 16A, no. 3; Reinhold Lingner, *Land-schaftsgestaltung* (Berlin: Aufbau-Verlag, 1952), 64; Reinhold Lingner, "Ist Deutschland von Naturkatastropen bedroht?" slide lecture, *Lichtbilder-Vorträge Nr. 2*, ed. Deutsches Friedens-komitee, n.d., NL Funcke, Akte 14. For the DEFA Wochenschau, see *Der Augenzeuge*, 1949/14, minutes 31–44. For newspaper articles, see F. Barkow, "Der sozialistische Plan zur Umgestaltung der Natur," *ND*, 25 November 1948, 4; "Frühling in Stalingrad," *ND*, 8 April 1949, 4; "Sowjetunion heute und morgen. Zum Jahrestag eines erstaunlichen Planes," *ND*, 29 October 1949, 3. Also see Otto Müller, *Die Umgestaltung der Natur in der Sowjetunion* (Berlin: Verlag der Nation, 1952). For Pniower's writings on landscape design protecting agricultural lands and on lumber production outside of forests, see NL Pniower, nos. 24, 44.

30. Rudolf Eichhorn, *Straßenbau und Straßen-Obstkultur* (Regensburg: Habbel, 1947); G. Raschke, *10 Jahre Lignikultur im Dienste des Flurholzanbaus*, ed. Forschungsstelle für Flurholzanbau der Lignikultur, Gesellschaft für Holzerzeugung außerhalb des Waldes e.V. (Reinbek: Ed. Wagner, 1956); Forstmeister Bernhardt, "Die wirtschaftliche Bedeutung des Flurholzanbaus an Straßen und Wegen," *Holz-Zentralblatt*, 12 November 1955, 1604; W. Koch, "Waldbäume an Straßen," *Natur und Landschaft* 28, no. 2 (1953): 22–24.

31. Lingner, "Wiederaufbau der Wälder und Parks," 10 July 1946, LAB, C Rep. 109, Nr. 174. On a larger scale, Lingner undertook a "landscape diagnosis" (Landschaftsdiagnose) for the restoration of East Germany's landscape. He sought to convince the SED of a plan for the transformation of nature in Germany, taking obvious clues from the Great Stalin Plan. Kurt Hueck, *Vorschläge für die Wiederbepflanzung der Grünanlagen und Schaffung von Wind-schutzpflanzungen auf pflanzensoziologischer Grundlage im Landschaftsraum von Groß-Berlin*, ed. Magistrat von Groß-Berlin (Berlin: 1948); Reinhold Lingner and Hans Georg Büchner, "Bepflanzungskonzeptionen für Städte," *Wissenschaftliche Zeitschrift der Hum-boldt-Universität zu Berlin, Mathematisch-naturwissenschaftliche Reihe* 17, no. 2 (1968): 215–30.

32. John Evelyn, *Fumifugium; or, The Inconveniencie of the Aer and Smoak of London . . .* (London: W. Godbid . . . , 1661); Otto Behre, *Das Klima von Berlin* (Berlin: Otto Salle, 1908), 146–47; A. Löbner, "Staubverteilung in einer Großstadt," *Die Umschau* 41, no. 21 (1937): 474–77; Heinrich Schulz, "Grünanlagen—die besten Staubfilter," *Volksheil* 15 (1938): 662.

33. Reisner, "Die Verunreinigung der Stadtluft," *Technisches Gemeindeblatt für Straßenbau, Landesplanung, Siedlungswesen* 42 (1939): 125–29 (127).

34. For the health hazards attributed to rubble dust, see A. Löbner, "Lufthygienische Fragen in kriegszerstörten Städten," *Gesundheits-Ingenieur* 68, no. 2 (1947): 48–50; D. Hennebo, *Staubfilterung durch Grünanlagen* (Berlin: VEB Verlag Technik, 1955), 22, 26–29; Hellmut Reifferscheid, "Staub in Trümmerstädten," *Die Umschau* 50, no. 18 (1950): 566–67.

35. Fritz Schumacher, *Probleme der Grosstadt vor und nach dem Kriege* (Cologne: E. A. See-mann, 1949), 126–27; Brigitte von Mangoldt, "Trümmerberge als grüne Oasen," *TS*, 15 July 1952, 4; Reinhold Lingner, "Bericht über bisherige Vorarbeiten am Aufbau Berlins," 23 May 1947, LAB, C Rep. 109, Nr. 154; "Besprechung mit den Gartenamtsleitern am 28. Juni 1945," C Rep. 109, Nr. 1046, 1945–49; Rolf Schwedler, "Der Wiederaufbau in West-Berlin," in *Die unzerstörbare Stadt* (Berlin: Carl Heymanns Verlag Köln-Berlin, 1953), 180–89 (181).

36. Lbg., "Kolleg über die Trümmerflora," *NZ*, 3 September 1948, 3; Karl Schmid, "Begrünung

von Trümmerbergen," *Das Gartenamt* 1, no. 12 (1952): 12–14 (14); Stephen Daniels, "Marxism, Culture, and the Duplicity of Landscape," in *New Models in Geography*, ed. Richard Peet and Nigel Thrift (London: Unwin Hyman, 1989), 2:196–220. For the planting on mining sites, see Ernst Günter, "Haldenbegrünung im Duisburger Raum," *Garten und Landschaft* 62, no. 7 (1952): 23–25; Hans-Joachim Berthold, "Erfahrungen bei der Aufforstung von Halden auf dem Gelände des westdeutschen Steinkohlenbergbaues," *Forschung und Beratung: Forstwirtschaft*, Auszüge aus Dissertationen und Forschungsaufträgen, *Schriftenreihe des Landesausschusses für landwirtschaftliche Forschung, Erziehung und Wirtschaftsberatung beim Ministerium für Ernährung, Landwirtschaft und Forsten Nordrhein-Westfalen*, no. 1 (1954): 146–50; Rudolf Heuson, *Praktische Kulturvorschläge für Kippen, Bruchfelder, Dünen und Ödländereien* (Berlin: J. Neumann, 1929). For the West German interest in climate and urban design, see Kurt Bullrich, "Durchlüftete Stadt," *Die neue Stadt* 2, nos. 8–9 (1948): 359–61 (359). Climate plans were drawn up for West Germany's heavily industrialized Ruhr region. See, for example, P. Rappaport, "Das Grün im Ruhrgebiet," *Garten und Landschaft* 62, no. 7 (1952): 21–23; Günter Hollweg, "Bioklimatische Erkenntnisse, angewendet auf den Großraum einer Industriestadt, Musterbeispiel Duisburg," *Garten und Landschaft* 64, no. 11 (1954): 6–8.

37. Günter Lehmann, "Wieviel Staub atmet der Berliner?" *TS*, 9 February 1951, 4; "Die Abteilung Wasser- und Lufthygiene im Robert-Koch-Institut in Berlin-Dahlem," BAB, R/154/130, Standort 51, Magazin M108, Reihe 89; von Mangoldt, "Trümmerberge"; Lingner, "Bericht über Vorarbeiten"; Ministerialrat Wedler, "Bepflanzbarkeit von Trümmerhalden," 20 May 1947, LAB, C Rep. 109, Nr. 154; Fritz Witte, "Grundsätze für die Trümmerhalden-Begrünung," *Das Gartenamt* 1, no. 12 (1952): 15–16; Fritz Witte, "Städteplanung, Grünplanung und Internationale Bauausstellung 1957 in Berlin," *Das Gartenamt* 6, no. 2 (1957): 54–58.

38. The protection of fertile topsoil was an important topic that had already been addressed before the war and regained attention in the postwar years. Hellmut Seier, "Vier Mark für eine Fuhre Laub," *TS*, 13 October 1950, 4; Irmgard Wolter, "Junges Grün," *TS*, 21 April 1954, 4; Gustav Rohde, "Kompostierung der Stadtabfälle," *Wissenschaftliche Zeitschrift der Humboldt-Universität zu Berlin, mathematisch-naturwissenschaftliche Reihe* 4, no. 5 (1954–55): 417–33. "Kompostierung als Weg aus der Düngernot," 26 June 1947, LAB, C Rep. 109, Nr. 154; "Verordnung zum Schutz der Muttererde vom 21.10.46 und 20.12.46—H Pla Grün I,3—für den Bereich der städt. Verwaltung," minutes of the meetings of district directors of gardens on 7 May 1947 and 21 May 1947, LAB, C Rep. 109, Nr. 1046; "Entwurf für eine Verordnung zum Schutz der Muttererde," LAB, C Rep. 102, Nr. 30; "Magistratsvorlage Nr. 180 für die Sitzung am Sonnabend, dem 13. April 1946," April 3, 1946, LAB, C Rep. 100-05, Nr. 770; Lingner, "Aufgaben und Ziele der Grünplanung," 11; Lingner, "Die Unterbringung unverwertbaren Trümmerschuttes in Berlin als Problem der Stadtplanung," July 1949, NL Greiner, 16 A, no. 3. Norbert Schindler, "Das Müllproblem," *Das Gartenamt* 1, no. 12 (1952): 5–11; Alwin Seifert, "Städtebau, gesunde und kranke Landschaft," in *Medizin und Städtebau*, ed. Paul Vogler and Erich Kühn (Munich, Berlin, and Vienna: Urban und Schwarzenberg, 1957), 91–96. Throughout the 1930s Seifert had promoted composting and the protection of topsoil, particularly with regard to the construction of the new motorways. Alwin Seifert, *Im Zeitalter des Lebendigen* (Planegg near Munich: Müllersche Verlagshandlung, 1943), 82–89.

39. Kurt Pöthig, "Über die Begrünung von Trümmerschuttflächen," *Garten und Landschaft* 61, no. 3 (1951): 1–3, 6; Fritz Witte, "Über die Trümmerberge und ihre Begrünung," *Garten und Landschaft* 62, no. 9 (1952): 8–9; Witte, "Trümmerhalden-Begrünung." Lange, "Tätigkeitsbericht des Hauptamtes für Grünplanung und Gartenbau für die Zeit vom 1.10. bis 31.12.1947," 6 January 1948; Reinhold Lingner, "Tätigkeitsbericht des Hauptamtes für Grünplanung und Gartenbau für die Zeit vom 1.1. bis 31.3.1948," 2 April 1948, LAB, C Rep. 109, Nr. 154.

40. For Heuson's relevant publications, see Heuson, *Praktische Kulturvorschläge*; Rudolf Heuson, *Bodenkultur der Zukunft* (Berlin: J. Neumann); Rudolf Heuson, *Die Kultivierung roher Mineralböden* (Berlin: Siebenreicher, 1947).

41. For the pioneering attempts to plant rubble areas in Kiel, see Max K. Schwarz, "Kiel—ein

Beispiel für die Begrünung von Trümmerschutt," *Garten und Landschaft* 61, no. 3 (1951): 4–5; Walter Baumgarten, "Abfallwirtschaft—Begrünung der Trümmerflächen—Gestaltungsfragen der Stadtlandschaft," *Baurundschau* 38, nos. 13–14 (1948): 332–36; Hans F. Kammeyer, *Begrünung enttrümmerter Flächen*, Mitteilung der Abteilung Garten- und Landschaftsgestaltung an der Versuchs- und Forschungsanstalt für Gartenbau Dresden-Pillnitz (Dresden-Pillnitz, 1951), 4; Schmid, "Begrünung von Trümmerbergen," 14; Camillo Schneider, "Begrünung von Flächen in ausgebombten Städten," *Gartenwelt* 49, no. 4 (1948): 59–60.

42. Eberhard Fink, "Die Gemeinschaft der Robinie als Motiv der Trümmerschuttbepflanzungen," *Das Gartenamt* 3, no. 1 (1954): 3; "Blick in die Fachpresse . . . Robinienstämme für Zaunpfähle," *Das Gartenamt* 2, no. 12 (1953): 247.

43. K. Wienke, "Die Methode der Direktbegrünung innerhalb der Stadt," *Deutsche Gartenarchitektur* 2, no. 2 (1961): 45–46; Kammeyer, *Begrünung enttrümmerter Flächen*, 8.

44. For documents on the Wild Grass Program, see Stadtbauamt des Magistrat von Groß-Berlin to Stadtbezirksämter, "Durchführung des Wildgrasprogramms," 2 January 1963, LAB, C Rep. 110-04, Nr. 10; LDB, NGA, Mitte, 0509, 0510, and 0514. For expenditures regarding the Wildgrasprogramm in Pankow, see documents in "Analysen 1966–1970, Teil II," NGABP. For an early evaluation and suggestions for rubble greening in Berlin, see G. Bickerich, "Gutachten über die Bepflanzung von Trümmergelände," NL Pniower, no. 19. Gottfried Funeck, "Die Entwicklung des Berliner Stadtgartenamtes und seine Auswirkung auf die städtebauliche Planung," in *Entwicklung der Volksparke*, ed. Kulturbund der Dt. Demokratischen Republik, Zentraler Fachausschuß Dendrologie u. Gartenarchitektur, Zentrales Parkaktiv/Joachim Berger and Detlef Karg (Weimar: Druckhaus Weimar, 1979), 21–32 (32); W. Weise and G. Funeck, "Erfahrungen bei der provisorischen Begrünung in der Praxis," *Deutsche Gartenarchitektur* 5, no. 3 (1964): 63–64. For the citizens' reaction to the program, see Max, "Gestrüpphain," *BZ am Abend*, 3 August 1962, n.p. [6].

45. Kammeyer, *Begrünung enttrümmerter Flächen*, 8; Hans F. Kammeyer, "Pflanzen für Trümmerflächen," *Der Landschaftsgärtner*, supplement of *Deutschen Gärtnerpost*, no. 34 (22 August 1952): 13–14. Also see Walter Pisternik, "Die Lehren aus der bisherigen Arbeit zur Verschönerung der Städte," in *Schriftenreihe Demokratischer Aufbau* (Berlin: VEB Deutscher Zentralverlag Berlin, 1954), 16; Kurt Irmscher, "Zur Trümmerflächenbegrünung," *Der Deutsche Gartenbau* 1, no. 7 (1954): 221–22.

46. D. Hennebo, *Staubfilterung durch Grünanlagen* (Berlin: VEB Verlag Technik, 1955). Later, after he had left East Berlin for West Germany, Hennebo also published on the dust-binding capacity of vegetation in West German journals. D. Hennebo, "Möglichkeiten und Grenzen der Staubfilterung durch Grünanlagen und Anpflanzungen," *Die neue Landschaft* 6, no. 11 (1961): 239–40.

47. Hennebo, *Staubfilterung*, 66. Besides Hennebo's dissertation, Pniower advised two other dissertations dealing with urban trees: Alfred Hoffmann, "Der Straßenbaum in der Großstadt unter besonderer Berücksichtigung der Berliner Verhältnisse" (doctoral diss., Humboldt University, Berlin, 1954); Harri Günther, "Über das Verhalten von Gehölzen unter großstädtischen Bedingungen, untersucht an einigen Gehölzarten in Berlin" (doctoral diss., Humboldt University, Berlin, 1959).

48. Wolfgang Böer, *Klimaforschung im Dienste des Städtebaus* (Berlin: Deutsche Bauakademie, 1954); Wolfgang Böer, "Richtlinien zur Berücksichtigung klimatischer Werte und Erkenntnisse im Städtebau," BAB, DH/2/21625, Standort 51, Magazin M304, Reihe 49. Böer also showed how forests, tree rows, and hedges slowed down wind and swirled the air. R. Lingner, "Grünflächen dienen dem Fortschritt," in *Beiträge über Fragen der Grünplanung in der DDR*, BAB, DH/2/21623, Standort 51, Magazin M304, Reihe 49.

49. Albert Kratzer, *Das Stadtklima* (Braunschweig: Friedr. Vieweg und Sohn, 1937), 126; Erich Kühn, "Hygiene und Klima in der Stadt," in *Medizin und Städtebau*, ed. Paul Vogler and Erich Kühn (Munich: Urban und Schwarzenberg, 1957), 2: 124–31 (130); Hennebo, *Staubfilterung*, 60; Georg Pniower, *Bodenreform und Gartenbau* (Berlin: Siebenreicher, 1948), 142–43.

50. "Niederschrift der Amtsleiterbesprechung am 9.2.1950," LAB, C Rep. 109, Nr. 1046. For Lingner's critique of the design procedure for the first building section of Stalinallee, see Reinhold Lingner, "Komplexe Projektierung verbilligt das Bauen," *Deutsche Architektur* 4, no. 2 (1955): 82–83.

51. Helmut Kruse, "Die *grüne* Strasse," *Sonntag*, 12 April 1953, 3. For the limited information available on Stalinallee's open-space design, see NL Greiner, no. 8. Lingner, "Komplexe Projektierung."

52. Herbert Nicolaus and Alexander Obeth, *Die Stalinallee: Geschichte einer deutschen Straße* (Berlin: Verlag für Bauwesen, 1997), 125–26, 143; Kurt Liebknecht, *Mein bewegtes Leben* (Berlin: VEB Verlag für Bauwesen, 1986), 147.

53. Josef Pertl, "Haupthochamt, Hauptplanungsamt, Dienstblatt Teil III Seite 79, ausgegeben 30.7.1941. 76, Pla VII A 2. An die Bezirksbürgermeister. Pflanzung von Straßenbäumen," LAB, A Rep. 037-08, Nr. 383; "Baumpflanzungen und Grünanlagen auf Stadtstraßen und Plätzen," *Das Gartenamt* 2, no. 2 (1953): 35–36; "Vorläufige Richtlinien für den Ausbau von Stadt-straßen (RASt)," supplement to *Verkehrsblatt des amerikanischen und britischen Besatzungs-gebietes* 1, nos. 7–8 (1947); Camillo Sitte, "Erläuterungen zu dem Bebauungsplane von Marienburg," *Der Städtebau* 1, no. 10 (1904): 141–45 (145).

54. Eberhard Fink, "Der Baum im Straßenbild der Stadt," *Das Gartenamt* 2, no. 4 (1953): 68–73. For Hoffmann's research project, see "Abschlußbericht zur Forschungsarbeit 'Grundlagen der Grünplanung,'" BAB, DH/2/23365, Standort 51, Magazin M 304, Reihe 51; A. Hoffmann with K. Kirschner, "Die Straßenbepflanzung in Städten in ihren Beziehungen zur Lufthygiene sowie zu den verkehrs- und stadttechnischen Verhältnissen," in *Fragen der Grünplanung im Städtebau*, ed. Johann Greiner and Alfred Hoffmann (Berlin: Henschel, 1955), 62–69. Parts of Hoffmann's dissertation were published in West German journals: Hoffmann, "Der Straßen-baum in der Großstadt unter besonderer Berücksichtigung der Berliner Verhältnisse," 52–53; Alfred Hoffmann, "Eine ökologische Betrachtung zum Problemkreis des Straßenbaumes," *Das Gartenamt* 4, no. 10 (1955): 182–86; A. Hoffmann, "Zur Ökologie des Strassenbaumes in den Städten," *Das Gartenamt* 5, no. 5 (1956): 14–17; A. Hoffmann, "Bäume in Stadtstrassen," *Das Gartenamt* 5, no. 5 (1956): 87–91.

55. "Einfluss der Landschaftsgestaltung auf die Windströmung," *Garten und Landschaft* 79, no. 8 (1969): 239; Victor Olgyay, "Windwirkung und Strömungsbilder," *Bauwelt* 60, no. 19 (1969): 657–65; Victor Olgyay, *Design with Climate* (Princeton: Princeton University Press, 1963), 100–101; Robert F. White, *Effects of Landscape Development on the Natural Ventilation of Buildings and Their Adjacent Areas*, Research Report 45 (College Station: Texas Engineering Experiment Station, 1954); Robert F. White, "Landscape Development and Natural Ventila-tion," *Landscape Architecture* 45, no. 2 (1955): 72–81. On the lack of German studies on the role of street trees in influencing air circulation and first assumptions, see Klaus Ermer, Eckhard Holtz, and Günter Hachmann, *Strassenbäume in innerstädtischen Bereichen* (Berlin: Technische Universität Berlin, 1974), 17–19. Kühn, "Hygiene und Klima in der Stadt," 128.

## Chapter 6. Greening Trees

1. See documents in LAB, C Rep. 109, Nr. 155.

2. Fritz Witte, "Die grüne Brücke nach Berlin," *Garten und Landschaft* 61, no. 2 (1951): 1–4 (2); Fritz Witte, "3 Jahre 'Hilfe durch Grün' im Aufbau Berlins," *Garten und Landschaft* 62, no. 9 (1952): 4–7; "Hilfe durch Grün," *Das Gartenamt* 2, no. 8 (1953): 149–50; Fritz Witte, "Zur 'Grünen Woche Berlin 1955,'" *Das Gartenamt* 4, no. 1 (1955): 1–5; *Hilfe durch Grün*, no. 2 (1953); G. Pniower, "'Hilfe durch Grün' auf der Bundesgartenschau Köln 1957," *Der Deutsche Gartenbau* 4, no. 8 (1957): 203–5.

3. For "sozialistische Stadtbegrünung," see Hinkefuss, "Exposé über die Lage der Stadtbe-grünung in Berlin," LAB, C Rep. 110-04, Nr. 6; LAB, C Rep. 110-04, Nr. 11. For the importance attributed to Lenné by East German landscape architects and planners, see Kurt Illner, "Die

Bedeutung Peter Josef Lennés für die Landschaftspflege," *Wissenschaftliche Zeitschrift Humboldt-Universität Berlin, Mathematisch-naturwissenschaftliche Reihe* 16, no. 3 (1967): 373–78; Reinhold Lingner, "Die Bedeutung Peter Josef Lennés für die moderne Garten-architektur," *Wissenschaftliche Zeitschrift Humboldt-Universität Berlin, Mathematisch-naturwissenschaftliche Reihe* 16, no. 3 (1967): 361–70.

4. Reinhold Lingner, "Komplexe Projektierung verbilligt das Bauen," *Deutsche Architektur* 4, no. 2 (1955): 82–83; Hans Georg Büchner, "Bäume in allen Straßen," *Landschaftsarchitektur* 15, no. 1 (1986): 10–12 (10). For the egalitarian attitude toward architecture and landscape architecture, see Hinkefuss, "Exposé über die Lage der Stadtbegrünung in Berlin"; E. B. Zalesskaja, "Park- und Erholungsanlagen in der Sowjetunion," *Deutsche Gartenarchitektur* 9, no. 1 (1968): 1–4 (1); "Junges Grün auf Berlins Straßen," *BZ*, 8 November 1951, 6; Johannes Greiner, "Stadtentwicklung und Großvegetation," NL Greiner, R16A, no. 2; D.-K. Gandert and H.-J. Albrecht, "Beiträge zur Gehölzkunde 1977," *Beiträge zur Gehölzkunde* (1977): 3–5 (3); H. Lichey, "Straßenbäume und Stadtverkehr," *Deutsche Gartenarchitektur* 3, no. 3 (1962): 78–79.

5. Rud. Kühn, "Ästhetik des Strassenbaumes," *Das Gartenamt* 6, no. 12 (1957): 248–50 (249); Harri Günther, "Über das Verhalten von Gehölzen unter großstädtischen Bedingungen, untersucht an einigen Gehölzarten in Berlin" (doctoral diss., Humboldt University, Berlin, 1959), 194; Hinkefuss, "Exposé über die Lage der Stadtbegrünung in Berlin."

6. Här, "Großreinemachen bis Ende Juni," *BZ*, 20 May 1952, 6. Whereas the planting of thirty-five hundred more trees was planned, the *Berliner Zeitung* reported forty-seven hundred were planted in May 1952; "Junges Grün." On the early tree-planting initiatives in East Berlin, also see *ND*, 11 June 1949, 6; "Grosser Filmball im November," *NZ*, 17 October 1953, 6; "Kleine Berliner Chronik," *BZ*, 4 July 1950, 6; "Lustgarten wird aufgeforstet," *BZ*, 7 April 1951, 6; "Berlins Straßen werden schattiger," *NZ*, 19 April 1952, 5; "Fünf Wochen Großreinemachen in Berlin," *NZ*, 20 May 1952, 5.

7. "Verordnung über die Verwendung des Derbholzes der Strassenbäume," *Gesetzesblatt der Deutschen Demokratischen Republik*, no. 38, 25 March 1953; "Anordnung zur Erfassung von Abfallholz aus den Betrieben und von den Baustellen, vom 3. Mai 1952," *Verordnungsblatt für Groß-Berlin* 8, part 1, no. 22 (3 June 1952), 226–27; documents in "Nicht zuzuord. Unterlagen N-Z," NGABP.

8. For the dissolution of the Hauptamt für Grünanlagen and the development of a new Department of Gardens in East Berlin, see Gottfried Funeck, "Die Entwicklung des Berliner Stadtgartenamtes und seine Auswirkung auf die städtebauliche Planung," in *Entwicklung der Volksparke*, ed. Kulturbund der Dt. Demokratischen Republik, Zentraler Fachausschuß Dendrologie u. Gartenarchitektur, Zentrales Parkaktiv/Joachim Berger and Detlef Karg (Weimar: Druckhaus Weimar, 1979), 21–32 (32); H. Lichey, "Stellungnahme zu Fragen der Neuordung des Grünflächenwesens der Stadt Groß-Berlin," 17 August 1959, LAB, C Rep. 110-04, Nr. 7; Hinkefuss to Stadtbaudirektor Gisske, "Exposé zur Lage und Struktur der Stadtbegrünung in Berlin," 28 April 1958; Hinkefuss, "Exposé über die Lage der Stadtbe-grünung in Berlin," LAB, C Rep. 110-04, Nr. 6. Also see Klaus-Dietrich Gandert, "Entwick-lungstendenzen bei der Bewirtschaftung kommunaler Grünanlagen, dargestellt am Beispiel Berlin," *Wissenschaftliche Zeitschrift der Humboldt-Universität, Mathematisch-naturwissen-schaftliche Reihe* 9, no. 5 (1959–60): 757–69; Hans Georg Büchner, "Die Entwicklung der Verwaltung des öffentlichen Grüns in Berlin-Ost 1948 bis 1990," in *Denkmalpflege und Gesellschaft*, ed. Thomas Drachenberg, Axel Klausmeier, Ralph Paschke, and Michael Rohde (Rostock: Hinstorff, 2010), 199–208, and in a more elaborate typescript from 1992 shared with the author.

9. Walter Funcke, "Der Mensch verändert das Antlitz der Erde," lecture presented at the Kulturbundhaus in Potsdam on 15 November 1950, NL Funcke, Akte 210.

10. *Verordnung über die Erhaltung, Pflege und den Schutz der Bäume—Baumschutzverordnung*, 28 May 1981, paragraph 7 (3). The 1981 ordinance replaced the first East Berlin tree ordinance

signed into law in 1976, which focused solely on the trees' protection, care, and replacement. See "Ordnung zum Schutz der Bäume auf dem Territorium der Hauptstadt der DDR Berlin—Baumschutzordnung," in *Verordnungsblatt für Groß-Berlin* 32, no. 10 (17 May 1976), 33–36. Magistrat von Berlin, Hauptstadt der DDR, Stadtgartenamt, "Baumschutzordnung," in *Bäume in der Stadt. Ordnung über die Erhaltung, die Pflege und den Schutz der Bäume auf dem Territorium der Hauptstadt der DDR, Berlin* (Berlin, 1982), 3–11.

11. "Aufgaben und Ziele des Arbeitsausschußes 'Stadtbegrünung' im Fachverband Bauwesen der Kammer der Technik, Bez. Berlin," LAB, C Rep. 110-04, Nr. 6; Dr. Hennebo and Dr. Zander, *Anleitungen zur Grundlagenforschung in Grünplanung und Gartenkunst* (Berlin: Bernhard Patzer, 1956).

12. Johann Greiner, *Grünanlagen für mehrgeschossige Wohnbauten* (Berlin: VEB Verlag für Bauwesen, 1966), 35–36.

13. For the liquidation of the city's nurseries and the future provision of all trees and shrubs by the VE Baumschulen, see Lichey, "Direktive für die Ausarbeitung des Perspektivplanes zur Entwicklung und Bewirtschaftung des Stadtgrüns bis 1970," LDB, NGA, Mitte, 0650. For poplar production in GDR forestry, see Hans-Friedrich Joachim, "Zur Pappel- und Weidenforschung und zum Anbau dieser schnellwüchsigen Baumarten," in *Umweltschutz in der DDR*, ed. Hermann Behrens and Jens Hoffmann (Munich: oekom, 2007), 2:107–26.

14. Walter Meissner and Hans-Joachim Dreger, "Bäume in Berlin, Hauptstadt der DDR," in *100 Jahre Arboretum (1879–1979)*, ed. Walter Vent (Berlin: Akad.-Verl., 1980), 83–92 (89–90).

15. H. Linke, "Großbaum im Schlauchcontainer," *Landschaftsarchitektur* 6, no. 1 (1977): 24–26; Harald Linke, "Baumnachwuchs für Neubauviertel," *ND*, 3 April 1975, 3; Harald Linke and Hans Prugger, "Verfahren zur Anzucht von Großbäumen," DDR Wirtschaftspatent A01g/186.925, filed 26 June 1975 and issued 5 June 1976; Wolfgang Fischer, "Beitrag zur Entwicklung eines Anzuchtsverfahrens von Stark- und Großbäumen im Container" (doctoral diss., Technical University Dresden, 1985), 130. Trees from abroad, for example, an elm tree variety resistant to Dutch elm disease from the United States, were imported and tested on their adaptability to the Berlin climate. See Joachim Kutz, "Neue Straßen- und Alleebaumselektionen sowie neue Sorten strauchartiger Gehölze für städtische Grünanlagen," *Landschaftsarchitektur* 17, no. 4 (1988): 117–18. The best tree crops produced in GDR nurseries were sold to West Germany to gain the highly sought-after foreign exchange.

16. Georg Pniower, "Über die Verpflanzung starker Gehölze," *Der Deutsche Gartenbau* 1, no. 1 (1954): 25–29; A. Hoffmann and K. Kirschner, "Arbeitsmethodik und Arbeitsorganisation beim Verpflanzen größerer Bäume," *Der Deutsche Gartenbau* 2, no. 2 (1955): 95–96.

17. Hans Georg Büchner, "Industriemäßige Produktion und ihre Auswirkung auf die Gestaltung der Grünanlagen," *Wissenschaftliche Zeitschrift der Humboldt-Universität zu Berlin, Mathematisch-naturwissenschaftliche Reihe* 17, no. 2 (1968): 207–9.

18. "Neue Technik," appendix in Felix Arnold and Harry Hannemann, "Wie soll die Betriebsparteiorganisation mit dem Plan, dem Vertragssystem und der Statistik arbeiten?" *Der Parteiarbeiter*, no. 4 (Berlin: Dietz, 1961): 134–41. "Perspektivplan 1965–1970"; Lichey, "Direktive für die Ausarbeitung des Perspektivplanes zur Entwicklung und Bewirtschaftung des Stadtgrüns bis 1970," LDB, NGA, Mitte, 0650. See comments by various district Departments of Gardens on operational standards in LDB, NGA, Mitte, 0404, Allgemeines.

19. H. Lüderitz, "Probleme mit den Bäumen an Stadtstraßen und Beschreibung eines Gerätes zum Ausbohren von Baumstubben," *Landschaftsarchitektur* 13, no. 1 (1984): 26–27. In 1946, many of the remaining tree stumps in Tiergarten were removed with the help of explosives. See Finanzamt für Liegenschaften des Magistrats der Stadt Berlin to Hauptamt für Grünplanungen, 27 February 1946, LAB, C Rep. 109, Nr. 318, 1945–48: Notprogramm zur Neugestaltung des Tiergartens. For the use of explosives for the removal of tree stumps and tree felling—possibly inspired by World War I—see Georg Bela Pniower, "Baumfällen durch Sprengung," *Der Deutsche Gartenarchitekt* 4, no. 10 (1927): 120; Georg B. Pniower, "Aus der Werkstatt des Gartengestalters," *Haus Hof Garten* 49, no. 42 (1927): 502–4.

20. Lichey, "NV 'Baumgruben auf Trümmerboden,'" 15 November 1971, LDB, NGA, Mitte, 0404, Allgemeines; Klaus-Dietrich Gandert, "Möglichkeiten und Forderungen für die rationelle Gestaltung von Vegetationsflächen in Grünanlagen," *Wissenschaftliche Zeitschrift der Humboldt-Universität zu Berlin, Mathematisch-naturwissenschaftliche Reihe* 17, no. 2 (1968): 185–91.

21. Gandert, "Rationelle Gestaltung von Vegetationsflächen."

22. K.-D. Gandert and K. Ostwald, "Zur Methode der Bepflanzung in Städten," *Deutsche Gartenarchitektur* 7, no. 3 (1966): 63–67; Greiner, *Grünanlagen*, 105–26. For a tree-planting concept for a new Berlin neighborhood, see O. Foth, "Gehölzkonzeption für den Wohnkomplex—Salvador-Allende-Viertel—in der Hauptstadt Berlin," *Landschaftsarchitektur* 10, no. 1 (1981): 12.

23. Reinhold Lingner and Hans Georg Büchner, "Bepflanzungskonzeptionen für Städte," *Wissenschaftliche Zeitschrift der Humboldt-Universität zu Berlin, Mathematisch-naturwissenschaftliche Reihe* 17, no. 2 (1968): 215–30. For Halle-Neustadt, see H. G. Büchner, "Bepflanzungskonzeption für Städte, dargestellt am Beispiel Halle-Neustadt," *Deutsche Gartenarchitektur* 8, no. 3 (1967): 61–62. Greiner, *Grünanlagen*, 101–26; Johann Greiner and Helmut Gelbrich, *Grünflächen der Stadt* (Berlin: VEB Verlag für Bauwesen, 1972), 171–78.

24. E. Zinn, "Die Bepflanzungskonzeption für Eisenhüttenstadt und die Entwicklung des Gehölzbestandes seit 1958," *Landschaftsarchitektur* 10, no. 1 (1981): 13–15; Andreas Seidel, "Erste sozialistische Stadt im Grünen," in *Eisenhüttenstadt: "Erste sozialistische Stadt Deutschlands*," ed. Arbeitsgruppe Stadtgeschichte (Berlin: be.bra, 1999), 70–87 (76–80); Elisabeth Knauer-Romani, *Eisenhüttenstadt und die Idealstadt des 20. Jahrhunderts* (Weimar: VDG, 2000), 51–67; Lingner and Büchner, "Bepflanzungskonzeptionen für Städte"; Kurt W. Leucht, *Die erste neue Stadt in der Deutschen Demokratischen Republik* (Berlin: VEB Verlag Technik, 1957), 51; Walter Funcke, "Stalinstadt—erste sozialistische Stadt im Grünen," *Deutsche Gärtnerpost* 5, no. 51 (1953): 3–4.

25. Lingner, "Gutachten zur Projektierung der Grünanlagen in Stalinstadt durch das Volkseigene Entwurfsbüro für Industriebahnbau Abt. Grünplanung, Chefarchitekt Walter Funcke," 22 September 1953, NL Funcke, Akte 14. For Funcke, see Susanne Karn, *Freiflächen- und Landschaftsplanung in der DDR: Am Beispiel von Werken des Landschaftsarchitekten Walter Funcke (1907–87)* (Münster: Lit, 2004), 38, 138–47, 265–67.

26. Gandert and Ostwald, "Zur Methode der Bepflanzung."

27. See *Richtlinie zur Ausarbeitung von Generalbebauungsplänen der Städte, Entwurf* (Berlin: Deutsche Bauakademie, Institut für Städtebau und Architektur, May 1966), 62; *Richtlinie zur Ausarbeitung von Generalbebauungsplänen der Städte . . . Nachdruck* (Weimar: Hochschule für Architektur und Bauwesen, Sektion Gebietsplanung und Städtebau, 1972), 1. G. Brause et al., *Planwerk Generalbebauung der Städte: Karten und Pläne* (Berlin: Deutsche Bauakademie, 1969). For the Urban Design Regulation and tree plantings, see "Prinzipien für die Planung und Gestaltung der Städte der DDR. Entwurf," 11; "'Grundsätze für die städtebauliche Planung und Gestaltung der Städte der Deutschen Demokratischen Republik,' in Auswertung der 4. Baukonferenz überarbeiteter Entwurf, 3.12.1965," 19–20; "Deutsche Bauakademie, Grundsätze der Planung und Gestaltung der Städte der DDR in der Periode des umfassenden Aufbaus des Sozialismus," February 1965, 21; "Entwurf. Feindisposition für die Grundsätze des Städtebaus (Aufgrund der Aufgabenstellung vom 23.12.65)," 29 December 1965 and 31 December 1965, BAB, DH/1, 18297, Standort 51, M 303, Reihe 80. Stefan Klotz, "Flora und Vegetation in der Stadt, ihre Spezifik und Indikatorfunktion," *Landschaftsarchitektur* 17, no. 4 (1988): 104–7.

28. The full conference title was "The Street Tree in the Metropolis—Its Use along Streets, Main Transportation Corridors, and in City Centers" ("Der Straßenbaum in der Großstadt—Seine Verwendung an Straßen, Magistralen und in Stadtzentren"). H. Lichey, "Der Straßenbaum in der Großstadt," *Deutsche Gartenarchitektur* 7, no. 2 (1966): 37–39 (38); "Niederschrift zum Erfahrungsaustausch 'Straßenbaum' Stadtgartenamt 27.1.1966," LDB, NGA, Mitte, 0410, Allgemeines; P. Rullmann, "Baumpflanzungen auf dem Alexanderplatz," *Deutsche Gartenarchitektur* 10, no. 2 (1969): 67.

29. Lichey, "Straßenbaum in der Großstadt," 37, 39; Ministerium für Verkehrswesen, Hauptverwaltung des Straßenwesens—Forschungsgemeinschaft Städtischer Verkehr, Arbeitsgruppe Stadtstraßen, ed., *Richtlinie für Stadtstraßen [RIST], April 1966* (Berlin: Deutsche Bauinformation, 1967), 50–51. For Lichey's interest in street trees, see W. Meißner and G. Funeck, "Stadtgartendirektor H. Lichey 60 Jahre," *Deutsche Gartenarchitektur* 11, no. 2 (1970): 49; Lichey, "Straßenbäume und Stadtverkehr"; H. Lichey, "Die Aufgaben der Berliner Stadtbegrünung," *Deutsche Gartenarchitektur* 5, no. 3 (1964): 50–51 (50).

30. Aufbauamt Gruppe Grünflächen to Fachinspektion Abt. Aufbau, "Ausführung von Gartenanlagen im Bezirk Mitte," 21 December 1950; Aufbauamt Gruppe Grünflächen to VVG Baumschulen L. Späth, "Ausführung von Gartenanlagen im Bezirk Mitte," 15 December 1950; Aufbauamt Gruppe Grünflächen to Amt für Grünplanung, "Bericht über die Bauvorhaben der Gruppe Grünflächen Mitte im Monat April," 5 May 1950, LDB, NGA, Mitte, 0409, Allgemeines.

31. Wolfang Reckling, "Erfassung des Straßenbaumbestandes Berlins, Hauptstadt der DDR; Maßnahmen zur Pflege und Erhaltung des Straßenbaumbestandes" (Diplomarbeit [master's thesis], Humboldt University, Berlin, 1974).

32. K.-D. Gandert, "Über die Verwendung von Bäumen in Berliner Grünanlagen," STUG 002–48, "Materialsammlung Baumschutz 1," pp. 31–33.

33. Harald Engler, *Wilfried Stallknecht und das industrielle Bauen. Ein Architektenleben in der DDR* (Berlin: Lukas, 2014), 31–37; Greiner and Gelbrich, *Grünflächen der Stadt*, 176–78; Alfred Wagner, "Von Bau- und Baumfreiheit," *BZ*, 17 December 1972, 1; Gandert, "Verwendung von Bäumen."

34. Sander to Funeck, 20 January 1981, "Magistratbeschluß 365/80 'Maßnahmen zur Erhöhung des Baumbestandes in der Hauptstadt der DDR—Berlin'"; Funeck to Sander, 23 February 1981, LDB, NGA, Mitte, 0647, Allgemeines. H. Matthes, "Die Bearbeitung einer Bepflanzungskonzeption für das Territorium der Hauptstadt der DDR/Berlin," *Landschaftsarchitektur* 10, no. 1 (1981): 9–10 (10); Hans Georg Büchner, "Zum geforderten und zum erreichten Niveau der Freiraumgestaltung in Berlin/Ost," *Landschaftsarchitektur* 20, no. 4 (1990): 109–10. For the preparation of new trees to be delivered to East Berlin by 1980, see Reckling, "Erfassung des Straßenbaumbestandes Berlins." For the phytosociological map of Berlin, see Hueck, "Vorschläge für die Wiederbepflanzung."

35. Greiner, *Grünanlagen*, 54; Büchner, "Bäume in allen Straßen," 12; Dorothea Krause, Wolfgang Krause, and Holger Fahrland, "Baum und Stadt—Durchgrünung des dicht-besiedelten Stadtbezirks Berlin-Prenzlauer Berg," *Architektur der DDR* 35, no. 10 (1986): 604–9.

36. Numbers vary in contemporary and secondary literature. See the brochure by Falk Trillitzsch, *Baum, Stadt, Strasse*, ed. Der Senator für Stadtentwicklung und Umweltschutz (Berlin: Gerike GmbH, 1985), 18; fa., "Eine Milliarde für das Grün in der Stadt," *TS*, 11 July 1976, 16; fa., "10 000 neue Straßenbäume für neun Millionen Mark," *TS*, 21 May 1976, 10; fa., "Überprüfung der Straßen, wo Lücken für Bäume sind," *TS*, 14 September 1975, 16; Herbert Krause, "Pflanzungen an Straßen," in *Berlin und seine Bauten*, part 11, *Gartenwesen* (Berlin, Munich, and Düsseldorf: Wilhelm Ernst und Sohn, 1972), 143–52 (148); Norbert Schindler, "Berliner Stadtgrün," lecture on 19 September 1972 in *100 Jahre Berliner Grün* (Berlin: Senator für Bau- und Wohnungswesen, 1970), 54; Rudolf Kühn, *Die Straßenbäume* (Hannover, Berlin, and Sarstedt: Patzer, 1961), 82; "Mehr als 200 000 Straßenbäume," *TS*, 23 April 1961, 11; Witte, "Zur 'Grünen Woche,'" 3.

37. "Platanen auf dem Kurfürstendamm," *TS*, 22 March 1950; "Platanen oder Linden?" *TS*, 23 January 1951, 4; *TS*, 5 April 1951, 4.

38. Bonatz to Hauptämter für Bau- und Wohnungswesen and Bezirksämter der Westsektoren, "Straßenbaumpflanzungen," 14 May 1949; Gartenamt to Tiefbauamt, 6 January 1949, LAB, A Rep. 044-08, Nr. 275.

39. Helmut Schildt, "Der Strassenbaum im neuen Städtebild," *Garten und Landschaft* 59, nos. 5–6 (1949): 17–19 (19).

40. Carl Hampel, *Stadtbäume* (Berlin: Paul Parey, 1893), 13–15; Joseph Stübben, *Der Städtebau*, 9. Halb-Band des Handbuches der Architektur, Vierter Teil (Darmstadt: Arnold Bergsträsser, 1890), 439–58, esp. 441–42; F. Zahn, "Der praktische und ästhetische Wert der Bäume und Vorgärten im Städtebau," *Die Gartenkunst* 8, no. 5 (1906): 75–76.

41. Camillo Sitte, "Greenery within the City," appendix 1, in George Collins and Christiane Crasemann Collins, *Camillo Sitte: The Birth of Modern City Planning* (New York: Rizzoli, 1986), 303–21 (309). Sitte published these ideas in a series of articles in 1900 in *Der Lotse* and subsequently as an appendix to his influential 1909 and 1922 editions of *Der Städtebau nach seinen künstlerischen Grundsätzen* (City Planning According to Artistic Principles).

42. For Erwin Barth, Walter von Engelhardt, and street trees, see Wilhelm Koch, "Der Baum im Straszen- und Städtebild," *Die Gartenkunst* 52, no. 9 (1939): 185–91, and W. v. Engelhardt, "Unter welchen Voraussetzungen und Bedingungen dient der Baum am besten der Schönheit des Straßenbildes?" *Die Gartenkunst* 34, no. 7 (1921): 79–81, 89–96. Magistrat to Bezirksamt Kreuzberg, Gartenamt, "Baumpflanzungen in Straßen mit Vorgärten," 6 September 1927, LAB, A Rep. 007, Nr. 68.

43. I. V. Dr. Maretzky to Bezirksbürgermeister des Verwaltungsbezirks Charlottenburg, "Entfernung von Bäumen," 6 April 1936; "Haupthochamt, Hauptplanungsamt, Dienstblatt Teil III Seite 79, ausgegeben 30.7.1941. 76, Pla VII A 2. An die Bezirksbürgermeister. Pflanzung von Straßenbäumen"; "Auszugsweise Abschrift der Rdvfg. des Obm. V. 8.10.38," LAB, A Rep. 037-08, Nr. 383. Franz Kolbrand, "Die deutsche Stadt im Grünen," *Die Gartenkunst* 49, no. 3 (1936): 44–47 (47); Alwin Seifert, "Die Hausgärten auf der Düsseldorfer Ausstellung," *Die Gartenkunst* 50, nos. 7–8 (1937): 170–75 (175); Josef Pertl and Michael Mappes, *Vorgärten: So oder so?* (Halle: Ewald Ebelt, 1938), 6–12. On the politicization of front yard and street design in 1930s Berlin, see Sonja Dümpelmann, "American System and Italian Beauty: Transatlantische Aspekte in der Park- und Freiraumplanung Anfang des 20. Jahrhunderts," *Die Gartenkunst* 18, no. 1 (2006): 119–42 (137–38).

44. Th. Falck, "Bäume an Stadtstraßen," *Bauamt und Gemeindebau* 22, no. 4 (1940): 32; Th. Falck, "Bäume an Stadtstraßen (Fortsetzung)," *Bauamt und Gemeindebau* 22, no. 5 (1940): 36.

45. Hans Bernhard Reichow, *Organische Stadtbaukunst* (Braunschweig: G. Westermann, 1948). Kühn, "Ästhetik des Strassenbaumes," 249.

46. Herbert Göner, "'Wie atmet die Stadt?' Was sagt der Städtebauer dazu?" *Gesundheits-Ingenieur* 55, no. 31 (1932): 373–75 (374). J. Goldmerstein and K. Stodieck, *Wie atmet die Stadt?* (Berlin: VDI-Verlag, 1931).

47. For East Germany, see Alfred Hoffmann, "Der Straßenbaum in der Großstadt unter besonderer Berücksichtigung der Berliner Verhältnisse" (doctoral diss., Humboldt University, Berlin, 1954), 44–50; Ingrid Fisch, K. Horn, and W. Muschter, "Über die hygienische Bedeutung von Grünflächen für das Stadtklima," *Deutsche Gartenarchitektur* 9, no. 2 (1968): 34; I. Ormos, "Bedeutung von Baumpflanzungen in Städten," *Deutsche Gartenarchitektur* 7, no. 2 (1966): 36–37 (37). For the surging interest in bioclimatological issues, see, for example, B. de Rudder and F. Linke, eds., *Biologie der Großstadt* (Dresden: Steinkopf, 1940). The volume collected papers presented at the Fourth Frankfurt Conference for the Collaboration between Medicine and the Natural Sciences, 9–10 May 1940.

48. On Hellpach's role in the development of environmental psychology and his attitudes toward Nazi racial ideologies, see Rudolf Miller, "Umweltpsychologische Aspekte im Werk von Willy Hellpach," in Willy Hellpach, *Beiträge zu Werk und Biographie*, ed. Walter Stallmeister and Helmut E. Lück (Frankfurt: Peter Lang, 1991), 1:88–89; Horst Gundlach, "Willy Hellpachs Sozial- und Völkerpsychologie unter dem Aspekt der Auseinandersetzung mit der Rassenideologie," in *Rassenmythos und Sozialwissenschaften in Deutschland*, ed. Carsten Klingemann (Opladen: Westdeutscher Verlag, 1986), 242–76.

49. Willy Hellpach, *Mensch und Volk der Großstadt* (Stuttgart: Ferdinand Enke, 1952), 49; Willy Hellpach, *Geopsyche: Die Menschenseele unterm Einfluß von Wetter, Klima, Boden und Landschaft* (Leipzig: Wilhelm Engelmann, 1939), 218–43, esp. 218–23. For references to Hellpach in street tree literature, see Hoffmann, "Der Straßenbaum in der Großstadt unter

besonderer Berücksichtigung der Berliner Verhältnisse," 44–46; "Grundlagen der Grünplanung, Begrünung von Straßen und Plätzen, Teilergebnis 3," 23–29, BAB, DH/2/23365,
Standort 51, Magazin M 304, Reihe 51; A. Hoffmann with K. Kirschner, "Die Straßenbe-
pflanzung in Städten in ihren Beziehungen zur Lufthygiene sowie zu den verkehrs- und
stadttechnischen Verhältnissen," in *Fragen der Grünplanung im Städtebau*, ed. Johann
Greiner and Alfred Hoffmann (Berlin: Henschel, 1955), 62–91 (63); Dietrich Hieronymus,
"Grossstadt und Garten," *Das Gartenamt* 2, no. 10 (1953): 189–92 (190); Di., "Stadtgrün—
Stadtmensch," *Das Gartenamt* 4, no. 3 (1955): 42–44; Martin Rock, "Zum Verhältnis
Mensch-Natur," *Das Gartenamt* 38, no. 12 (1989): 739–42.

50. See G. Pinkenburg, *Der Lärm in den Städten und seine Verhinderung* (Jena: Gustav Fischer,
1903); W. Zeller and H. Kaesser, "Über die Aufgaben unserer neuen Zeitschrift," *Lärmbekämpfung* 1, no. 1 (1956): n.p.

51. F. J. Meister and W. Ruhrberg, "Der Einfluß von Grünanlagen auf den Verkehrsgeräuschpe-
gel," *VDI Zeitschrift* 97, no. 30 (1955): 1063–67; F. J. Meister and W. Ruhrberg, "Der Einfluß
von Grünanlagen auf die Ausbreitung von Geräuschen," *Lärmbekämpfung* 1, no. 1 (1956):
5–11; R. Dittmann, "Öffentliche Grünanlagen, ein Faktor zur Lärmbekämpfung," *Das Gartenamt* 7, no. 8 (1958): 175–80; Viktor v. Medem, "Lärm- und Staubschutzpflanzungen in der
Industriestadt aus der Sicht des Landschaftsgärtners," *Die neue Landschaft*, no. 9 (1961): 193–97.

52. A. Moles, *Physique et technique du bruit* (Paris: Dunod, 1952), 131; Gerhard Beck, *Pflanzen als
Mittel zur Lärmbekämpfung* (Hannover and Berlin: Patzer, 1965); Gerhard Beck, "Untersuchung über Planungsgrundlagen für eine Lärmbekämpfung im Freiraum mit Experimenten
zum artspezifischen Lärmminderungsvermögen verschiedener Baum- und Straucharten"
(doctoral diss., Technical University Berlin, 1965); also see Herbert Fischer, "Lärmschutz
durch Bepflanzung," *Baum-Zeitung* 5, no. 3 (1971): 42.

53. Ulrich Ruge, "Erfahrungen mit dem öffentlichen Grün in Städten," *Das Gartenamt* 24, no. 2
(1975): 93–96; "Ideal sind 100 Bäume pro Kilometer," *TS*, 27 February 1975, 12; Norbert
Schindler, "Leitlinien für das Berliner Stadtgrün," *Das Gartenamt* 24, no. 7 (1975): 431–35;
Herbert Wichmann, "Verpflanzung von Grossgehölzen," *Das Gartenamt* 26, no. 4 (1977):
226–35 (228); Der Senator für Stadtentwicklung und Umweltschutz, *Berliner Grün in Zahlen*
(Berlin: Verwaltungsdruckerei Berlin, 1987).

54. Hoffmann, "Der Straßenbaum in der Großstadt unter besonderer Berücksichtigung der
Berliner Verhältnisse," 52–53; Hoffmann with Kirschner, "Straßenbepflanzung in Städten,"
78–88; Aloys Bernatzky, "Bäume in der Stadt," *Garten und Landschaft* 54, no. 10 (1974):
700–706; Aloys Bernatzky, "Bäume in der Stadt," *Garten und Landschaft* 54, no. 11 (1974):
638–41; Aloys Bernatzky, *Tree Ecology and Preservation* (Amsterdam and New York: Elsevier,
1978), 86–93. East Berlin garden director Lichey stressed the need for further research on trees
and traffic guidance in the early 1960s. See Lichey, "Straßenbäume und Stadtverkehr." Ruge,
"Erfahrungen mit dem öffentlichen Grün."

55. Max Bromme, "Sind Alleen und Einzelbäume an der Straße in ihrer Beziehung zum
Verkehr, zur Landschaft und zum Ortsbild künftig lebensberechtigt?" *Mitteilungen der
deutschen Dendrologischen Gesellschaft* 49 (1937): 165–73; Rudolf Kühn, *Die Straßenbäume*
(Hannover, Berlin, and Sarstedt: Patzer, 1961), 16–27.

## Chapter 7. Shades of Red

1. "Gehölze und Stauden. Salzversuch," in *Bundesgartenschau Berlin 1985, Offizieller Ausstel-
lungskatalog*, ed. Bundesgartenschau Berlin 1985 GmbH (Berlin: Thormann und Goetsch,
1985), 33; fa., "Große Bäume wuchsen nach der Verpflanzung nicht weiter," *TS*, 24 December 1976, 8; fa., "Baum-Veteran soll auf Wanderschaft gehen," *TS*, 23 October 1976, 8; fa.,
"Die ersten fünf Linden auf Wanderschaft zur Gartenschau," *TS*, 13 October 1976, 9; Falk
Trillitzsch and Edelgard Jost, "Dying Trees as Exhibition Pieces," *Garten und Landschaft* 95,
no. 4 (1985): 60; "Der Straßenbaum," in *Bundesgartenschau Berlin 1985*, 123.

2. Tsp., "Start des grünen Baumpaten-Busses," *TS*, 19 March 1976, 11; Interessengemeinschaft Berliner Kunsthändler, *Achte Frühjahrsmesse Berliner Galerien 1976, Akademie der Künste, 11.–14. März 1976* (Berlin: Kupijai und Prochnow, 1976). On Wargin's tours and happenings in West German cities, see Ben Wargin, *Nenn mich nicht Künstler* (Berlin: Ch. Links, 2015), 112–15. we., "Was Bäumen zugemutet wird," *TS*, 10 March 1976, 9.

3. we., "Kurfürstendamm soll mit Ginkgos geschmückt werden," *TS*, 15 April 1978, 8; L. H., "Wargin pflanzt heute Bäume auf dem Teufelsberg," *TS*, 29 November 1978, 12; on Wargin and his tree-planting campaigns and tree activism, also see Wargin, *Nenn mich nicht Künstler*, 98–100, 120–22, 141.

4. erk., "Im Wildwuchs-Gefecht steht Udo Jürgens Wargin bei," *TS*, 17 September 1980, 12; Tsp., "Ben Wargins Pflanzenwelt zum wiederholten Mal zerstört," *TS*, 13 June 1981, 8; Wargin, *Nenn mich nicht Künstler*, 109.

5. For a cultural history of the ginkgo, see Peter Crane, *Ginkgo* (New Haven and London: Yale University Press, 2013). Wargin, *Nenn mich nicht Künstler*, 114–15. Although Wargin has claimed that Goethe's ginkgo poem left him uninspired, he quoted it in Ben Wargin, *Poesie der Strasse oder wie ein Jahr vergeht* (Berlin: Nicolai, 1983), n.p.

6. See Claudia Mesch, *Modern Art at the Berlin Wall* (London and New York: Tauris, 2008), 17, 50, 95–102.

7. Manfred Butzmann, "Straßengärtner," *Form und Zweck* 15, no. 1 (1983): 39–40.

8. For Christo and Jeanne Claude's "revelation through concealment," see David Bourdon, *Christo* (New York: Harry N. Abrams, 1970), 9. "'Der Baum muß weg,'" *Das Gartenamt* 34, no. 1 (1986): 5; "Junggärtner lassen Bäume über Nacht verschwinden," *Augsburger Allgemeine Zeitung*, 9 September 1985, 20.

9. *The Tree Show: The Tree as a Metaphor in a Selection of Contemporary Art* (Boston: Massachusetts College of Art, 1987); Daniela Dahn, *Kunst und Kohle* (Darmstadt: Hermann Luchterhand, 1987), 233–34; Mesch, *Modern Art*, 164–65, 192, 197–200, 226.

10. Joseph Beuys, in Johannes Stüttgen, *Beschreibung eines Kunstwerks, 7000 Eichen, Ein Arbeitspapier der Free International University [FIU]* (Bielefeld: WICO-Druck, 1982), 7.

11. Much has been written about Joseph Beuys's *7,000 Oaks*. See Stiftung 7000 Eichen, ed., *30 Jahre. Joseph Beuys 7000 Eichen* (Cologne: König, 2012); Reinhard Zimmermann, *Kunst und Ökologie im Christentum. Die "7000 Eichen" von Joseph Beuys* (Wiesbaden: Dr. Ludwig Reichert, 1994); Christine Drößler, *Stadtverwaldung statt Stadtverwaltung* (Frankfurt: FB 3, WBE Methodologie der J.-W.-Goethe-Universität, 1990); Fernando Groener and Rose-Marie Kandler, eds., *7000 Eichen–Joseph Beuys* (Cologne: König, 1987); Karl Heinrich Hülbusch and Norbert Scholz, eds., *Joseph Beuys — 7000 Eichen zur documenta 7 in Kassel* (Kassel: Kasseler, 1984).

12. fa, "Trockener Sommer und Tausalz bedrohen Baumbestand der Straßen," *TS*, 12 November 1971, 10; Tsp., "Tausalz greift Beton und Bäume an," *TS*, 11 February 1972, 9; L. H., "Mitten im Sommer welkendes Laub an den Straßenbäumen," *TS*, 4 August 1972, 9; "Bei Verwendung von Auftaumitteln an die Straßenbäume denken," *TS*, 17 December 1972, 15. See also Tsp., "27 Straßenbäume werden gefällt," *TS*, 15 December 1973, 10; L. H., "Mit dem Unkraut werden auch Bäume vernichtet," *TS*, 23 April 1977, 8; L. H., "Die Tausalzregelung kommt für viele Bäume zu spät," *TS*, 24 November 1979, 10; Tsp., "Für generelles Tausalzverbot Beschluß der Berliner CDU," *TS*, 27 November 1979, 12; Tsp., "Eine exakte Kontrolle über die gestreute Salzmenge fehlt," *TS*, 8 January 1980, 10; Tsp., "Lüder beziffert Tausalzschaden an Bäumen auf 750 Millionen Mark," *TS*, 7 December 1979, 15. Klaus Esche for "Initiative zur Rettung der Berliner Straßenbäume to Senator für Bau- und Wohnungswesen et al.," open letter, August 1979, private collection Jürgen Hübner-Kosney "Aktion Tausalzstopp."

13. L. H., "Bevorzugt sollten Eichen, Robinien und Schnurbäume gepflanzt werden," *TS*, 16 June 1974, 16; "Straßenbäume in Berlin," *Das Gartenamt* 24, no. 6 (1975): 350; Hans-Otfried Leh, "Schäden an Bäumen durch Auftausalze," *Baum-Zeitung* 6, no. 2 (1972): 22–23; Hans-Otfried Leh, "Untersuchungen über die Auswirkungen der Anwendung von Natriumchlorid

als Auftaumittel auf die Straßenbäume in Berlin," *Nachrichtenblatt des Deutschen Pflanzen-schutzdienstes* 25, no. 11 (1973): 163–70; H.-O. Leh, "Schäden an Strassenbäumen durch Auftausalze," *Gesunde Pflanzen* 23, no. 11 (1971): 217–20; H.-O. Leh, "Untersuchungen über die Auswirkungen des tausalzfreien Winterdienstes auf den Gesundheitszustand der Straßen-bäume in Berlin," *Nachrichtenblatt des Deutschen Pflanzenschutzdienstes* 42, no. 9 (1990): 134–42; H.-O. Leh, "Auswirkungen innerstädtischer Streßfaktoren auf Straßenbäume," *Gesunde Pflanzen* 44, no. 9 (1992): 283–91; Kaspar Klaffke, "Winterschäden an Strassenbäu-men und Verkehrsgrünflächen," *Das Gartenamt* 29, no. 11 (1980): 699–700; Peter Kiermeier, "Zur Problematik stadtfester Gehölze: 11. Gleditsia triacanthos—ein Baum wie eine Straußen-feder," *Das Gartenamt* 32, no. 9 (1983): 547–51; Carola Wünsche, "Pflege, Erhaltung und Erweiterung des Straßenbaumbestandes im Stadtbezirk Berlin-Prenzlauer Berg" (Diplomar-beit [master's thesis], Humboldt University, Berlin, 1983), 63–65.

14. fa., "Robinien und Roteichen trotzen dem Tausalz," *TS*, 8 April 1975, 10; L. H., "Bäume in Kübeln für Straßen ohne natürlichen Lebensraum," *TS*, 12 October 1977, 10; ock, "Die Bäume in den Pflanzenkübeln wirken erst in einigen Jahren," *TS*, 3 June 1979, 15; Tsp., "Straßenbäume sollen vom gepökelten Schnee befreit werden," *TS*, 9 February 1979, 8; L. H., "Für die Bäume das Salz aus der Erde herausspülen," *TS*, 17 February 1979, 8; Tsp., "Bürger können Bäume vor dem Tod durch gepökelten Schnee retten," *TS*, 1 March 1979, 15; "Die gemeinsame Baumhilfe fiel ins Regenwasser," *TS*, 6 March 1979, 12; Klaffke, "Winterschäden."

15. UBA, "Streusalz soll umweltfreundlicher werden," *Das Gartenamt* 29, no. 10 (1980): 659; Hermann Seiberth, "Neue Ansätze im Naturschutz einer Grosstadt," *Das Gartenamt* 29, no. 5 (1980): 344–53; Kaspar Klaffke, "Auswirkungen des Streusalzes (NaCl) auf die Pflanzen," *Das Gartenamt* 30, no. 11 (1981): 805–9; *Der Baumschutz*, no. 3 (1977): 6; L. H., "Trotz günstiger Witterung vertrocknete Straßenbäume," *TS*, 12 August 1984, 8; Tap, "Der Salzverzicht lohnt sich für die Straßenbäume," *TS*, 19 February 1988, 12.

16. Wolfang Reckling, "Erfassung des Straßenbaumbestandes Berlins, Hauptstadt der DDR; Maßnahmen zur Pflege und Erhaltung des Straßenbaumbestandes" (Diplomarbeit [master's thesis], Humboldt University, Berlin, 1974); "Lauge-Schäden und Schutz," *NZ*, 25 February 1965, 5; Friedrich Lohrmann, "Chemie contra Glatteis," *ND*, 1 December 1973, 13.

17. Reckling, "Erfassung des Straßenbaumbestandes," 91a.

18. Holger Brandt, "Straßenbäume—Städtische Sorgenkinder!" *Arche Nova* 3 (1989), reprinted in *Arche Nova: Opposition in der DDR*, ed. Carlo Jordan and Hans Michael Kloth (Berlin: Basisdruck GmbH, 1995): 330–32 (331); "Maßnahmen zur Erhöhung des Baumbestandes in der Hauptstadt der DDR, Berlin," STUG 002-48, "Materialsammlung Baumschutz 1," p. 105; For the use of concrete planting containers, see, for example, "Jahresanalyse 1987," in "Analysen 1985–1993, Teil IV," NGABP.

19. Hartmut Tauchitz, "Satzungen zum Schutze des Baumbestandes," *Das Gartenamt* 30, no. 5 (1981): 377–79; Jürgen Milchert, "Zehn Thesen zur Zukunft des Stadtgrüns," *Das Gartenamt* 33, no. 10 (1984): 675–85 (685); *Der Baumschutz*, no. 3 (1977): 7.

20. "Satzung der Gemeinschaft zum Schutze des Berliner Baumbestandes e.V.," *Der Baum-schutz*, no. 2 (1976): 14–16.

21. *Der Baumschutz*, no. 5 (1979): 20; *Der Baumschutz*, no. 3 (1977): 10; *Der Baumschutz*, no. 2 (1976): 7–8 (8); Tsp., "Uferbäume müssen fallen," *TS*, 12 July 1974, 10; —erk, "Baumschutz-gemeinschaft klagt über Beschädigung 100 Jahre alter Bäume," *TS*, 9 July 1975, 12; Tsp., "Sorge um die Eichen an der Postdamer Chaussee in Zehlendorf," *TS*, 16 April 1976, 12; L. H., "Bäume sollen zum Naturdenkmal erklärt werden," *TS*, 10 February 1977, 12; cms, "Bürgerprotest unter Bäumen," *TS*, 25 February 1977, 11; Tsp., "Zweifel an Tausalz-Kontrolle," *TS*, 8 December 1979, 10; Tsp., "Ohne Tausalz auf begrenztem Raum in den nächsten Winter," *TS*, 10 January 1980, 12; typescript "Aktion Tausalzstopp," PT; also see documents in private collection of Jürgen Hübner-Kosney, "Aktion Tausalzstopp."

22. *Der Baumschutz*, no. 6 (1979): 2–9; *Der Baumschutz*, no. 8 (1981): 1–9; Michael Teske, "Der Alarmplan steht: Berlin ist für den Smog gerüstet," *Der Abend*, 20 September 1978, 16. Also see the newspaper article collection "Berliner Luft–Smog," STUG 167-5.

23. See *Der Baumschutz*, no. 8 (1981): 10–12; Giesbert Lange, "Zwanzig Jahre Baumschutz-gemeinschaft Berlin—ein Grund zum Feiern?!" *Baumschutz*, no. 15 (1992): 2.

24. bk, "Wunsch nach Bürgerhilfe beim Schutz der Straßenbäume," *TS*, 10 August 1989, 12; *Berliner Luftzeitung* no. 28 (1995): 6; *Baumschutz-Sondernummer* 19 (1995): 8.

25. The information about the amount of protected natural monuments in Prenzlauer Berg differs slightly in Rat des Stadtbezirks Berlin Prenzlauer-Berg, Gartenamt, *Bäume und Naturdenkmale im Stadtbezirk Berlin-Prenzlauer Berg—Ein Naturführer* (Berlin: Nationales Druckhaus Berlin, 1987), written by W. Krause and C. Wünsche, and in Walter Meissner and Hans-Joachim Dreger, "Bäume in Berlin, Hauptstadt der DDR," in *100 Jahre Arboretum (1879–1979)*, ed. Walter Vent (Berlin: Akad.-Verl., 1980), 83–92 (85). Wolfgang Krause, Ursula Rändel, and Reinhild Zagrodnik, *Platz für Natur im Prenzlauer Berg*, ed. Bezirksamt Prenzlauer Berg von Berlin (Berlin: Argon, 1995), 14–15; Reckling, "Erfassung des Straßenbaumbestandes," 29–31. Herbert Pohl and Wolf Dietrich Werner, "Analysen und Vorstellungen," *Form und Zweck* 15, no. 1 (1983): 8–14; Brian Ladd, "Local Responses in Berlin to Urban Decay and the Demise of the German Democratic Republic," in *Composing Urban History and the Constitution of Civic Identities*, ed. John J. Czaplicka and Blair A. Ruble (2003), 263–84 (267).

26. Alexander Haeder, "Hinterhofgestaltung im Altbaugebiet," *Bildende Kunst*, no. 7 (1985): 304–5. In the years leading up to the city's 750-year-anniversary celebrations in 1987, facades and small stores along Greifswalder Straße, Schönhauser Allee, and Husemannstraße as well as the courtyards of the buildings along Husemannstraße were restored, reconstructed, and replanted to appear once again in turn-of-the-century guise, the time when these streets had been bustling commercial centers of a working-class neighborhood. For the urban government's restoration and reconstruction activities, see Dorothea Krause, Uwe Klasen, and Wolfgang Penzel, "Rekonstrution im Stil der Jahrhundertwende," *Architektur der DDR* 36, no. 10 (1987): 14–21; Brian Ladd, "Local Responses."

27. Krause, Rändel, and Zagrodnik, *Platz für Natur*, 14–15; Rat des Stadtbezirks Berlin Prenzlauer-Berg, Gartenamt, *Bäume und Naturdenkmale*; Dorothea Krause, Wolfgang Krause, and Holger Fahrland, "Baum und Stadt—Durchgrünung des dicht-besiedelten Stadtbezirks Berlin-Prenzlauer Berg," *Architektur der DDR* 35, no. 10 (1986): 604–9; Dahn, *Kunst und Kohle*, 74; Stadtbezirksversammlung Kreisausschuß der Nationalen Front Berlin-Prenzlauer Berg, "Schöner unsere Hauptstadt Berlin-Mach mit! Programm 1982 des Stadtbezirks Berlin-Prenzlauer Berg," brochure held in Museum Pankow-Archiv, Chronik PB, box 5, 8.1 NAW PB "Mach mit." Wünsche, "Pflege, Erhaltung und Erweiterung," 2, 5–6, 71–73. The number of planted trees varies in different sources.

28. Krause, Krause, and Fahrland, "Baum und Stadt"; "Bericht und Planweiterführung der Baumpflanzung und Abrechnung 1985," LAB, C Rep. 134-02-02, Nr. 1345. For tree-care contracts and surveys, see LAB, C Rep. 112, Nr. 365. Holger Brandt, "Straßenbaumzählung—eine Inventur des Sterbens?" *Arche Nova* 3 (1989), reprinted in Jordan and Kloth, *Arche Nova*, 332–35. For the street tree–planting campaign in Schwerin and the role of tree-planting campaigns in oppositional movements, see Ehrhart Neubert, *Geschichte der Opposition in der DDR 1949–1989* (Berlin: Ch. Links, 1998), 451–52.

29. See Mary Fulbrook, *The People's State: East German Society from Hitler to Honecker* (New Haven and London: Yale University Press, 2005), esp. 235–68. *Ergänzte Konzeption für die volkswirtschaftliche Masseninitiative (Wettbewerb der Städte und Gemeinden) im Jahre 1971* (Berlin: Nationalrat der Nationalen Front, Sekretariat, ca. 1971).

30. Amtsleiter Kozlik to Bezirksabt. Aufbau, "Tätigkeit, Stand und Zweck des Ref. Grünflächen der Abteilung Aufbau," 13 December 1952, LAB, C Rep. 134-06, Nr. 54; Magistrat von Gross-Berlin to Hauptamt Stadtplanung—Koll. Käding, "Magistratsvorlage Nr. 968 zur

Beschlussfassung für die Sitzung am 15. Mai 1952. Massnahmen zur Verbesserung der Strassenordnung und des Strassenbildes," 12 May 1952, LAB, C Rep. 110-04, Nr. 7; *Verordnungsblatt für Groß-Berlin*, vol. 8, part 1, no. 25, 12 June 1952.

31. E. Jaenisch, "Gedanken und Erfahrungen über den Abschluß von Pflegeverträgen für Grünanlagen mit der Bevölkerung in Berlin," *Deutsche Gartenarchitektur* 2, no. 4 (1961): 99–102.

32. Stadtgartendirektor Dr. Lichey, "Direktive für die Ausarbeitung des Perspektivplanes zur Entwicklung und Bewirtschaftung des Stadtgrüns bis 1970," 17 December 1964, LAB, C Rep. 110-04, Nr. 8; "Perspektivplan 1965–1970," LDB, NGA, Mitte, 0650; H. Lichey, "Die Aufgaben der Berliner Stadtbegrünung," *Deutsche Gartenarchitektur* 5, no. 3 (1964): 50–51 (50); M. Pawlow, "Die Gesellschaften zum Schutz der Natur und zur Begrünung von Siedlungen in der UdSSR," *Deutsche Gartenarchitektur* 2, no. 4 (1961): 103–4; Dr. Lichey to Mr. Max Keppel, 8 January 1962, LAB, C Rep. 110-04, Nr. 4; A. Aluf, *The Development of Socialist Methods and Forms of Labour* (Moscow: Co-operative Publishing Society of Foreign Workers in the U.S.S.R., 1932), 11.

33. See *Stadtordnung zur Gewährleistung der Ordnung, Sauberkeit und Hygiene in der Hauptstadt der DDR, Berlin*, 17 November 1969; Gottfried Funeck, "Die Entwicklung des Berliner Stadtgartenamtes und seine Auswirkung auf die städtebauliche Planung," in *Entwicklung der Volksparke*, ed. Kulturbund der Dt. Demokratischen Republik, Zentraler Fachausschuß Dendrologie u. Gartenarchitektur, Zentrales Parkaktiv/Joachim Berger and Detlef Karg (Weimar: Druckhaus Weimar, 1979), 21–32 (32); R. Buchmann, "Tagung des Nationalrates—Schöner unsere Städte und Gemeinden," *Landschaftsarchitektur* 5, no. 3 (1976): 71–72; Manfred Fiedler, in *Inhalte und Ziele der Umweltpolitik nach dem XI. Parteitag der SED und die weitere Förderung der Bürgerinitiative "Schöner unsere Städte und Gemeinden—Mach mit!"* Erfahrungsaustausch des Ministeriums für Umweltschutz und Wasserwirtschaft . . . , Dachwig, 1986, 43–53, NL Greiner.

34. *Verordnung über die Erhaltung, Pflege und den Schutz der Bäume—Baumschutzverordnung*, 28 May 1981, paragraph 3 (2); *Verordnung zum Schutze des Baumbestandes in Berlin (Baumschutzverordnung—BaumSchVo)*, 11 January 1982; Freimut Wenzel, "Maßnahmen zum Schutz vorhandener Straßenbäume," *Die Straße* 22, no. 5 (1982): 168–72.

35. Walter Pisternik, *Die Lehren aus der bisherigen Arbeit zur Verschönerung der Städte*, Schriftenreihe Demokratischer Aufbau, ed. Hauptabteilung Örtliche Organe des Staates beim Ministerpräsidenten der Regierung der Deutschen Demokratischen Republik (Berlin: VEB Deutscher Zentralverlag Berlin, 1954), 12, 16; Reinhold Lingner et al., "Richtlinien für die Verschönerung unserer Städte durch Grünanlagen im Rahmen des Nationalen Aufbauwerkes," in Pisternik, *Verschönerung der Städte*, 19–24 (19, 22); Reinhold Lingner, "Gutachten zur Projektierung der Grünanlagen in Stalinstadt durch das Volkseigene Entwurfsbüro für Industriebahnbau Abt. Grünplanung, Chefarchitekt Walter Funcke," 22 September 1953, NL Funcke; E. Jaenisch, "Gedanken und Erfahrungen über den Abschluß von Pflegeverträgen für Grünanlagen mit der Bevölkerung in Berlin," *Deutsche Gartenarchitektur* 2, no. 4 (1961): 99–102. For a listing of care contracts and the expenditures they saved the Pankow district Department of Gardens in the late 1960s, see "Analysen 1966–1970, Teil II," NGABP.

36. Leiterin des Gartenamtes Zagrodnik, "Einschätzung des Frühjahrsputzes und der Baumpflanzung am 30.3. und 6.4.1974 im Stadtbezirk Prenzlauer Berg," 10 April 1974, LDB, NGA, Mitte, 0404, Allgemeines. "Entwurf Programm des Stadtbezirks Berlin-Prenzlauer Berg im Wettbewerb 'Schöner unsere sozialistische Hauptstadt Berlin—Mach mit' 1974 zu Ehren des 25. Jahrestages der DDR, November 28, 1973," Museum Pankow-Archiv, Chronik PB, box 5, 8.1 NAW PB "Mach mit." "Bericht und Planweiterführung der Baumpflanzung und Abrechnung 1985," LAB, C Rep. 134-02-02, Nr. 1345. "Jahresanalyse 1981," p. 3, and "Jahresanalyse 1983," p. 2, in "Analysen 1971–1985, Teil III"; "Jahresanalyse 1985," p. 1, and "Jahresanalyse 1988," p. 2, in "Analysen 1986–1993, Teil IV," NGABP.

37. Brandt, "Straßenbäume—Städtische Sorgenkinder!" 331; Hans Georg Büchner, "Zum

geforderten und zum erreichten Niveau der Freiraumgestaltung in Berlin/Ost," *Landschaftsarchitektur* 20, no. 4 (1990): 109–10; Hans Georg Büchner, "Die Entwicklung der Verwaltung des öffentlichen Grüns in Berlin-Ost 1948 bis 1990," in *Denkmalpflege und Gesellschaft*, ed. Thomas Drachenberg, Axel Klausmeier, Ralph Paschke, and Michael Rohde (Rostock: Hinstorff, 2010), 199–208, and in a more elaborate manuscript from 1992 shared with the author. Neubert, *Geschichte der Opposition*, 451–52; Michael Beleites, "Die unabhängige Umweltbewegung in der DDR," in *Umweltschutz in der DDR*, ed. Hermann Behrens and Jens Hoffmann (Munich: oekom, 2007), 3:179–224 (188–89); Hannelore Oehring, "Den Baumbestand erhalten und erweitern," *Landschaftsarchitektur* 11, no. 2 (1982): 52. Wolfgang Krause in a conversation with the author in Berlin on 14 March 2017.

38. Werner Nohl, "Über die Erlebniswirksamkeit von Bäumen," *Mitteilungen der deutschen Dendrologischen Gesellschaft* 67 (1974): 104–27.

39. Dieter Boeminghaus, "Der Baum an Straßen in der freien Landschaft und seine Bedeutung für die Wahrnehmung" (doctoral diss., Rheinisch-Westfälische Technical University, Aachen, 1974); Dieter Boeminghaus, "Der Baum an Landstrassen als informationspsychologische Grösse für den Autofahrer," *Das Gartenamt* 23, no. 10 (1974): 563–74; Klaus Ermer, Eckhard Holtz, and Günter Hachmann, *Strassenbäume in innerstädtischen Bereichen* (Berlin: Technische Universität Berlin, 1974), 32–37.

40. Michael Maurer, "Technische Massnahmen zur Erhaltung und Pflege alter Bäume. Erfahrungen aus den USA," *Garten und Landschaft* 60, no. 11 (1950): 13–14; Michael Maurer, "Über die Wertbestimmung von Bäumen—Erfahrungen aus den USA," *Garten und Landschaft* 61, no. 4 (1951): 8–9 (9); SDW, "Baumveteranen werden operiert," *Das Gartenamt* 2, no. 5 (1953): 98; Michael Maurer, "Von der Tätigkeit des Baumchirurgen," *Garten und Landschaft* 64, no. 3 (1954): 7–9 (7); Michael Maurer, "Von der Tätigkeit des Baumchirurgen," *Garten und Landschaft* 66, no. 4 (1956): 103; Michael Maurer, "Baumchirurgie in den USA," *Baum-Zeitung* 1, no. 2 (1967): 30–32; Karl Peßler, "Baumpflege und Baumchirurgie," *Garten und Landschaft* 84, no. 6 (1974): 317–22; Redaktion Baum-Zeitung, "Retter der Bäume. Michael Maurer zum 70. Geburtstag," *Baum-Zeitung* 9, no. 2 (1975): 32; "Bundesverdienstkreuz für Michael Maurer," *Baum-Zeitung* 9, no. 4 (1975): 56; "Michael Maurer," *Baum-Zeitung* 16, no. 2 (1982): 18–20; Werner Hoffmann, "Eine Stadt schützt ihre Bäume," *Garten und Landschaft* 66, no. 3 (1956): 65–67 (67).

41. Ephraim Porter Felt, *Our Shade Trees* (New York: Orange Judd, 1942), 76–81; Ephraim Porter Felt, *Our Shade Trees* (New York: Orange Judd, 1938), 36–39; E. Porter Felt, "Why Trees Act as They Do," *Horticulture* 17, no. 3 (1939): 45; E. P. Felt, "How a Tree May Be Valued," *American City* 42, no. 3 (1930): 102; Ephraim Porter Felt, *Shelter Trees in War and Peace* (New York: Orange Judd, 1943), 42–43, 44, 62–67. For the various methods employed by American tree experts to evaluate trees in the 1970s, see Gordon S. King, "Plant Material Evaluation," *Journal of Arboriculture* 3, no. 4 (1977): 61–64.

42. Werner Hoffmann, "Die Baumschutzverordnung und ihre Bedeutung für die Erhaltung des Baumbestandes im Land Hamburg," *Das Gartenamt* 9, no. 9 (1960): 219–22; Werner Hoffmann, "Über die Wertbestimmung von Bäumen," *Die neue Landschaft* 6, no. 7 (1961): 147–48; Heinrich Wawrik, "Wertberechnung von Strassen- und sonstigen Zierbäumen," *Das Gartenamt* 18, no. 1 (1969): 26–28; Erich Ostermeyer, "Wieviel ist ein Baum wert?" *Das Gartenamt* 18, no. 1 (1969): 28–30; Günther Heydenreich, "Wertberechnung von Bäumen," *Garten und Landschaft* 79, no. 8 (1969): 140; Grünflächenamt Mannheim, "Wertberechnung von Bäumen," *Baum-Zeitung* 2, no. 4 (1968): 57–59.

43. "Die Baumwertrichtlinien des Deutschen Städtetages," *Garten und Landschaft* 82, no. 3 (1972): 87; Heinrich Wawrik, "Entwurf von Richtlinien für die Wertberechnung von Bäumen," *Das Gartenamt* 21, no. 5 (1971): 230–34; Wolfgang Gellrich, "Wertberechnung von Bäumen und Schadensersatz," *Das Gartenamt* 21, no. 7 (1971): 318–24; Heinrich Wawrik, "Endgültige Formulierung der Baumwertrichtlinien des Deutschen Städtetages," *Das Gartenamt* 24, no. 5 (1975): 271–72; Werner Koch, "Hinweise zur Baumwertermittlung nach

dem Sachwertverfahren," *Das Gartenamt* 21, no. 11 (1972): 622–27; H. Wawrik, "Baumwerte. Endgültige Formulierung der Baumwertrichtlinien des Deutschen Städtetags," *Neue Landschaft* 17, no. 4 (1972): 192–95. On the Kastanienbaumurteil, see Werner Koch, "Der Berliner Kastanienbaumfall ist nun endgültig abgeschlossen," *Das Gartenamt* 25, no. 10 (1976): 626–28; Werner Koch, "Schadensberechnung für Gehölze," *Das Gartenamt* 25, no. 1 (1976): 58–62; Werner Koch, "Kalkulationsfragen bei der Sachwertermittlung für Gehölze und bei Gehölzschäden," *Das Gartenamt* 39, no. 2 (1990): 86–90; Helge Breloer, "Das Kastanienbaumurteil und andere Urteile zu Bäumen," in *Beiträge zur Rosskastanie, LWF Wissen 48,* ed. Alexandra Wauer and Hildegard Klessig (Freising: Lerchl Druck, 2005), 52–57.

44. Klaus Ermer, "Verfahren zur Wertberechnung von Strassenbäumen," *Das Gartenamt* 23, no. 10 (1974): 574–77; Ermer, Holtz, and Hachmann, *Strassenbäume in innerstädtischen Bereichen,* 40–63; Werner Koch, "Baumwertrichtlinien des Deutschen Städtetages," *Garten und Landschaft* 95, no. 3 (1985): 4; Werner Koch, "Baumwertrichtlinien des Deutschen Städtetages fortgeschrieben," *Das Gartenamt* 34, no. 1 (1985): 38–39; Franz Otto, "Die Bedeutung von Gehölzwerten bei der Anwendung von Baumschutzregelungen," *Das Gartenamt* 35, no. 1 (1986): 10–14; *Das Gartenamt* 38, no. 3 (1989); Werner Koch, "Kalkulationsfragen bei der Sachwertermittlung für Gehölze und bei Gehölzschäden," *Das Gartenamt* 39, no. 2 (1990): 86–90; Heinrich Wawrik and Johannes von Malek, "Die Schadstufenbestimmung bei Bäumen," *Das Gartenamt* 41, no. 11 (1992): 774–76; Helge Breloer, "Baumwertermittlung," *Stadt und Grün* 44, no. 11 (1995): 801–2; R. Schultz, "Wertermittlung nach der Methode Koch," *Stadt und Grün* 47, no. 7 (1998): 506–8.

45. W. Neumann, "Was ist ein Baum wert?" *Naturschutzarbeit in Berlin und Brandenburg* 9, no. 1 (1973): 2–3; "Baumschutz in Frankfurt/Oder," *Baum-Zeitung* 8, no. 3 (1974): 41; Schröder, "Stellungnahme zum 1. Entwurf 'Beschluß zur Sicherung des Baumbestandes,'" 25 January 1974; "Richtlinie für die Bewertung von Bäumen, 15.12.1976," STUG 002-01, "Arbeitsgruppe Baumschutzordnung," pp. 80–81, 240–49.

46. W. Zipperling, "Schutz und Pflege der Bäume in der Großstadt," *Deutsche Gartenarchitektur* 7, no. 2 (1966): 40; H. Linke and F. Klein, "Bäume und Baumbewertung," *Landschaftsarchitektur* 2, no. 3 (1973): 87; Friedbert Klein, "Grundlagen für die ökonomische Bewertung von Freiräumen—dargestellt am Vegetationselement Baum" (doctoral diss., Technical University Dresden, 1975); F. Klein, "Die Baumschutzordnung—eine Information zum Sachstand," *Landschaftsarchitektur* 4, no. 4 (1975): 121; "1. Entwurf der Ordnung zum Schutz der Bäume in Städten und Gemeinden—Musterbaumschutzordnung—29.8.1974"; "Richtlinie für die Bewertung von Bäumen, 19.3.1976," STUG 002-01, "Arbeitsgruppe Baumschutzordnung," pp. 20–61, 171–79.

47. Klaus-Dietrich Gandert, "Ziele und Möglichkeiten zur Erhaltung und Pflege von Bäumen," *Beiträge zur Gehölzkunde* (1977): 44–51. See Magistrat für Groß-Berlin, *Verordnungsblatt für Groß-Berlin,* vol. 32, no. 10, 17 May 1976; Wolfgang Reckling, "Schutz den Bäumen!" *Gärtnerpost* 29, no. 13 (1977): 16. Günter Hoffmann, "Bäume außerhalb des Waldes erhalten und schützen," in *Sonderheft Baumschutz 1982,* issued by the Kulturbund der DDR, Bezirksvorstand Erfurt; K.-D. Gandert to H. Thomasius, 23 December 1980, STUG 002-01, "Arbeitsgruppe Baumschutzordnung," pp. 311–12.

48. Ewald Müller, "Die Baum-Endoskopie," *Das Gartenamt* 31, no. 12 (1982): 742–43; Hans W. Eschborn, "Baum-Endoskopie," *Das Gartenamt* 33, no. 2 (1984): 112; Ewald Müller, "Baum-Endoskopie," *Das Gartenamt* 33, no. 12 (1984): 828–32; H. Tauchnitz, A. Habermehl, V. Schwartz, and H.-W. Ridder, "Computertomographische Untersuchungen an Stammscheiben von Bäumen der Gattungen Aesculus, Quercus, Betula und Tilia," *Das Gartenamt* 39, no. 2 (1990): 69–80; "Baum-Seminar in Minden—ein großer Erfolg!" *Baum-Zeitung* 14, no. 4 (1980): 50; Aloys Bernatzky, "Das mobile Computer-Tomographie-Gerät (Prof. Habermehl) zur Erkennung von Faulstellen in Baumstämmen," *Baum-Zeitung* 15, no. 1 (1981): 5–6; Günter Sinn, "Standsicherheit von Bäumen und Möglichkeiten der statischen Berechnung," *Das Gartenamt* 32, no. 9 (1983): 556–64; Günter Sinn, "Statische Berechnungen zur Standsi-

cherheit von Bäumen," *Das Gartenamt* 33, no. 2 (1984): 114–15; Günter Sinn, "Standsicherheitsuntersuchungen von Bäumen," *Das Gartenamt* 33, no. 9 (1984): 593–94; Günter Sinn, "Ergebnis einer Bruchsicherheitsberechnung überlastiger Äste," *Das Gartenamt* 35, no. 8 (1986): 462–65; Günter Sinn, "Entscheidungshilfen zur Beurteilung der Stand- und Bruchsicherheit von Bäumen," *Das Gartenamt* 39, no. 2 (1990): 67–68; Horst Ehsen, "7. Osnabrücker Baumpflegetage am 3. und 4. Oktober 1989—Wurzeln, Wasserhaushalt und Wundbehandlung," *Das Gartenamt* 39, no. 1 (1990): 35–39 (39); Lothar Wessolly, "Eine neue zerstörungsfreie Messmethode zur Ermittlung der Bruchsicherheit geschädigter Bäume: Elastomethode," *Das Gartenamt* 37, no. 12 (1988): 768–74.

49. Günter Sinn, "Sachstand der Baumstatik," *Das Gartenamt* 37, no. 3 (1988): 151–56. On the Sonderforschungsbereich 230, also see Konzepte SFB 230: "Tragende Lebewesen," *Arbeitshefte des Sonderforschungsbereiches* 230, no. 9 (December 1985); Frei Otto et al., *Natürliche Konstruktionen* (Stuttgart: Deutsche Verlags-Anstalt, 1982), 7–9, 38–45, 92–93.

50. Heinrich Wawrik, "Der Stadtstrassenbaum zwischen ökologischer Nische und Kunstbaum," *Das Gartenamt* 39, no. 9 (1990): 581–84.

51. Heinrich Wawrik, "Baumpflegelehrgänge in Heidelberg seit 1972," *Das Gartenamt* 27, no. 12 (1978): 771–74; Heiner Wawrik, "'Baumpfleger' ein neuer Fortbildungsberuf für Gärtner," *Das Gartenamt* 30, no. 2 (1981): 115–19; Heiner Wawrik, "Heidelberger Baumseminare," *Das Gartenamt* 33, no. 4 (1984): 558; Horst Schmidt, "Strassenbäume," *Das Gartenamt* 35, no. 8 (1986): 457–58; "Beeinflussen die forstlichen Untersuchungsergebnisse von Shigo die Wert- und Schadenermittlung von Stadt- und Alleebäumen?" *Baum-Zeitung* 20, no. 1 (1986): 20–22; "Baumpflege-Gespräch," *Das Gartenamt* 36, no. 2 (1987): 102–4.

52. *Der Baumschutz*, no. 2 (1976): 7–8 (8); L. H., "Kranke Bäume werden in Farbe aus der Luft gefilmt," *TS*, 8 March 1979, 12; rast, "Untersuchung der Baumschäden verzögert sich um ein Jahr," *TS*, 2 July 1985, 12; Michael Fietz and Bernd Meißner, "Zur Auswertung von Color-Infrarot-Luftbildern (Berlin/West 1979) bei der Vegetationskartierung," *Berliner geowissenschaftliche Abhandlungen*, series C3, vol. 3 (1984): 23–42.

53. The first German cities to use infrared photography for the detection of tree damage were Freiburg (1972), Hamburg (1974), Kassel (1977), and Wiesbaden (1978). Rigobert Monard, "Das Infrarotluftbild," *Das Gartenamt* 21, no. 1 (1972): 17–19; A. Kadro and H. Kenneweg, "'Das Baumsterben' auf dem Farb-Infrarotluftbild," *Das Gartenamt* 22, no. 3 (1973): 149–57; H. Kenneweg, "Das Infrarot-Luftbild und seine Anwendung bei der Inventur des städtischen Grüns und der Ermittlung seines Vitalitätszustandes," in *Inventur des städtischen Grüns und Ermittlung seines Vitalitätszustandes durch Fernerkundung, Pilotprojekt des Hessischen Ministers für Landesentwicklung, Umwelt, Landwirtschaft und Forsten sowie der Landeshauptstadt Wiesbaden* (Wiesbaden, 1979), 28–38; H.-J. Liesecke, "16. Arbeitstagung der Ständigen Konferenz der Gartenbauamtsleiter beim Deutschen Städtetag," *Das Gartenamt* 24, no. 1 (1975): 5–7. For an early East German reference to the use of aerial infrared photography in the management of urban trees, see Johann Greiner and Helmut Gelbrich, *Grünflächen der Stadt* (Berlin: Verlag für Bauwesen, 1972), 172; Wolfgang Fischer, "Möglichkeiten zur Ermittlung des Bestandes und des Zustandes der Vegetation in städtischen Teilgebieten," *Landschaftsarchitektur* 17, no. 4 (1988): 110–12; Zentrale Fachgruppe "Landschaftsarchitektur" im Bund der Architekten der DDR, "Pflanzungen in der Stadt—Standpunkte," Seminar der ZFG, May 1988, Jena, NL Greiner, R 16, no. 37. For the use of Spektrozonalfilm, see Michael Fietz, "Art- und schadensbedingtes Abbildungsverhalten von Berliner Strassenbäumen auf Colorinfrarot-Luftbildern" (doctoral diss., Free University of Berlin, 1992), 34.

54. "Zwei Drittel aller Straßenbäume in Neukölln erkrankt," *TS*, 29 January 1984, 16; Tsp., "Der alte Baumbestand an Straßen am stärksten geschädigt," *TS*, 15 February 1984, 10; Ulrich Förster, "Zur Auswertung von Colorinfrarot-Luftbildern," *Garten und Landschaft* 98, no. 7 (1988): 41–43; "Systematische Auswertung der neuen Infrarot-Baumfotos," *TS*, 2 September 1984, 16; Michael Fietz, "Terrestrische Infrarot-Fotografie," *Das Gartenamt* 33, no. 9 (1984): 579–85 (585); rast, "Untersuchung der Baumschäden."

55. Hartmut Kenneweg, "Objektive Kennziffern für die Grünplanung in Stadtgebieten aus Infrarot-Farbluftbildern," *Landschaft und Stadt* 7, no. 1 (1975): 35–43 (37); Hartmut Kenneweg, "Luftbilder für die Gewinnung von Informationen über Vegetationsbestände in Ballungsräumen," *Allgemeine Forstzeitschrift* 30, nos. 1–2 (1975): 23–27; Kadro and Kenneweg, "Das Baumsterben auf dem Farb-Infrarotluftbild."

56. Michael Fietz, "Berliner Strassenbäume im Color-Infrarotluftbild," *Berliner geowissenschaftliche Abhandlungen*, series A, vol. 47 (1983): 179–88; "Zwei Drittel aller Straßenbäume in Neukölln erkrankt," *TS*, 29 January 1984, 16; Michael Fietz, "Lebensbedingungen von Strassenbäumen in Berlin-Neukölln," *Berliner geowissenschaftliche Abhandlungen*, series C3, vol. 3 (1984): 53–64; Fietz, "Art- und schadensbedingtes Abbildungsverhalten," 9–11, 76–82, 189–90.

57. Ulrich Förster, "Vitalitätsbestimmung von Strassenbäumen. Ergebnisse der Color-Infrarot-Befliegung in Berlin (West) 1979," *Berliner geowissenschaftliche Abhandlungen*, series C3, vol. 3 (1984): 43–51; Ulrich Förster, "Strassenbäume in Berlin," *Das Gartenamt* 34, no. 4 (1985): 301–3.

58. Ernst Kürsten, "Das Strassenbaumkataster," *Das Gartenamt* 33, no. 4 (1984): 245–52; S. Sommer, "Erfassung des Straßenbaumbestandes—Grundlage für Erhaltung und Rekonstruktion," *Landschaftsarchitektur* 4, no. 4 (1975): 113–14; Siegfried Sommer, "Zur Auswertung der Straßenbaumerfassung im Stadtgebiet von Dresden," *50 Jahre Landschaftsarchitektur, Schriftenreihe der Sektion Architektur, Technische Universität Dresden*, no. 14 (1979): 149–54; Harald Linke and Horst Burggraf, "Sektion Architektur, Übersichtsbericht zur Forschung 1975," Universitätsarchiv Dresden, S 18, no. 96. For street trees in Dresden, also see Siefried Sommer and Stefan Löbel, "Dresdner Straßenbäume. Vom 16. Jahrhundert bis zum Ende des Zweiten Weltkrieges (Erster Teil)," in *Dresdner Geschichtsbuch* 17, ed. Stadtmuseum Dresden (Altenburg: DZA, 2012), 165–94; Stefan Löbel and Siegfried Sommer, "Dresdner Straßenbäume, Vom Ende des Zweiten Weltkrieges bis zum Jahr 2012," in *Dresdner Geschichtsbuch* 18, ed. Stadtmuseum Dresden (Altenburg: DZA, 2013), 193–223.

59. Klaus Weber, "Erste Straßenbäume dieses Jahrgangs sind im Boden," *ND*, 16 February 1989, 8; ADN, "Pankower Straßenbäume werden elektronisch erfaßt," *BZ*, 25–26 February 1989, 8; use, "Bäume werden gezählt," *NZ*, 25 February 1989, 6; H. Wermer, "Bäume werden gezählt," *Der Morgen*, 21 February 1989, 8; Bernd Schilling, "Der böse Traum vom Baum," *"Extra"-Pankow*, no. 2 (February 1991). For Pankow's street tree survey with the help of index cards beginning in 1967, see "Jahresanalyse 1967 des Gartenamtes Berlin-Pankow, 8 February 1968," p. 18; "Jahresanalyse 1968 des Gartenamtes Berlin-Pankow, 7 February 1969," p. 15; "Jahresanalyse 1970 des Gartenamtes Berlin-Pankow, 4 February 1970," p. 14; "Analysen 1966–1970, Teil II"; "Analyse der Tätigkeit des Gartenamtes Berlin-Pankow im Jahre 1972," p. 4; "Analysen 1971–1984, Teil III," NGABP.

60. Bernd Meißner, "Fernerkundung bei der Grünplanung für die Stadtregion Berlin (West)," *Berliner geowissenschaftliche Abhandlungen*, series C3, vol. 3 (1984): 5–21; Gottfried Borys, "Karten zum Strassenbaumkataster—Berlin (West)—1:2000," *Berliner geowissenschaftliche Abhandlungen*, series C3, vol. 3 (1984): 101–20.

61. Erhard Mahler, "Schwerpunkte der künftigen Grünpolitik Berlins," *Das Gartenamt* 34, no. 4 (1985): 233–41 (236); Förster, "Strassenbäume in Berlin."

## Chapter 8. Unity and Variety

1. Adolf Arndt, "Die Demokratie als Bauherr," *Bauwelt* 52, no. 1 (1961): 7–13; Eva Schweitzer, "Baum oder Säule," *TS*, 18 June 1996, 9; Oliver Elser, "Dekorierter Schuppen," *Frankfurter Rundschau*, June 19, 1996, 8; Axel Schultes, "The New Chancellory," *Journal of Architecture* 2 (Autumn 1997): 269–82; Heinrich Wefing, *Kulisse der Macht* (Stuttgart and Munich: Deutsche Verlags-Anstalt, 2001); Michael Z. Wise, *Capital Dilemma* (New York: Princeton Architectural Press, 1998), 65–79.

2. Wolfgang Endler, "Grenzenloses Grün statt grausames Grau," *Der Baumschutz* (1990):
4; Hannelore Petroschka, "Bäume für den Grenzstreifen," *BZ*, 23 March 1992, 9; "7000
japanische Kirschbäume für Berlin und Brandenburg," *TS*, 7 April 1995, 14; Sakura Campaign
Komitee, *Sakura Campaign* (Berlin: Medialis, 2004); "Wie wär's mit Grün?" *BZ*, 22 December
ber 1989, 3; "Fahrraddemo für einen Mauerpark," *ND*, 3 April 1990, 8; Helmut Gelbrich,
"Ein Stadtpark quer durch Berlin?" *Landschaftsarchitektur* 19, no. 3 (1990): 78–79.

3. Peter Böttcher, "Spielcasino oder Ginkgo-Hain?" *BZ*, 21 March 1990, 8; "Grüner unsere
Grenzen," *NZ*, 22 March 1990, 12; *TS*, 14 September 1990, 13. In August 1994, Wargin
organized a tree-watering campaign with the television moderator Arno Müller of RTL.
Museum Pankow-Archiv, 3. Natur und Umwelt, 3.01 Umweltschutz u. Pflege, 3.01.b Baum-
bestandsschutz u. Pflege, Allgemein, "BSR und Baumpate Ben Wargin bitten die Berliner:
'Unsere Straßenbäume brauchen Wasser—helfen Sie mit.'" Ben Wargin, *Nenn mich nicht
Künstler* (Berlin: Ch. Links, 2015), 169–73; Ben Wargin, *Ausstellung "Die Wüste ist in uns"*
(Berlin: Baumpaten e.V. Europa, 1992).

4. Wargin, *Nenn mich nicht Künstler*, 189–93.

5. Only twenty trees were reported to be planted. See "Linden bis zum Pariser Platz," *TS*, 8
February 1990, 14; Funeck, "Planauslegungsverfahren in der Straße Unter den Linden
zwischen Otto-Grotewohl-Straße und Pariser Platz," 14 February 1990, LDB, NGA, Mitte,
no. 0843.

6. R. Rittmeyer, report "Seelenachse," UDL, Inv. Nr. 7907; Max Heinrich, report "Preisfrage?"
UDL, Inv. Nr. 7952.

7. M. Friedeberg, "Über die Umgestaltung der Straße 'Unter den Linden' in Berlin," *Deutsche
Bauzeitung* 22, no. 27 (1888): 182–83; "Über die Umgestaltung der Straße 'Unter den Linden'
zu Berlin," *Zeitschrift für Gartenbau und Gartenkunst*, no. 49 (1897): 295; Theodor Goecke,
"Berliner Plätze und Prachtstrassen," *Der Städtebau* 1, no. 11 (1904): 157–60 (160).

8. R. Rittmeyer, "Seelenachse," UDL, Inv. Nr. 7907–12; N. N., "Alte und Neue Zeit," UDL, Inv.
Nr. 7913–14; Alexander Klein and Ernst Serck, "Monumentalstraße Unter den Linden," UDL,
Inv. Nr. 7960–75; N.N., "Neues Ortsgesetz," UDL, Inv. Nr. 19383–84; Franziska Bollerey,
*Cornelis van Eesteren. Urbanismus zwischen 'de Stijl' und C.I.A.M.* (Brunswick and Wies-
baden: Viehweg, 1999), 66–82, 204–8. For the announcement of the competition and its prize
winners, see *Wasmuths Monatshefte für Baukunst und Städtebau* 9, no. 5 (1925): 218; and
*Wasmuths Monatshefte für Baukunst und Städtebau* 9, no. 11 (1925): 495. For reports on the
competition results, see "Ideenwettbewerb zur Umgestaltung der Straße 'Unter den Linden,'"
*Deutsche Bauzeitung* 59, no. 98 (1925): 89–91.

9. "Entscheidung des Preisgerichts im Wettbewerb 'Wie soll Berlins Hauptstrasse "Unter
den Linden" sich im Laufe des 20. Jahrhunderts gestalten?'" *Wasmuths Monatshefte für
Baukunst und Städtebau* 9, no. 11 (1925): 497–98; "Ergebnisse des 'Linden-' Wettbewerbs,"
*Wasmuths Monatshefte für Baukunst und Städtebau* 10, no. 2 (1926): 61–76; "Zum 'Linden'-
Wettbewerb," *Der Städtebau*, nos. 5–8 (1925): 67–68; "Ergebnisse des 'Linden-' Wettbewerbs,"
*Der Städtebau*, no. 2 (1926): 24–32; "Ergebnisse des 'Linden-' Wettbewerbs," *Der Städtebau*,
nos. 9–10 (1926): 160–68; "Ergebnisse des 'Linden-' Wettbewerbs," *Der Städtebau*, no. 11
(1926): 176–84.

10. "Erste Anordnung zu der Reichspolizeiverordnung über das Auftreten der Juden in der
Öffentlichkeit vom 3. Dezember 1938," *Amtsblatt für den Landespolizeibezirk Berlin*, 5 Decem-
ber 1938, 335. For the redesign in the 1930s, see Wilhelm Bösselmann, "Die Umgestaltung der
Strasse 'Unter den Linden' Berlin," *Zentralblatt der Bauverwaltung* 56, no. 35 (1936): 1050–55;
Martin, "Neupflanzung der Straße Unter den Linden," 8 January 1935; "Gutachten des
Tiergartendirektors Timm v. 24.7.34 über die Beseitigung der Bäume in der Strasse Unter
den Linden anlässlich des Tunnelbaus der Reichsbahn," LAB, A Rep. 007, Nr. 50.

11. For the trees and the architectural and cultural history of Unter den Linden, see Hartmut
Balder, Gysbert Krüger, and Haile Noé, "'Unter den Linden,' Probleme einer Berliner
Prachtallee im Wandel der Zeit," *Das Gartenamt* 43, no. 1 (1994): 35–39; Hartmut Balder,

Kerstin Ehlebracht, and Erhard Mahler, *Strassenbäume Planen, Pflanzen, Pflegen* (Berlin: Patzer, 1997), 32–37; Rudolf Kühn, *Die Straßenbäume* (Berlin: Patzer, 1961), 73–76; Claus Siebenborn, *Unter den Linden* (Berlin: Oswald Arnold, 1949); Birgit Verwiebe, ed., *Unter den Linden* (Berlin: G-und-H, 1997); Helmut Engel, Laurenz Demps, and Werner Rackwitz, *Die Linden* (Berlin: Staatsbibliothek zu Berlin, 1997); Winfried Löschburg, *Unter den Linden* (Berlin: Ch. Links, 1991). Information on how many trees survived World War II varies. See, for example, Heinz Dannemann, "'Unter den Linden in Berlin'–alte Allee in neuer Gestalt," *Garten und Landschaft* 68, no. 5 (1958): 121.

12. For the pedestrianization plans of Unter den Linden, see Bundesminister für Wohnungsbau, Bonn, and Senator für Bau- und Wohnungswesen, Berlin, eds., *Berlin. Ergebnis des Internationalen städtebaulichen Ideenwettbewerbs Hauptstadt Berlin* (Stuttgart: Karl Krämer, 1960), 117. For the 1964 replanting and redesign of the central promenade, see LDB, NGA, Mitte, 0707 and 0761; "Zur bisherigen Diskussion über die Straße Unter den Linden," *Deutsche Architektur* 16, no. 1 (1967): 8–9; Hans Gericke, "Aufbau der Straße Unter den Linden," *Deutsche Architektur* 11, no. 11 (1962): 635–41.

13. Stephan Strauss, Sigrid Eckmann, and Petra Hennig, "Studien zur Aufwertung des Mittelraumes Unter den Linden," Institut für Städtebau und Architektur, November 1988, IRS.

14. See Luftbild & Vegetation GbR, "Strassenbaumleitplan für die Berliner Innenstadt (Hauptstraßen)," final report prepared in 1993 by order of Senatsverwaltung für Stadtentwicklung u. Umweltschutz, Abt. III., 133.

15. See St. Olbrich, "Tilia tomentosa und Tilia alba spectabilis," *Die Gartenkunst* 4, no. 2 (1902): 30–31; O. Schulze, "Welche Linden eignen sich zu Alleebäumen?" *Die Gartenkunst* 11, no. 7 (1905): 189–90; C. Heicke, *Die Baumpflanzungen in Straßen der Städte* (Neudamm: J. Neumann, 1896), 20; Georg Thiem, *Der Alleebaum* (Stuttgart: Ulmer, 1906), 3; Karl Hampel, *Stadtbäume* (Berlin: Parey, 1893), 12; Aug. Hoffmann, *Hygienische und soziale Betätigung deutscher Städte auf dem Gebiete des Gartenbaus* (Düsseldorf: August Babel, 1904), 45; Kurt Grottewitz, "Veränderungen in der deutschen Baumwelt," *Mitteilungen der Deutschen Dendrologischen Gesellschaft* 14 (1905): 107–11.

16. rm, "Sie blühen!" *Berliner Lokal-Anzeiger*, evening edition, 17 July 1941, n.p.

17. Ludwig Kroeber, "Volkstümliche Arzneipflanzen in alter und neuer Betrachtung," *Heil- und Gewürz-Pflanzen* 9, no. 1 (1926): 17–23 (17). For the use of linden fruit, seed, wood, and bark in the production of oil, and for the collection of linden flowers and their use as infusions substituting for tea during wartime, see Udo Dammer, "Kriegsnutzung unserer Gehölze," *Mitteilungen der Deutschen Dendrologischen Gesellschaft* 24 (1915): 167–70; Kriegsausschuß für Öle und Fette, "Öl- und Fettgewinnung aus Gehölzen," *Mitteilungen der Deutschen Dendrologischen Gesellschaft* 26 (1917): 137–42; Ernst Lehmann, "Die Botanik im Kriege," in *Deutsche Naturwissenschaft Technik u. Erfindung im Weltkriege*, ed. Bastian Schmid (Munich and Leipzig: Otto Nemnich, 1919), 553–83 (562–63, 572–75); Wilhelm Borgmann, "Die Forstwirtschaft im Kriege," in *Deutsche Naturwissenschaft Technik u. Erfindung im Weltkriege*, ed. Bastian Schmid (Munich and Leipzig: Otto Nemnich, 1919), 913–43 (932–33, 935); A. Tschirch, "Kriegsbotanik," *Berichte der Deutschen Pharmazeutischen Gesellschaft* 26, nos. 7–8 (1916): 326–52; H. Thoms, "Unsere Nahrungsmittel während des Krieges," *Berichte der Deutschen Pharmazeutischen Gesellschaft* 26, no. 4 (1916): 154–79; H. Thoms and Hugo Michaelis, "Die Linde als Fettlieferant," *Berichte der Deutschen Pharmazeutischen Gesellschaft* 26, no. 4 (1916): 185–92; H. Roß, "Pflanzt Holzgewächse welche Arzneidrogen liefern," *Heil- und Gewürzpflanzen* 1, nos. 10–11 (1917–18): 270–77; "Kleinere Mitteilungen," *Heil- und Gewürzpflanzen* 1, no. 12 (1918): 301; Gustav Edel, "Die Arzneipflanzensammlung im Schulbezirk Saulgau im Sommer und Herbst 1916 und 1917," *Heil- und Gewürzpflanzen* 1, no. 9 (1918): 243–45; "Sammeln von Arzneipflanzen hinter der deutschen Front," *Heil- und Gewürzpflanzen* 2, no. 3 (1918): 58–59; "Lindenblütensammeln in Würzburg," *Heil- und Gewürzpflanzen* 2, no. 3 (1918): 59.

18. See communication between the Magistrate and Kreuzberg district department, 9 March

1923, 27 April 1923, LAB, A Rep. 007, Nr. 68; "Gesellschaft für Anbau und Verwertung von Heilpflanzen in Berlin," *Heil- und Gewürzpflanzen* 3, no. 7 (1919–20): 174. During both world wars, children and women were enlisted to collect plants that provided fruit with a high vitamin C content and leaves that could replace imported tea and be used in infusions. Besides contributing to the country's autarky and public health, the Nazi initiative to procure "food out of the woods" (*Ernährung aus dem Walde*) was also considered an activity that could forge close ties with the native homeland and draw attention to its necessary protection. See *Handreichung für das Sachgebiet "Ernährung aus dem Walde" (RAW)* (Dresden: Nationalsozialistischer Lehrerbund, Gauwaltung Sachsen, Abteilung Erziehung und Unterricht, 1940); *Unsere Ernährung aus dem Walde* (Lüneburg: Dt. Frauenwerk, Gaustelle Ost-Hannover, 1943).

19. Helmut Donath, "Vergiftungen blütenbesuchender Insekten durch den Nektar von Silberlinden," *Landschaftsarchitektur* 17, no. 2 (1988): 58; "Ein Umwelt-Krimi: Der Fall Silberlinde," *ND*, 2–3 December 1989, 16.

20. Günter Madel, "Vergiftungen von Hummeln durch den Nektar der Silberlinde *Tilia tomentosa Moench*," *Bonner Zoologische Beiträge* 28, nos. 1–2 (1977): 149–53.

21. Wolfgang Blenau, "Giftwirkung fremdländischer Lindenarten auf einheimische blütensuchende Insekten," Jahresarbeit, Humboldt University, Berlin, 1989–90, NL Gandert, ZFA Dendrologie und Gartenarchitektur, folder 002-28. For an early warning about the toxicity of silver linden flowers for bees in the East German specialist press, see Hans-Joachim Winkelmann, "Die Linden als Straßenbaum," *Der Deutsche Gartenbau* 2, no. 2 (1955): 55–56. For the discussion about native versus non-native trees in East and West Germany, see Peter A. Schmidt, "Fremdländische Gehölze in der Stadt, ja oder nein?" *Beiträge zur Gehölzkunde* (1991): 28–34; Ingo Kowarik, "Einheimisch oder nichteinheimisch?" *Garten und Landschaft* 99, no. 5 (1989): 15–18; Ingo Kowarik, "Ökosystemorientierte Gehölzartenwahl für Grünflächen," *Das Gartenamt* 35, no. 9 (1986): 524–32.

22. Letters between K.-D. Gandert and B. Surholt, dated between 1990 and 1997, NL Gandert, ZFA Dendrologie und Gartenarchitektur, folder 002-28; Werner Mühlen, Volkhard Riedel, Thomas Baal, and Bernhard Surholt, "Insektensterben unter blühenden Linden," *Natur und Landschaft* 69, no. 3 (1994): 95–100; Thomas Baal, Bernd Denker, Werner Mühlen, and Berhard Surholt, "Die Ursachen des Massensterbens von Hummeln unter spätblühenden Linden," *Natur und Landschaft* 69, no. 9 (1994): 412–18; Berhard Surholt and Thomas Baal, "Die Bedeutung blühender Silberlinden für Insekten im Hochsommer," *Natur und Landschaft* 70, no. 6 (1995): 252–58; Bernhard Surholt und Werner Mühlen, "Blühende Silberlinden. Heimtückische Insektenfallen oder wertvolle Nahrungsquellen?" *Beiträge zur Gehölzkunde* (1997): 4–17.

23. Surholt to Gandert, 17 August 1997, NL Gandert, ZFA Dendrologie und Gartenarchitektur, folder 002-28.

24. erk, "Silberlinden—tödlicher Irrtum," *TS*, 20 February 1990, 12; *TS*, 23 March 1990, 11; "Silberlinden killen keine Hummeln," *Das Gartenamt* 38, no. 12 (1989): 729–30.

25. For the preference of silver linden in Berlin, see Harri Günther, "Über das Verhalten von Gehölzen unter großstädtischen Bedingungen, untersucht an einigen Gehölzarten in Berlin" (doctoral diss., Humboldt University, Berlin, 1959), 168–76. For 'Pallida,' see L. Späth Baumschule, ed., *Späth-Buch 1720–1930* (Berlin: Berlin-Baumschulenweg, 1930), 310. For 'Greenspire,' see William Flemer III, Linden Tree, United States Plant Patent 2,086, filed 7 November 1960 and issued 5 September 1961, https://www.uspto.gov/. For the more recent use of linden trees along Berlin's streets, see species list for Unter den Linden, 2017, courtesy Senatsverwaltung für Stadtentwicklung und Umwelt, Verkehr und Klimaschutz Berlin; and e-mail communication with Jürgen Götte, Bezirksamt Mitte, Straßen- und Grünflächenamt, Berlin, 16 February 2017. Information on Berlin's street trees can also be retrieved at http://fbinter.stadt-berlin.de/fb/index.jsp, accessed 10 February 2018.

26. Erhard Mahler, "Alleepflanzung als Tradition—Straßenbäume in Berlin," *Landschafts-*

*architektur* 23, no. 3 (1993): 20–23. In 1995, *Der Tagesspiegel* reported 396,000 street trees in the city. See "Für die Straßenbäume ist schon Herbst," *TS*, 8 August 1995, 8.

27. Georg Pniower, *Bodenreform und Gartenbau* (Berlin: Siebenreicher, 1948), 127–28, 134–36, 139–40, 142–43; Georg Pniower, "Über die Entwicklungsgeschichte und landeskulturelle Bedeutung der Dendrologie," in *Gehölzkunde und Landeskultur: Referate der Ersten Zentralen Tagung für Dendrologie in Dresden-Pillnitz am 29. bis 31. August 1953* (Leipzig and Jena: Urania, 1954), 13–141 (129–34); Günther, "Verhalten von Gehölzen," 191–94.

28. For green open-space planning, see Fritz Witte, "Zur Grünplanung in Berlin," *Garten und Landschaft* 67, no. 10 (1957): 267–69. For the urban design competition, see Bundesminister für Wohnungsbau, Bonn, and Senator für Bau- und Wohnungswesen, Berlin, eds., *Berlin. Planungsgrundlagen für den städtebaulichen Ideenwettbewerb "Hauptstadt Berlin"* (Berlin: Lindemann und Lüdecke, 1957), and Bundesminister für Wohnungsbau, Bonn, and Senator für Bau- und Wohnungswesen, Berlin, eds., *Berlin. Ergebnis des Internationalen städtebaulichen Indeenwettbewerbs "Hauptstadt Berlin"* (Stuttgart: Karl Krämer, 1960); Berlinische Galerie e.V., ed., *Hauptstadt Berlin* (Berlin: Gebrüder Hartmann GmbH, 1990); Wolfgang Schäche, "Der Hauptstadtwettbewerb 1957/58. Entwürfe gegen die Vergangenheit," in *Stadt der Architektur, Architektur der Stadt*, ed. Thorsten Scheer, Josef Paul Kleihues, and Paul Kahlfeld (Berlin: Nicolai, 2000), 251–57; Werner Durth, Jörn Düwel, and Niels Gutschow, *Architektur und Städtebau der DDR*, vol. 1 (Frankfurt: Campus, 1998).

29. Schwöbel, Senatsverwaltung für Stadtentwicklung und Umwelt, "Vermerk, 25.03.2004: Entwicklung des Straßenbaumbestandes in Berlin" (courtesy Senatsverwaltung für Stadtentwicklung und Umwelt, Verkehr und Klimaschutz Berlin); Michael Fietz, *Art- und schadensbedingtes Abbildungsverhalten von Berliner Straßenbäumen auf Colorinfrarot-Luftbildern, Berliner geowissenschaftliche Abhandlungen*, series D 188, vol. 2 (1992), 54–55, 57–58.

30. Carola Wünsche, "Pflege, Erhaltung und Erweiterung des Straßenbaumbestandes im Stadtbezirk Berlin-Prenzlauer Berg" (Diplomarbeit [master's thesis], Humboldt University, Berlin, 1983), 2, 25–26; Mahler, "Alleepflanzung."

31. H. Balder and G. Krüger, "Vitalität des Öffentlichen Grüns—ein Ost-/Westvergleich," *Gesunde Pflanzen* 44, no. 9 (1992): 291–95; Dirk Dujesiefken, "Baumpflege in der DDR," *Das Gartenamt* 39, no. 9 (1990): 585–87.

32. Krause to Dr. Joachim, 29 March 1988, STUG 002-43, "Überarbeitung des Standards 'Anlagen des Straßenverkehrs; Terminologie,'" p. 16; Stefan Kuschel, "Immer mehr Bäume ersticken," *TS*, 19 August 1996, 10.

33. L. Fintelmann, *Über Baumpflanzungen in den Städten* (Breslau: Kern's, 1877), 14; Hampel, *Stadtbäume*, 59–66; Kny, "Einfluss des Leuchtgases auf die Baumvegetation," *Botanische Zeitung* 29, no. 50 (1871): 852–54; Kny, "Einfluss des Leuchtgases auf die Baumvegetation," *Botanische Zeitung* 29, no. 51 (1871): 867–69; Gartendirector Meyer, "Die Einwirkung des Leuchtgases auf die Baumzucht," *Berlin und seine Entwicklung, Städtisches Jahrbuch* 6 (1872), 91–94; "Die Baumnoth in großen Städten," *Berlin und seine Entwicklung, Städtisches Jahrbuch* 3 (1869): 96–106; Richard Böckh, ed., *Statistisches Jahrbuch der Stadt Berlin, Statistik des Jahres 1882* (Berlin: P. Stankiewicz' Buchdruckerei, 1884), 109.

34. *Baumschutz*, no. 19 (1995): 2–3, 9–10; *Baumschutz*, no. 14 (1991): 8; "Baum-Boom statt Bau-Boom," *TS*, 13 May 1995, 10; gih, "Bauarbeiten gefährden alte Lindenallee," *TS*, 19 August 1993, 10; Fritz, "Berlins schönste Allee bleibt erhalten," *TS*, 20 September 1993, 11.

35. Mahler, "Alleepflanzung."

36. Hampel, *Stadtbäume*, 58.

37. Christian van Lessen, "Das geht unter die Rinde," *TS*, 18 May 2008, 9.

38. Herbert Krause, "Pflanzungen an Straßen," in *Berlin und seine Bauten*, part 11, *Gartenwesen*, ed. Architekten- und Ingenieurverein (Berlin: Wilhelm Ernst und Sohn, 1972), 143–52; Mächtig, "Berlin," in Hoffmann, *Hygienische und soziale Betätigung*, 73–103 (98–99); Hoffmann, *Hygienische und soziale Betätigung*, 45–46; *Die Gartenwelt* 13, no. 37 (1909): 444.

## Epilogue

1. Chauncey D. Leake, "Social Aspects of Air Pollution," in *Proceedings, National Conference on Air Pollution, Washington, D.C., November 18–20, 1958* (Washington, DC: U.S. Government Printing Office, 1959), 20–24 (23); Richard A. Wolff, "Planting Trees to Cleanse Air," *NYT*, 9 December 1960, 30; "Trees May Protect Earth from Disastrous Pollution Heating," 17 November 1958, press release by Science Service, CAP.

2. For the turn-of-the-century attention to trees' ozone production, see Heinrich Pudor, "Der Volkspark von Gross-Berlin," *Die Gartenkunst* 10, no. 1 (1908): 9–12; Lothar Abel, *Die Baumpflanzungen in der Stadt und auf dem Lande* (Vienna: Georg Paul Faesy, 1882), 34–36; L. Fintelmann, *Ueber Baumpflanzungen in den Städten* (Breslau: F. U. Kern's, 1877), 9. For the medicinal properties attributed to ozone in combating tuberculosis and pulmonary disease, see B. Bandelier and O. Roepke, *Die Klinik der Tuberkulose* (Würzburg: Curt Kabitzsch, 1912), 116–17, 163, 196, 220.

3. Dave Kendal and Mark McDonnell, "Adapting Urban Forests to Climate Change: Potential Consequences for Management, Urban Ecosystems and the Urban Public," *Citygreen* 8 (2014): 130–37; Mark J. McDonnell and Amy K. Hahs, "Adaptation and Adaptedness of Organisms to Urban Environments," *Annual Review of Ecology, Evolution, and Systematics* 46 (2015): 261–80.

4. Johann Christoph Gewecke and Gunter Hägele, "Mehr tun für Bäume!" *Das Gartenamt* 26, no. 5 (1977): 311–12, 321–23.

5. Andrew Jackson Downing, in *The Horticulturalist and Journal of Rural Art and Rural Taste* 1, no. 9 (1847): 393–97; Smith Riley, "Native Atmosphere from Trees," *Parks & Recreation* 8, no. 6 (1925): 554–62 (556–57); Wayne C. Holsworth, "What of the Poplars?" *Parks & Recreation* 6, no. 6 (1923): 519–20.

6. Memorandum by Philip J. Cruise, "Trees Planted—Spring 1942," 8 May 1942, DPR, Manhattan, 1942, folder 36; memorandum by David Schweizer, "Street Tree Planting—Capital Budget Program," 29 April 1946, DPR, Administration, 1946, folder 54.

7. *Projekt Stadtgrün 2021, Abschlussbericht zum Forschungsvorhaben KL/08/02* (Veitshöchheim: Bayerische Landesanstalt für Weinbau und Gartenbau, 2012); Philipp Schönfeld, Susanne Böll, and Klaus Körber, *Weitere Baumarten im Projekt Stadtgrün 2021* (Veitshöchheim: Bayerische Landesanstalt für Weinbau und Gartenbau, 2016); street tree lists issued by GALK in 1978, 1980, 1983, 1986, 1991, 1995, 2001, and 2006, published in *Das Gartenamt* and *Stadt und Grün*. For a street tree project in the United States, see T. Davis Sydnor, "Ohio's Shade Tree Evaluation Project: In Search of a Well Adapted Tree," *Arboricultural Journal* 8 (1984): 115–22.

8. "A Report on the Street Trees of Hartford, Conn.," *Park and Cemetery and Landscape Gardening* 13, no. 3 (1903): 39–40; E. P. Felt and S. W. Bromley, "Shade Tree Problems," *Sixth National Shade Tree Conference, Proceedings of Annual Meeting* (1930), 13–23 (13).

9. William Solotaroff, "Progress in Municipal Shade Tree Control," *Park and Cemetery and Landscape Gardening* 21, no. 11 (1912): 751–52 (751); "Free Distribution of Trees by Department of Public Grounds in Boston," *Park and Cemetery and Landscape Gardening* 15, no. 11 (1906): 458–59; Henry M. Hyde, "Chicago Lining All Its Streets with Shade Trees," *CDT*, 14 May 1914, 13; J. H. Prost, "City Forestry in Chicago," *American City* 4, no. 6 (1911): 277–81.

10. "Trees," 26 August 1956; 15 March 1953, NYCPPR. Stuart Constable to John A. Mulcahy, 11 September 1959, DPR, Manhattan, 1959, folder 23.

11. Charles Abrams, "Memorandum in Support of a Bill to Amend the City Charter to Authorize the Planting of Trees by Borough Presidents," CAP; memorandum by W. H. Latham, "WPA Tree Planting," 2 October 1939, DPR, Brooklyn, 1939, folder 6; memorandum by Philip J. Cruise, "Street Trees," 4 September 1941, DPR, Manhattan, 1941, folder 57; documents in DPR, Administration, 1943, folder 28; Administration, 1939, folders 46, 47.

12. Documents in LAB, A Rep. 036-08, Nr. 127; LAB, A Rep. 036-08, Nr. 128; Carl Hampel, *Stadtbäume* (Berlin: Parey, 1893), 32–33.

13. David K. Randall, "Maybe Only God Can Make a Tree, but Only People Can Put a Price on It," *NYT*, 18 April 2007, B3; Jill Jonnes, "What Is a Tree Worth?" *Wilson Quarterly* 35 (Winter 2011): 34–41; www.itreetools.org, accessed 4 June 2017. For trees' "biological-technical functions," see K.-D. Gandert and K. Ostwald, "Zur Methode der Bepflanzung in Städten," *Deutsche Gartenarchitektur* 7, no. 3 (1966): 63–67.

14. "A Tree Is a Tree, but Not in City," *NYT*, 21 August 1966, 66; Peter Kiermeier, "Entwicklung von neuen Strassenbaumsorten in den USA," *Das Gartenamt* 30, no. 2 (1981): 92–106; Ruth S. Foster and Joan Blaine, "Urban Tree Survival: Trees in the Sidewalk," *Journal of Arboriculture* 4, no. 1 (1978): 14–17; Gary Moll, "The State of Our Urban Forest," *American Forests* 95, nos. 11–12 (1989): 61–64; Bob Skiera and Gary Moll, "The Sad State of City Trees," *American Forests* 98, nos. 3–4 (1992): 61–64; Marianne Holden, "New York Street Trees Are Endangered," *NYT*, 22 May 1985, A26; Elisabeth Bumiller, "Thousands of Trees? Millions? Volunteers Begin Count of New York's Street Trees," *NYT*, 29 July 1995, 21.

15. For street trees' thriftiness, see *Bulletin of the Tree Planting Association of New York City*, September 1913, 3; J. Victor, "Noch einmal: Der Strassenbaum," *Das Gartenamt* 12, no. 10 (1963): 272–74.

# INDEX

Page numbers in *italics* refer to illustrations.